WERTHEIM PUBLICATIONS
IN INDUSTRIAL RELATIONS

Established in 1923 by the family of the late Jacob Wertheim "for the support of original research in the field of industrial cooperation . . ."

JAMES J. HEALY   DEREK C. BOK

E. R. LIVERNASH   JOHN T. DUNLOP,

GEORGE S. HOMANS   *Chairman*

HD 6515
B86 B

THE BUTCHER WORKMEN
A STUDY OF UNIONIZATION

DAVID BRODY

63841

ST. JOSEPH'S UNIVERSITY   STX
HD 6515 .B86b
The butcher workmen;

3 9353 00087 3222

HARVARD UNIVERSITY PRESS

CAMBRIDGE, MASSACHUSETTS

1964

© 1964 by the President and Fellows of Harvard College

All rights reserved

Distributed in Great Britain by Oxford University Press, London

Library of Congress catalog card number 64–21240

Printed in the United States of America

TO MY PARENTS

TO MY PARENTS

FOREWORD

This volume is the story of one of our most interesting international unions. Few unions faced greater adversity from managements in an earlier day than in the packing plants, and few unions have organized a more diverse group of workers. The present membership of 350,000 or more includes major representation in retail butcher shops and chain stores, meat packing plants, food-processing plants, the fur and leather industries, and among migratory agricultural workers and sheep shearers. Few unions have survived, much less prospered, after two nationwide strikes which saw unionism virtually eliminated from the meat packing plants. The union has adapted its policies and its structure to rapid changes in technology, significant shifts in product market competition and geographical location, to the changing role of government, to new policies of management, and to new rivals and marked changes in the federation level of our labor movement.

Throughout its long history, the *Amalgamated Meat Cutters and Butcher Workmen* has been characterized by a distinctive spirit. It has long included both craft and industrial groups; it was sympathetic to the cause of industrial unionism and the purposes of John L. Lewis in the 1930's, and yet it never deserted the AFL; it has accommodated both local and national collective bargaining; it has maintained a progressive spirit without ideological tinge; it has grown in size and resources without losing its drive; it has amalgamated with a variety of other unions without loss of its historical character; and it has had strong leadership and effective administration at the international level and still preserved local vitality.

It is fortunate that Professor David Brody, a historian with special interests in labor and social developments, was available to make this study of an international union with such rich experiences. Indeed, his training and interests are particularly well suited to the task of presenting a full picture of the complex and shifting ingredients which comprise the history of the *Butcher Workmen*.

Among the many themes of this volume which will be of widespread current interest is the account of the process of growth in membership and expansion in scope or jurisdiction of the union. The union has grown through a variety of measures and methods: It has been imaginative in its approach to workers; it has worked cooperatively with other unions; it has sought to influence employers; it has used economic pressure; and it has amalgamated with other international unions to improve its strategic position. The volume also provides the basis for an appraisal of the role of the government in affecting union growth.

This volume is the eighth in a continuing series on labor-management history. It was written with full access to all records and materials of the international union and with the cooperation of its officers. It was the explicit understanding that "the standards of scholarship and the interpretation of events would, of course, be the responsibility solely of those scholars working on the project." This relationship was essential since the union, along with other labor and management organizations and foundations, made a financial contribution to Harvard University toward the larger labor-management history project of which this is a part.

The Butcher Workmen enhances our understanding of the labor movement, and the growth and adaptability of its institutions under changing economic and political conditions; it also provides insights into the operation of collective bargaining and the limits and effects of governmental intervention in industrial relations.

John T. Dunlop

AUTHOR'S PREFACE

At the opening of the twentieth century, trade unions occupied a minor place in American life. Only a few important industries—railroading, coal mining, construction—felt the impact of strong unions. The organizing struggle, persisting for many years, quickened during the New Deal era. By mid-century, the issue was no longer in doubt. The labor movement had overborn the resistance of most major industries. Trade unions claimed a membership of upwards of sixteen million, and through collective bargaining fixed the terms of employment in a broad sector of the American economy. Here, by any measure, was a central change in the nation's life. Yet we lack a precise understanding of how this labor achievement came about. Recent labor history has still to be examined with the focus on the unionization process.

That is what this book proposes to do for the American meat trade. This field seemed suited for an investigation of unionization. One segment— meat packing—had the characteristics of mass-production industry: giant firms, national markets, the production economies of scale, a concentrated labor force of semiskilled and unskilled workmen. Here could be traced the play of forces that, after joining together abortively twice earlier, finally led to mass-production unionism following the emergence of the CIO. Meat retailing, on the other hand, in many ways typified the economic sector favorable to AFL unions of the old line: a multitude of small employers, a local market, skilled labor. And the chain-store revolution, striking in the later 1920's, illustrated the effect of industry change on the pattern of unionization.

The trade unions also differed in significant ways. The Amalgamated

Meat Cutters and Butcher Workmen of North America functioned within the framework of the AFL. Not only could one follow the progress of the Amalgamated among the skilled retail butchers; one could also plot over half a century the maximum effort of an AFL union in a mass-production field. The CIO vehicle in meat packing, the United Packinghouse Workers of America, demonstrated the new approach to mass production that accompanied the split in the labor movement in the 1930's. To the industrial union came the larger successes in meat packing in the great organizing years of the New Deal era. The triumphant UPWA, however, could not sustain its effectiveness after World War II. The fortunes of the rival unions unexpectedly shifted. The diversity, decentralized structure, and organizing tactics of the Amalgamated worked to greater advantage in exploiting the opportunities of the 1950's. There were, as it turned out, liabilities to the unionism molded by the needs of the 1930's.

This is an inquiry into unionization. By that term I mean more than the recruitment of workingmen into labor organizations. The process ran its full course only when it became irreversible. To reach this critical point, a union had to do more than build membership: it also needed to achieve internal order and genuine collective bargaining. Lacking these, the Amalgamated had seen great organizing gains disintegrate in the past. Complex process that it was, unionization unfolded in different ways. It had, for example, a more eventful and complicated history in meat packing than in the retail trade. This partly accounts for the unequal apportionment of space in this book. Collective bargaining did not come simultaneously with the organization of the packinghouse workers. For the major packers made recognition their real line of resistance to trade unionism. What was more, the internal problems accompanying packinghouse organization were always immense, involving in the 1930's the birth of a new national union. And the ensuing disruptive rivalry abated only slowly in the face of hard experience.

One determinant of the scope of this book, therefore, was the nature of unionization. The causes constituted the other. I wanted to know why unionization proceeded as it did in the meat industry. This involved, first, a focus on the trade unions, for they were the unionizing vehicles. Their strategies, above all, claimed my attention. Strategy, in turn, hinged on the operative conditions: management policy, industry characteristics, the role of government and law, the state of the butcher workmen themselves. These added up, at any given point, to the prospects for unionization. The total became really favorable, of course, only during the later years of the New Deal and during World War II. The extent to which the prospect became a reality depended on the union response. Unionization sprang from the interaction of union effort and larger conditions. In this book I have tried to take both into full account.

I have, by the same token, excluded information that did not bear directly on my problem. The conventional categories of labor history therefore receive only partial treatment here.

This volume does claim thoroughness as an exploration of the process and causes of unionization. I hope that the reader leaves it with some understanding of how unionization occurred among the butcher workmen and, by an extension of that, how it proceeded for the whole of American industry. For the forces acting on the meat field were at work everywhere. The Amalgamated and the UPWA reflected main strains of American trade unionism in the twentieth century. And the industry characteristics in meat packing and distribution were shared by other economic sectors in process of unionization. The experience of the butcher workmen had wider applicability. That is not, of course, the justification for this book. The events related here, it need not be emphasized, were in themselves of large importance in recent American history.

I have incurred heavy debts during my work on this book. The Labor-Management History Project at Harvard University employed me as a research associate in 1959–61 during the initial stages of work and covered all the incidental expenses of research and writing. That is, as every historian will know, an unusually comfortable arrangement for one of our craft. Professor John T. Dunlop, the head of the project, invited me to write a history of the Amalgamated Meat Cutters and Butcher Workmen of North America. Then he left me free to develop the subject as I pleased, listened to my ideas, permitted me to appropriate his, read and criticized the manuscript and, from start to finish, encouraged me in my labors. The Amalgamated gave me unrestricted access to its files and treated me with every courtesy while I was working at its headquarters. I am particularly grateful to Secretary-Treasurer Patrick E. Gorman, who patiently answered my questions and saved me from a number of errors by his careful reading of an earlier draft of the manuscript. The United Packinghouse Workers of America also generously opened to me the records of the Packinghouse Workers Organizing Committee. Mr. Irwin Nack, who organized those records in the course of work on his Columbia doctoral dissertation on the PWOC, shared freely with me his ideas on the subject. I invariably received courtesy and assistance from the staffs of the National Archives, the State

Historical Society of Wisconsin, the New York Public Library, the Harvard University Library and the Columbia University Library. My wife Susan is responsible for the felicity of my household, and also for the index.

D. B.

New York City
April 25, 1964

CONTENTS

CONTENTS

CHAPTER 1

THE MODERNIZATION
OF THE MEAT TRADE

In the final decades of the nineteenth century, the American butcher work-man entered a new industrial world. The meat trade was rapidly taking its modern form. Changing structure and technology had, among other effects, major consequences for labor in the meat industry. By 1900 the industrial development had run its course, revealed its full impact on the butcher work-men, and laid the basis for unionization.

The meat industry had been dispersed and undifferentiated. Except for pork packing, slaughtering and processing remained a local business that was carried on largely in conjunction with retailing. This pattern altered after 1875, and meat packing began to concentrate in a few centers in the middle west. The locational shift transformed the industry.

The direction of national growth created the vital conditions for a cen-tralized industry. Urban population was multiplying in the cities of the east and the old northwest, and livestock production meanwhile was mov-ing westward to the corn belt and the grazing ranges of the Great Plains. A network of railroads bridged the distances between producing and con-suming regions. Initially, the live animals were shipped to the urban centers. Stockyards sprang up, first in Chicago in 1865, then in Kansas City in 1871, and later at other railroad termini, to centralize the distribution of live-stock. Slaughtering took place at the end of the journey—a system that wasted both car space and animals during shipment and fostered the un-sanitary and inefficient operations of local slaughterhouses. However, no other arrangement was feasible without further technological advances.

The chief obstacle centered around the distribution of fresh meat. Cured products, not facing the problem of spoilage in transit, had long had the advantages of centralized production. Some 475,000 hogs were slaughtered at Cincinnati—"Porkopolis"—in the winter season of 1848. Natural ice refrigeration gradually made pork packing a year-round operation. By 1875, Chicago houses were processing over two million hogs annually, but fresh meat shipments to eastern cities awaited the application of refrigeration to transportation. There were numerous experiments to perfect a refrigerator car. George H. Hammond of Detroit had shipped dressed beef as early as 1871. The first major commercial success came to Gustavus F. Swift, a Massachusetts cattle dealer who arrived in Chicago in 1875. His efforts, hampered by very slight resources, finally resulted in an efficient car and, by the end of the decade, the start of the dressed-beef trade. Meat canning, coming into general use by 1880, further encouraged centralized slaughtering by providing an outlet for inferior grades of cattle. A government official was certain by 1884 "that the old method of shipping live cattle in open cars will be wholly superseded." The advantages of central operations were evident from the start.[1]

The dressed-beef trade nevertheless had to overcome bitter opposition in the early years. The Vanderbilt and Pennsylvania lines, well equipped to handle livestock and interested in the eastern stockyards, discouraged the new business. The packers were forced to build their own refrigerator cars and at first to ship by the Grand Trunk line through Canada. Local butchers saw their own trade threatened by the western beef. Organizations were formed, prejudices enflamed, and, finally, laws passed in a number of states prohibiting the sale of out-of-state meat. (These were quickly set aside as interferences with interstate commerce.) The packers forced their way into the local markets with ruthless price-cutting tactics, and, as an eastern slaughterer reported in 1895, put "the New York butchers at the mercy of the Western packing houses."[2] The number of cattle killed in Chicago increased from 224,309 in 1875 to 2,223,971 in 1890.

A half-dozen firms meanwhile emerged to dominate the industry. Swift & Co. and Armour & Co. were the largest, followed at a distance by Nelson Morris & Co., Cudahy Packing, and Schwarzschild and Sulzberger (later Wilson & Co.). The National Packing Company in March 1903 temporarily combined many of the lesser properties, including the Hammond and Fowler interests. There were a number of independents such as Kingan, Jacob Dold, T. M. Sinclair, and others. But the giants engrossed the bulk of the trade. The Big Six slaughtered 89.5 per cent of the Federally inspected cattle east of the Rockies in 1903. Competition, moreover, was never very keen among the packers. Shipment pools had existed since the mid-1880's, as had price agreements on some meats.[3]

Notwithstanding this limitation of competition, the meat trade was always uncertain. Despite repeated accusations by stock men and retailers, the packers could not effectively control the price either of livestock or wholesale meat. Philip Armour was, in fact, noted for his speculations in pork. Unlike in the steel industry, firms did not integrate down to the raw materials, except for Morris to a small degree. The high volume of business in relation to capital added another element of risk. The packers had to operate with substantial short-term borrowing and inventories in relation to their fixed assets.[4] Sulzberger and Sons (formerly Schwarzschild and Sulzberger) failed in 1915, followed by Armour & Co. after World War I, because of the sharp drop in the market value of its inventory.[5] In both cases, control passed into the hands of bankers. The character of the meat industry never permitted the kind of stability that existed, say, in steel after 1901.

Economy was, from the outset, the objective of the packers. Theirs was not a new or improved product, as in the case of steel, oil, or electricity; in fact, western beef was long considered inferior to home-killed meat. The concentration of slaughtering in the western centers was designed rather to reduce the costs of serving an existing market. "We are here to make money," Armour told a Senate committee in 1889. "I wish I could make more . . . but if I had not been a little inventive and enterprising I do not believe I would make any at all. I know I couldn't in the old-fashioned way." [6]

Distribution savings—the initial objective of the packers—were disappointing; the average costs of freight, shrinkage, icing, and branch-house operation proved to be high, exceeding the plant expenses per head of cattle in the early 1900's by 25 per cent.[7] The important economies came instead from large-scale production. Local slaughtering, even in extensive urban markets, had been carried on in small units with little effort made to use the wastes (except in feeding the "slaughter-house hogs") or to raise the efficiency of operations. A movement was spreading for central local slaughterhouses, and pork packers had of course made significant progress. But by-product use and low-cost operation were perfected only with the growth of the dressed-meat industry. "The modern packinghouse eminently exemplifies scientific, commercial methods," observed Armour in 1900. "It has logically displaced the smaller slaughterhouses by the application of economies only obtainable in extensive operations." [8]

By-product changes advanced on two lines. First, the packers began to take over the work from the satellite rendering, glue, and fertilizer works around their plants. In 1877 Armour was the first packer to manufacture fertilizer from offal in his own plant. Second, there was a continuing extension of by-product use. Oleomargarine, requiring much experimentation, stimulated progress in this direction. Beginning with Nelson Morris in 1886, the packers employed staffs of chemists who developed a succession of such products as compound lards, fertilizers, glues, soaps, and pharmaceuticals. Their importance in the meat business was apparent in the books of three major packers for 1903–4: cattle by-products, including hides, came to one-fourth the net value of the beef.[9]

Labor savings constituted the other important economy in the western packinghouses. Machinery took over much of the by-product, sausage, and canning operations. Friction hoists, overhead conveyers, endless chains, chutes, and moving benches eliminated the manual handling of materials. The Hammond plant at St. Joseph, completed in 1904, seemed to culminate this development. "The beef is not moved or lifted a single inch by human force from the time the steer falls on the killing beds . . . until the dressed carcass finally arrives but a few feet in front of the refrigerator car." Without the modern equipment, surmised the *National Provisioner,* ten times as many men would be required to handle the huge volume of the plant. The killing and cutting tasks, however, were not in themselves susceptible to mechanization. Only the hog-scraping machine, first patented in 1876 and gradually improved, supplanted an important manual procedure on the killing floors.[10]

Assembly—or rather, disassembly—line operation was the basic achievement here. Division of labor had existed in the pork packinghouses of Cincinnati since before the Civil War. But now it was carried forward to the most minute degree. The packers, as John R. Commons observed, surveyed and laid off the animal like a map, apportioning to each workman a tiny part of the total operation. There were, for example, 78 different occupations in the beef-killing gang of 157 men at one Chicago plant. Two knockers and one sticker killed all the 1050 head of cattle handled by the gang in a ten-hour day. The division of labor had the advantages, first, of putting two thirds of the work into the hands of common labor and, second, of greatly improving the quality of the delicate splitting and skinning tasks. Output per man was also enormously increased. This resulted in part from the greater proficiency that came with specialization and, more important, from assembly-line methods. Moving rails and benches not only saved "the men the time taken in passing the hogs along the rail," but also prevented "the slowest man from regulating the speed of the gang." Management, not the men, set the work pace on the cutting line.[11]

Labor policies were in accord with the pattern of economy. There were further savings to be gained by the proper use of labor under the conveyer system. "If you need to turn out a little more," said a Swift superintendent, "you speed up the conveyers a little and the men speed up to keep pace." An intense competitive spirit also was fostered between gangs and plants. The Armour plant at Chicago held the record for a day's slaughtering— 3219 head of cattle—until the company's Kansas City plant killed 3249 on October 26, 1900. A reminiscent poem caught the spirit of "How They Used to Kill Cattle in the Old House (Swift's) in the Summer of 1896."

> All right, Herman; start that knocking,
>   Roll them out upon the floor;
> Get them ready for the headers,
>   Shackle on and let 'em roar.
> Get them on the rails and roll 'em
>   To their places—let them drop;
> Keep the heads and feet from piling,
>   Do not let the killing stop.
>
>   .   .   .   .
>
> My, O my, but how we're sailing,
>   We must now be nearly done;
> Ten hours killing eighteen hundred,
>   Now the boys are singing "Home, Sweet Home."

In 1884, five splitters in one gang dealt with 80 cattle an hour; ten years later, four splitters in the packinghouse were handling 120 per hour, almost doubling the output per man. "That marvelous speed and dexterity so much admired by visitors," charged Socialist A. M. Simons in 1899, "is simply inhumanly hard work." [12]

The system of compensation underwent a significant change. Until the mid-1880's, a day wage prevailed in Chicago. The cattle butchers in 1886 tried to get steady time, that is, a weekly salary which only a few skilled men then had. Instead, the packers forced the acceptance of an hourly rate. A great advantage of Chicago over the country towns, the *National Provisioner* explained in 1894, was the availability at the Union Stock Yards of "all the labor you want, for which you pay by the hour, and only use the labor as long as necessary." [13] The practice developed of not compensating for time lost during machinery breakdowns. Far more valuable was the ability under an hourly system of scheduling work to maximize economy. The bulk of livestock shipments normally arrived in the early part of the week. The packers expanded their capacity to permit the slaughter of all animals on the day of purchase. There was therefore very intense work during the first two or three days, but almost invariably short time at the end of the week. The arrangement, while concentrating the work and saving feeding and storage charges, gave the workmen an irregular and uncertain schedule. During the peak fall and winter months, the work week was usually close to fifty hours; in the slack season it often fell to thirty. The regular employees of the beef department in one plant averaged about two thirds of full time from 1886 to 1904.

The effect on earnings was disastrous. In Chicago, cattle splitters and floorsmen—the aristocrats of the packinghouses—earned 45 cents an hour in 1900, 5 cents more than in the 1880's. At the other end of the scale, laborers received 16½ cents an hour, ½ cent less than in 1890. These rates were not out of line with comparable occupations in the north central states. But short time shrank weekly earnings. Splitters in one Chicago plant averaged $16.15 a week in 1900. Unskilled men at 16½ cents an hour made $6.27 a week.[14] In addition, a segment of the work force—one fifth in 1910—suffered from seasonal unemployment each year. This bottom group the packers considered casual labor employed on a temporary basis.[15]

The low earnings for common labor tied into the changing character of packinghouse labor. Bohemians, Slovaks, and Poles jumped from 17½ per cent to over 40 per cent of the labor force of one major company in ten years after 1896. The Immigration Commission found that in 1909 eastern Europeans made up 43 per cent of the labor force in the packing industry in the four major western centers. The immigrants flooded the labor market and, for a variety of reasons, willingly accepted the common-labor earnings in packing. Simultaneously, an increasing number of women found a place in the packinghouses at wages well below the unskilled male rate. They increased from 2 per cent of the total labor force in 1890, to 4 in 1900, to 6 in 1905; in Chicago, 11.1 per cent were women in 1905.[16]

The labor policies of the major packers did not extend beyond their cost sheets. Plant and living conditions received practically no attention. According to government investigators, women worked in chilled rooms "without any ventilation whatever, depending entirely upon artificial light. The floors were wet and soggy . . . In a few cases even drippings from the refrigerator rooms above trickled through the ceilings upon the heads of the workers." Toilet facilities were primitive, dressing rooms and cafeterias nonexistent. Men ate at their work benches or hurried over to the saloons at "Whiskey Point" next to the Chicago yards. "The whole situation as we

saw it in these huge establishments," concluded the inspectors, "tends necessarily and inevitably to the moral degradation of thousands of workers . . .
under conditions that are entirely unnecessary and unpardonable." These
pervading hardships contributed to the poor health of the packinghouse workers. Rheumatism was widespread. Tuberculosis and pneumonia accounted for
20 per cent of the deaths in the Chicago stockyards district from 1894 to 1900.
The packinghouses were also dangerous places. Thirteen men died in the
Swift plant in Chicago from June 1907 to December 1910. No safety programs existed in the early period, nor any plant hospital in the Chicago
yards until Swift & Co. opened its medical department in 1900.[17] Occasional acts of kindness (or paternalism) did occur in the huge plants of
western centers. At Kansas City, for instance, Schwarzschild and Sulzberger in 1900 agreed to pay employees, instead of by check, in nonnegoti-
able receipts drawn on a nearby bank which would stay open on pay night,
thus freeing the men from the tyranny of the saloonkeeper. Armour & Co.
gave its employees turkeys for Christmas in 1901. Gustavus Swift urged
employees to buy stock in the company.[18] Beyond such insignificant acts,
however, the major packers left their employees to their own devices.

The fact was that the early years provided small opportunity for any
other policy. The packers were engaged in an intense drive for economy.
The average killing and dressing labor costs of three major firms in 1903
indicated the measure of success here: 61.49 cents per head of cattle.[19] The
industry was also fighting for acceptance by the nation's consumers, and
the leading firms were undergoing enormous expansion. Swift & Co., for
instance, grew from a single plant with 1600 hands in 1886 to seven giant
plants with a work force of over 23,000 in 1903. The struggle for empire left
little room for the problems of the workmen.

The concentration of slaughtering operations in the western centers precipitated a quieter revolution in the towns and cities of the country. The
local meat trade never entirely lost its slaughtering function. Home-killed
meat retained some importance in small towns which drew a share of their
supply from nearby livestock sources. City firms continued to cater to the
upper-class and Kosher trade; the four largest East Coast cities received
over a million head of live cattle in 1900—approximately a fourth of the
number slaughtered in the five chief western centers. Nevertheless, local
slaughtering rapidly declined in importance with the rise of the western
industry.[20] The local meat trade handled wholesale distribution, which was
engrossed mainly by the packers' branch houses and "peddler" cars. Some

intermediate processes—for instance, sausage making or hotel supply—still had local importance. But the major function remaining in the local meat industry involved the retail trade.

During the era of local slaughtering, retailing had not been sharply divided from the killing and wholesaling stages of the meat trade. In larger cities, it was true, there was an extensive marginal group of small neighborhood markets. But ordinarily all the stages of the meat trade, sometimes including sausage making and rendering, were united in a single enterprise. The retail trade was largely in the hands of firms which slaughtered their own cattle, either on the premises or in public abattoirs. A Des Moines retailer turned down an Armour salesman because "the sort of meat he said he would give I cannot tell anything about. I want to see the meat alive, anyway." That, butchers felt, was the proper way of doing business. "I sell no meat unless I see it killed," asserted a Pittsburgh butcher.[21]

Cheap western beef, however, swiftly undermined that upright position. The butchering side of the meat business became unprofitable. A Kansas City dealer was asked by a Congressional committee in 1889:

Q. You used to kill your own beef cattle? *A.* Yes, sir.
Q. Now you buy from the dressed beef men? *A.* Yes, sir; because I can buy it cheaper from them.

Similarly, the view of a New Hampshire dealer ("one of the old time retail butchers of Keene [who] has slaughtered his own stock for the past 27 years"):

In New England Western dressed beef can be handled more profitably than local slaughtered stuff. I figure that my own cattle stand me from 4 to 4½ cents dressed weight, including tallow and hide. I can have Western beef put into my box at from 5 to 5½ cents. That is a loss rather than a profit in local slaughtering. I have patrons who want native cattle, however, and I slaughter to satisfy the demand.[22]

Some meat men narrowed their operation to wholesaling, usually entering the employ or, especially in the early period, partnership with a western packer. Others in the local trade became retailers. One consequence of these shifts was the clear separation of retailing from other stages of the trade. Marketmen developed bitter opposition to wholesalers and slaughterers who persisted in selling to consumers.[23] The line between wholesale and retail, hitherto unclear, now became sharply drawn.

Western beef tended to broaden the access to the retail meat business. Earlier, the trade had been engrossed by men of experience and some capital.

There had been among butchers, particularly the Germans, a strong craft consciousness, reflected in such organizations as the Bloomingdale Germania Butcher Guard No. 1 of New York or the Butchers' Union No. 1 of Louisville. Annual butchers' parades and drills of the trade had been held, followed by balls at which the craftsmen "shook legs until daylight." The master butchers had always been substantial businessmen in the community. The divorce of slaughtering and retailing opened the trade to new men. "There are a good many shinners in the business," grumbled one Des Moines butcher in 1889, "but there are not many butchers . . . I do not call a man a butcher who will start a meat market. Most of them just buy their meat here . . . But for regular butchers there are not more than ten or a dozen in the town, men I call butchers, who served their time to it." [24] Entrance into the field was also encouraged by the widening of the market and the feasibility of small-scale operation. The packers, anxious to expand their business, drummed up trade among groups who had earlier not been regular consumers of fresh meat. At the same time, shopkeepers were able to buy dressed meat in small quantities. The local butcher business increasingly became another form of small-shop retailing, carried on frequently in combination with a grocery.

The first thorough investigation of the retail meat trade was undertaken by the Department of Agriculture in 1919. The most striking fact was the proliferation of stores: one per 821 inhabitants in the 33 urban districts surveyed. The ratio varied widely: Terre Haute (population 66,083) had a store for every 444 people; Eau Claire, Wisconsin (population 20,906) had only ten shops, one for 2091 people. There were relatively few substantial businesses. In seven completely canvassed cities, little more than a tenth of the meat markets had an annual sale of $50,000 or more. Half sold under $20,000 a year, a volume which could sustain only a one-man operation. Meat-grocery combinations had become very widespread, predominating particularly in New England.[25] The meat industry revolution, concentrating slaughtering operations, had the opposite effect at the local level.

Functioning in a crowded and competitive field, the retailers tried to mitigate their individual weaknesses by common action. Protective associations attempted to prevent price-cutting, to eliminate "dead beats," and to fight the trading-stamp evil. They also sought to exclude competition from department stores ("an institution that threatens to rob every man, young or old, of the chance . . . the Constitution grants and guarantees, that of successfully entering into business on one's own account"), meat peddlers, packers, and wholesalers. Some associations pooled resources to establish

rendering and icemaking plants—for example, in New York City, Rochester, and Watertown, New York.[26] These actions could bring only slight
leverage against the broader economic forces confronting the marketmen.
Many, however, secured prosperity by building up a clientele through personal attention, credit, and delivery service. (Only 25 per cent of the meat
markets in the 1919 survey were cash-and-carry.) The retail butcher,
handicapped in the economic struggle, often survived because of his ability
to serve his customers.

But the basic condition for success was a tight operation. The margin
between profit and loss was always narrow. According to the accounts of
206 meat markets for 1919, average net profits were 2.32 per cent of sales.
The cost of meat alone came to 80.95 per cent of receipts. From the remainder, the proprietor had to cover the charges of rent, wages, advertis-
ing, wrapping, interest, and insurance. His return was determined by the
extent to which these were trimmed down. "Running expenses," the *National Provisioner* told its retail audience repeatedly, "are the small eaters
which nibble off the margin of profit and often put the shop balance on the
other side." The successful butcher shop could be, in its fashion, as much a
model of economy as the great packinghouse.[27]

The retail development fixed the labor situation at the local level of the
meat trade. Old-style butchers continued to find a place wherever local
slaughterhouses persisted, although the larger of these began to adopt the
methods of the western plants. Still, the cattle and sheep butchers in New
York houses remained closer to journeymen butchers than did the specialists in the Chicago packinghouses. The great local demand, however,
was also for specialists—that is, meat cutters for the retail markets. As a
"Veteran Butcher" pointed out, "we do not need butchers any more, for
in our day the benchman need not know how to kill and dress." The division of labor was thus extended to the retail field, transforming the all-
round butcher into a meat cutter. This narrower calling should have been
equally demanding. "It takes years of study and practical experience," wrote
a California workman. And the meat cutter also "must be a gentleman and
must possess the patience of Job . . . and also present a neat, tidy appearance, he must be fairly educated . . . a good salesman." [28] People to fit this
description were always at a premium.

But, in fact, a noticeable decline occurred in the level of labor in the retail field. Marketmen repeatedly complained of the dearth of "good, eco-
nomical meat cutters." Technical competence, employers claimed, lessened
as butchering became exclusively retail work. The fault rested largely with

the entrance of inexperienced proprietors and grocers into the meat business. Shopkeepers, observed the *National Provisioner,* were likely to hire help "of a cheap order," assuming that an ordinary clerk could serve behind the meat counter. The consequence was an abundance "of hackers and sawyers masquerading under the title of experienced meat cutters," but few "competent and trustworthy" men. Many master butchers found it necessary to get "a likely boy" and teach him the trade. Thus, the character of the retail labor force was highly mixed in the early years, ranging from mere clerks to skilled cutters.[29]

One result was a diverse wage pattern. This was also fostered by the economics of meat retailing. The proprietor necessarily paid close attention to the wage bill because it was much the largest of his operating expenses —generally, two thirds—and he worked on a narrow margin. "Everything should therefore be done to economize in this respect and to make every dollar paid for wages produce an adequate return," a market expert told retail butchers. "It is a mistaken policy to pay the same amount to clerks of widely different experience and training . . . Successful merchants . . . do not pay their help the same rate regardless of their help to the business." Wages in St. Paul, the Minnesota Bureau of Labor reported in 1904, ranged from $9.60 to $18.00 a week, averaging $14.08. The going rate for competent meat cutters in California was $2.50 a day in 1902.[30] The shop butcher, while not a highly paid artisan, received an adequate wage and, in view of the steady work, annual earnings as high as the skilled men in the packinghouses.

The work schedule raised greater problems. The meat cutter's day began before dawn and ended at seven or later in the evening. Markets were generally open until midnight on Saturday and again on Sunday morning. One proprietor estimated that his employees averaged 88 hours a week. This schedule was standard near the turn of the century. The appearance of reform sentiment gave some hope for relief in the later 1890's. Proprietors, no less than their men, were victims of the unending hours. Sunday and early closing movements started in many places. Chicago marketmen, for instance, posted this notice in January 1900:

We, the undersigned grocers and butchers of the Northwest Side, have agreed to close our places of business on Sunday and ask your cooperation, as we believe we should have at least one day of rest out of the week.[31]

Such agreements were among the chief objectives of boss butchers' associations. (The Buffalo body was called the Retail Butchers' Early Closing and

Business Association.) Local ordinances and state laws, hitherto ignored, were now enforced. Still, the initial gains were modest. The reduction of hours was small: generally, early and Sunday closing did not come together, and, even in the optimum case, the work week would still be at least 70 hours. Moreover, the gains were not always, or even usually, permanent. In a competitive situation, recalcitrants or backsliders forced a general resumption of earlier conditions in many places. Most meat cutters continued to have inordinately long hours.

The country received an efficient system of meat production and distribution from the industrial transformation. But the impact on the butcher workmen seemed less beneficial. Old ways were disrupted, and new and disturbing terms of employment introduced. It was inescapable that one labor response should take the form of union organization.

CHAPTER 2

THE UNION IMPULSE

Trade unionism flickered and grew in the formative period of the modern meat industry. Concrete results were small; labor organization counted for next to nothing in the industry before 1900. Yet the record of false starts and defeats revealed the drift toward collective action by the butcher workmen. And there also emerged at that time the key instrument for unionization. A national butchers' union was established: The Amalgamated Meat Cutters and Butcher Workmen of North America.

The rising sense of grievance could not be mistaken in meat packing. Labor troubles plagued the western stockyards. During the eight-hour campaign of 1886, the unrest broke into the open. Without organization or outside leadership, the Chicago packinghouse workers on May 3, 1886 answered the call for a general strike by the Federation of Organized Trades and Labor Unions. When the packers restored the ten-hour day in the autumn, two more strikes occurred in the Chicago yards, both rank and file in character. The men, it was true, had by then formed unions under the Knights of Labor, but the strike decisions on October 8 and again in early November came directly from the packinghouse workers. They seemed to T. P. Barry, the Knights' official in charge, "a thick-headed, hot-headed mob." Involving over 9000 men, the Chicago strikes failed because of the packers' resistance and the ineptitude of the Knights' officialdom.[1] The mass uprising, short-lived as it was, exposed the latent pressures at work in the stockyards.

The second round of major strife came in the summer of 1894. By then, the discontent had deepened under the impact of continuing industry changes and of the depression. The Pullman boycott was collapsing in the second week of July when the American Railway Union made its desperation call for a sympathetic strike. The Chicago butchers walked out, again spontaneously, followed after a time by the men in other major centers (except Kansas City). The strike was fiercely and stubbornly waged and, unlike that of 1886, was attended by considerable violence. Troops had to be brought into Chicago and South Omaha. The overt issue involved the restoration of wage rates prevailing before the panic of 1893, but the spontaneity and dimensions of the strike clearly expressed deeper grievances.[2]

A series of lesser strikes punctuated the 1890's. Chicago sheep butchers struck unsuccessfully in November to prevent the introduction of the ring system, which would permit a division of labor in sheep killing. Generally, however, the strikes were touched off by wage or hour cuts. Reductions in 1893 caused walkouts in Kansas City, and to a lesser degree in Hammond and East St. Louis. Kansas City was again the scene of disturbances in 1896. Chicago hog butchers struck in November 1896 because the winter rate during the peak of the livestock flow, customarily an increase of a dollar a day, was only 75 cents that year.[3] Unrest in 1899, occasioned by the return of prosperity, resulted in much trouble in the Chicago stockyards.

Labor organization also had an erratic start in the packinghouses. There had been some ephemeral unions in the 1870's. John T. Joyce, a unionist whose long career spanned the entire history of modern meat packing, participated in a Knights of Labor local assembly of mostly packinghouse workers in 1881, and was among those trying to form an organization of cattle butchers in the Swift plant in November 1884 after a cut of 50 cents a day. Significant unions materialized in later years in the wake of strikes. The eight-hour struggle in the Chicago yards, for example, led to the formation of Knights of Labor assemblies. "Without organization the men had gained eight hours," observed the local Knights' journal, "and they had organized for the purpose of perpetuating the advantage." Butchers' unions similarly emerged from the 1894 troubles, as well as from minor strikes. These unions did not as a rule long outlast the crises from which they sprang, although some vestiges remained. After the 1886 strike an informal group of cattle butchers, calling themselves the Blackthorn Club, continued to exist in the Armour plant in Chicago, and activists tended to keep in contact and to form the nuclei of later union attempts. But very little progress was made toward permanent organization.[4]

The rapid expansion of the industry hindered this development. Plants were being erected at Kansas City, East St. Louis, South Omaha, and elsewhere during the late 1880's and the 1890's. Either voluntarily or by transfer, packinghouse workers moved to the new centers in considerable number. Mobility was not merely geographical. It was significant that some of the early strike leaders eventually became company officials or became successful in other fields. Slavic immigrants and women were meanwhile finding a larger place in the plants. The labor force lacked the stability that was a prerequisite for labor organization.

The narrow division of labor provided another obstacle. The strategic position of the skilled butchers was being undermined. The pork departments had always required a lesser degree of skill and had been under centralized operation for a long time. And here employers exhibited least restraint. The return to ten hours in October 1886 affected only the pork workers, not the Armour beef department until its men came out in sympathy, nor Swift and Morris (then primarily beef companies) until after the failure of the first strike.[5] By the 1890's, however, the cattle operations had also been drastically altered by the division of labor. That was one lesson of the 1894 fight. The skilled men believed that replacements would not be found, but some expert butchers were brought in from branch houses, and other men were hired from the army of unemployed. The *National Provisioner* reported in mid-August that the Chicago houses were "doing more killing than at this time last year, and that the new men are readily becoming efficient." The superintendent of the Swift plant at South Omaha assured the defeated strikers, when they demanded the dismissal of the strikebreakers, that he could fill every man's place in the plant within three days.[6] The assumption of a scarcity of skills, essential for craft unionism, was becoming obsolete in the packing industry.

Yet the craft instinct remained strong among the butchers. They invariably unionized along narrow craft lines in the 1880's and 1890's. Eight or more Knights of Labor assemblies existed in the Chicago yards in 1886, interplant in scope, including Cattle Butchers (No. 7802), Sheep Butchers (No. 8332), Hog Butchers (No. 8271), and Beef Carriers (No. 8738). As the minutes of the cattle butchers' assembly revealed, the cleavage was deep between skilled men and common labor. Similar divisions also existed in the packinghouse unions of the 1890's. The bylaws of the Sheep Butchers Protective Union of New York City, for instance, limited membership to "Practical Sheep Butchers, who shall be vouched for by three members in good standing." These constituted an examining board before which candidates had to demonstrate their ability; ten negative votes from the total membership sufficed for rejection. Butchers in the western centers shared this restrictive view. "We do not take helpers in our Union and we are preventing them from learning," wrote a St. Joseph man. "There is [sic] enough cattle butchers in the field now . . . and by not breaking in any more, so much the better for ourselves. We do not take in every Tom, Dick and Harry." [7] As the division of labor proceeded, this exclusiveness contributed heavily to the undermining of early packinghouse organization. [8]

The employers also were an obstacle. Trade unions unquestionably disrupted the employee relations desired by the packers. During the 1896 presidential campaign, for instance, a Kansas City federal labor union of packinghouse workers denounced the Armour Packing Company for its "outrageous assumption of authority, subversive of the rights of citizenship . . . by intimidating voters and asking them to violate the secrecy of the ballot by declaring for whom they intended to vote." Gustavus Swift, who put a high premium on worker loyalty, closely examined his foremen after the 1886 troubles to determine the reasons for their inability to hold their men. "When he had finished," his son Louis F. Swift later recalled, "they had a new comprehension of the value of working with men in the way that wins loyalty." During the November 1886 strike, one packer asserted that he would rehire "only such men as recognize the fact that their first allegiance is to us, their employers, and not to any system of tyranny in the guise of a labor organization." Besides undermining loyalty and discipline, unions also tended to infringe on managerial rights. The cattle butchers' union in Kansas City, for instance, demanded in 1893 the dismissal of several nonunion butchers in the Armour plant. The union struck on meeting refusal, as it did again several months later because five steady-time butchers were not included in general wage cuts. [9] These invasions of their authority, jealously held, necessarily imbued the packers with an underlying bias against labor unions.

No rigid hostility, however, existed in the early period of the industry. It was true that in the heat of the 1886 controversy the Packers' Association resolved to hire only nonunion men and to employ yellow-dog contracts. The resolution, however, was withdrawn after a few days, with the proviso: "we will exercise the right to employ and discharge whom we please, and conduct the business . . . according to our best interest." Thereafter, the packers pursued a pragmatic policy on organized labor. The biographers of P. D. Armour put the matter clearly: "He preferred to deal in a free labor market, but he was not an open shop fanatic and when he had to do business with labor unions he knew how to do it." [10]

The economics of meat packing, particularly in the formative period, provided strong reasons for avoiding labor troubles. The packers accepted unions where that seemed the more profitable course. Unions of coopers, machinists, stationary engineers, and other auxiliary mechanics were at times firmly established in the stockyards, and the packers also were willing on occasion to deal with organizations of butcher workmen. In the summer of 1886, for instance, they offered a one-year contract on the basis of nine hours' pay (instead of ten) for eight hours' work. Again, they moved prudently when strong organization developed in Kansas City in the mid-1890's. The unions acted in concert through the Industrial Council (the central labor body), and made effective use of the boycott weapon. Consequently, when a strike of female meat trimmers broke out at the Swift plant, company officials met with a committee from the Anchor Federal Labor Union (covering the unskilled packinghouse workers) and signed an agreement revoking the wage cut and paying the women twenty dollars each for lost time. Concessions ended other disputes, including an agreement by Armour officials with Beef Boners' Union 6151 not only to pay Chicago rates, but also to give union members "the preference." Plant representatives, urging their friendly intentions, assured the Kansas City unions of their willingness to consider reasonable grievances and requests.[11]

The balance between tolerance and opposition, however, almost invariably tipped over to opposition. For, from the packers' standpoint, the unions quickly became unreasonable. The turning point at Kansas City, for instance, came in May 1896 when Stationary Firemen's Union 6406 demanded an eight-hour day at $2.00 instead of twelve hours at $2.25. Led by the Armour Packing Company, the local companies refused. "Every packer realizes if the firemen win," explained the *National Provisioner*, "that labor of all descriptions will have demands, that the now tiny rivulet will become a flood." The Industrial Council ordered a boycott, and Armour retaliated with the dismissal of 1100 union employees and an anti-boycott injunction. The controversy ended in a compromise after two weeks. Thereafter Armour began quietly to eliminate the unionists. By August, the Industrial Council concluded that the plot to "disunionize" all the Kansas City plants "is being secretly and persistently accomplished." When the council resumed the boycott, endorsed this time by the AFL Executive Council, the break with Armour Packing Company was complete.[12] Each confrontation with labor unions in the early period led the packers to the same ultimate decision: that it was cheaper to fight than to appease.

Ample means were at hand to combat unionism. The packers commanded vast financial and physical resources. They were, moreover, united on labor matters. The 1886 troubles, in which Swift had played a lone hand outside the Packers' Association, had revealed the advantages of concerted action. Nor did the companies hesitate to employ their great power. Strikebreakers and

Pinkertons flooded the stockyards during labor troubles. The blacklist was openly used; after every conflict the activists found themselves shut out of the industry. Libby, McNeill and Libby, sued by a number of canning department girls for damages caused by an alleged blacklist, did not bother to deny the charge. The company simply argued that the blacklist was a legal device, and the court upheld its position. Mary McDowell, coming to the "back of the Yards" in 1894 to head the University of Chicago Settlement, saw the consequences of company repression. After the 1894 strike, she recalled, "the community was left . . . without courage or self-confidence. At my invitation to discuss social questions, they would invariably answer, 'We dare not; we would lose our jobs.' They seemed to us unmanly and without self-respect." [13] The unequal struggle had apparently subdued the packinghouse workers.

The union cause thus made little headway in the expanding packing centers. Mass movements there had been in moments of crisis, but there had evolved little permanent organization. A few small Knights of Labor assemblies had a fairly continuous existence in the stockyards of Chicago and Kansas City during the 1880's. Starting with Gompers' visit to Kansas City in 1889, AFL packinghouse locals there were in regular operation, except during the hard times of 1893-4. AFL unions of butcher workmen had come and gone in the other centers, except Chicago, where the unions had been independent. At the close of 1896, little more than 500 paid-up members belonged to the AFL packinghouse unions, plus an undetermined number in federal labor unions.[14] Meat packing was virtually unorganized.

The drift toward collective action took an entirely different turn in the retail trade. Here there were no explosive strikes and sudden spurts of union activity. Only in rare instances did discontent over conditions lead spontaneously to organization, for instance, in the Kosher butcher shops of New York's East Side in 1900. "These poor devils," reported a trade journal, were agitating for *a reduction to fifteen hours a day*.[15] Notwithstanding the general complaint about lengthy hours, the terms of employment did not normally lead to union activity in America's butcher shops.

The absence of a strong union impulse had more fundamental causes than accommodation to working conditions. The lives of meat cutters were spent in an immediate, and often close, relationship with the proprietor, who not uncommonly also boarded his single men. Contact with fellow workmen, except in large markets, was on the other hand minimal. The meat cutter also understood the problems of the business as packinghouse workers could not. (He had, for instance, to know the cost of meat and

gross profit margin in order to cut properly.) Nor did shop butchers perceive, as one Chicago workman complained, "the true principles of their
Class and Unionism." More often, indeed, they were incipient capitalists,
holding a reasonable expectation of eventual proprietorship. A faithful employee could hope, after due course, to be taken into the business. This was
sometimes the only way of holding a valuable man, for "as soon as meat
cutters . . . become useful, they usually take their services elsewhere, or if
ambitious and with sufficient money, start in business on their own account." The frequency of the latter course was evident in the personal
columns of the trade journals. Butcher clerks, when asked by interviewers
what they would do if they had $500, made clear their ambitions toward
proprietorship.[16] Finally, craft consciousness appeared weak: for one thing,
the labor force was diluted by the influx of inexperienced men, and, in addition, meat cutting was a recent occupation and in a new environment.
Trade unionism did not spring naturally from the situation in butcher shops.

Organizations of another sort did emerge spontaneously to meet mutual
protective and social needs, and also sometimes as a counterpart to employer
bodies. Local employee associations appeared in many places, for example,
the Provision Clerks' Benefit Association of New Bedford or the Shop
Butchers' Benevolent Association of Jersey City. Like their employers, the
shopmen held social functions; but benefit provisions were more important.
The Chicago Grocers' and Butchers' Clerks, organized in 1886, paid five
dollars a week for twenty weeks of illness and a hundred dollars for fu-
nerals.[17] The benefit associations were also drawn into the agitation for
shorter hours. In the later period this concern tended to stimulate employee
organization, as it did also among the proprietors. Sometimes, employer and
employee bodies were allies in the reform effort. The Retail Butchers' and
Grocers' Association of Chicago urged the clerks' organization to put pressure on shopkeepers who resisted the employers' Sunday closing campaign
in 1900.[18]

The protective associations were not trade unions. Their function was
not to negotiate with employers nor to employ economic pressure. Thus the
Chicago Grocers' and Butchers' Clerks' Association: "We do not believe in
strikes, but think a man should be paid wages according to his worth to his
employers." When they did edge into the area of working conditions,
notably hours, the associations resorted to petition, not negotiation. They
ordinarily were considered useful institutions by employers. The Benchmen's Association of New York always had the editorial support of the trade
journals, and officials of employers' bodies attended its conventions; several

proprietors, in fact, were present at the organizing meeting of the Essex County branch.[19]

Trade unionists saw the Benchmen's Association as a "bosses' organization," designed to divert the men from true unionism.[20] The fact was, however, that the pacific, noneconomic character of the protective associations was shared somewhat by the butchers' unions. Social and beneficial motives played a big part in forming trade unions; the evidence of early butchers' unions, indeed, consists chiefly of reports of their picnics and socials in trade journals. Nor was the distinction between protective associations and trade unions very pronounced. Had the former assumed bargaining functions, they would have been indistinguishable from contemporary butchers' unions. This move, however, appeared to be a rare phenomenon. Early butchers' unions sprang from another, an external, source.

The broader labor movement was the major stimulus of the occasional unions in the retail trade. The Knights of Labor organized butchers into separate local assemblies and, in smaller towns, into mixed assemblies. Butchers' Assembly 6341 in Washington, D.C. was still flourishing in 1895. The United Hebrew Trades sponsored a Kosher butchers' union in New York in 1895, and again at later dates. The greatest impetus came from the American Federation of Labor. Despite the lack of resources, the AFL operated as the organizing agency in fields outside the jurisdiction of existing internationals. Among the first local AFL trade and federal unions were some journeymen butchers' unions. Two were recorded in April and May 1888, and by the end of 1892 there were twelve on the AFL books.[21]

These pioneer unions appeared generally under certain favoring conditions. Foremost was the existence of a vigorous local movement. The city central bodies in such localities as Syracuse and Duluth played a crucial role, providing leadership, advice and meeting halls for a new union. Equally important, they gave an initial leverage against the proprietors. For the good opinion of the labor movement was highly valued by tradesmen in workingmen's communities. When, for example, the Trades and Labor Assembly of Akron in 1887 requested the meat dealers not to buy from Armour, one proprietor explained, "we were satisfied that all those who bought it would be boycotted . . . Our city being made up mostly of the laboring class and they being well organized, we concluded it was best for us to discontinue the use of Armour's meats." [22] Ethnic ties were also important, particularly because there were among the immigrant arrivals many skilled butchers. Unions of German, Bohemian, and Hebrew butchers existed in Chicago and New York. Finally, unions tended to appear where there remained a strong craft consciousness among shop butchers. It was significant that the early trade organizations were called "journeymen" butchers' unions.

Initial employer-union relations were as a rule amicable. Proprietors often failed to perceive the consequences of trade unionism and sometimes, even the division between labor and capital. A representative of the National Retail Butchers' Protective Association at one point actually queried Gompers about the possibility of affiliating with the AFL.[23] Butchers' unions did not immediately excite employer antagonism; for hours, not wages or conditions, were the first subject of negotiation. The unions sometimes succeeded where reforming employers had failed. A California trade journal reported: "In our opinion, the best and most effective way to compel Sunday closing—do as was done in San Francisco—hand it up to the Labor Union. What agitation and intimidation failed to do, the Butchers' Labor Union accomplished." Meat cutters' unions, unlike employee beneficial associations, had the ability to force recalcitrants into line. The Lewis Brothers Meat Market in Watertown, New York, was boycotted by the trades assembly in 1895 for refusing to close at 8:00 P.M. and to affiliate with the proprietors' association. The Utica butchers' union assisted the employers by visiting the mayor to protest retailing of meat by wholesalers. Of course, strained relations did develop on occasion. Toledo meat dealers, who favored long hours, agreed to Sunday closing in 1893 only after the union threatened to strike. Unions became an embarrassment during the frequent abandonments of shorter-hour movements. The Butchers' and Sausage Makers' Union of St. Louis waged a long and bitter fight against the firm of Anton Loux for refusing to sign its contract. But amity generally reigned. Employer resistance was not the obstacle to the organization of the meat cutters.[24]

Few localities were permanently organized before 1900. Only one or two of the dozen butchers' unions affiliated with the AFL in 1892 survived the panic of 1893. At the end of 1896, only Syracuse, Utica, Wheeling, Duluth, and a few other places had butchers' unions. For quite different reasons, the retail workers of the meat industry succeeded no better than did their packinghouse brothers. "Probably there is no class of wage earners that has been as slow in the matter of organizing as the butcher workmen," concluded one of their early union leaders in 1897.[25] The formidable task awaited the formation of a national union.

In December 1896, the American Federation of Labor had on its rolls five retail butchers' locals, three packinghouse locals in Kansas City (two others were listed but were in fact extinct), two in New York City, one in

Hammond, Indiana, and another in Marine, Texas. From these few scattered organizations emerged a national union. The formation of the Amalgamated Meat Cutters and Butcher Workmen of North America marked the real start of continuous labor organization in the meat industry. For the national body was the creator, not the consequence, of unionism among butcher workmen.

The immediate impulse came from the AFL. The executive council had been directed back in 1886 "to organize local Trade Unions and connect them with the Federation, until such time as there are a sufficient number to form a National or International Union, when it shall be the duty of the President of the Federation to see that such organization is formed." Gompers fulfilled these instructions boldly. He felt that "it was better to have the National Union started and set on its way to progress, though weak at first, than to have a fragmentary number of local unions . . . without any central head or common concert of action among them." A number of national (actually, international) unions were thus begun on Gompers' initiative, usually through the device of a conference coinciding with an AFL convention or executive council meeting.[26] In the late autumn of 1896, letters went out to the butchers' unions to send delegates to the forthcoming annual convention at Cincinnati in December.

Four men appeared: Homer D. Call of Meat Cutters' Union 5969, Syracuse, New York; George Byer of Sheep Butchers' Union 6496, Kansas City; and John F. Hart of Butcher Workmen's Union 6598 of Utica, New York. (Delegates from bicycle workers' unions were attending the convention for the same purpose.) John F. O'Sullivan, a union official from Boston, was assigned to advise the butchers. The group met in the anteroom of the Odd Fellows' Hall, the site of the convention, to consider the advisability of forming a national butchers' union.

The need seemed clear. Industrial changes had rendered local unionism entirely ineffectual in meat packing. Labor organization there had to cover every center and, given that, had to act in concert. The national scope of the dressed-meat market, for one thing, made wage rates interdependent throughout the industry. In 1886 Philip D. Armour enunciated the logic: ten hours' pay for eight hours' work was impossible while Cincinnati and Kansas City remained on the ten-hour day. The skilled labor market was also industrywide. Particularly during hard times, workmen were readily available from other centers (as employers invariably pointed out) who would be glad to work for less. By the early 1890's, finally, all the major packers were engaged in multi-plant operations. In the event of a strike, key men could be imported from other plants, as was done in Chicago in 1894. Or, alternately, the packer could expand production at other houses, as G. H. Hammond did in 1893 when its Hammond, Indiana, plant was struck. A local organization was consequently helpless: "You may close down their business in any one locality and they simply divert that business to other localities and move right along, without a ripple." [27]

The Chicago cattle butchers had tried to rectify this weakness when they struck in the summer of 1894. They secured a charter from the State of Illinois establishing the Journeyman Butchers' National Union of America with John C. Taylor as president. Two-man delegations were dispatched from the executive board to spread the strike to other centers, succeeding in South Omaha and East St. Louis. But the strike failed, and the organizations disappeared. Nor does it appear that, aside from the coordination of simultaneous strikes, the national union actually established any authority and machinery. Nevertheless, the necessity was plain.[28]

A national union had a weaker imperative in the retail trade. Competition here was entirely local. The labor market was somewhat wider, and meat cutters' unions were sometimes troubled by job-seeking strangers. Yet the threat was comparatively mild. Before 1900, meat cutters' unions did not ordinarily negotiate wage scales. Nor were there frequent strikes to give opportunities to outsiders. A sentiment nevertheless existed favorable to a national union in the local trade. The Duluth local, for instance, had asked the AFL in 1891 for the location of all affiliated butchers' unions because it planned a convention to form a national union. In 1893, again, there were reports of an effort by the butchers of Toledo, New York City, and Milwaukee to start a "joint organization." [29] Although unsuccessful, these attempts indicated the inclination of meat cutters and others in the local trade.

Their impulse toward national organization arose, in part, from the general trend in this direction. If other trades, why not butchers? National affiliation would also provide the benefits of financial aid and advice for the isolated locals. There was another, subtler influence at work. The local trade seemed to be in the clutches of the Meat Trust. Men still engaged in local slaughtering operations, of course, felt in a very direct way the impact of the western industry. But meat cutters also seemed injured, hearing as they did the complaints of their employers.

What is true of the slaughterhouse butcher is also true of the meat cutter, for the retail meat dealer of today exists only at the sufferance of the packers and the wholesale price is always kept so close to the retail price there is only a meager living in the profits therefrom.

A strong national union would somehow counteract the forces, national in scope, controlling the local trade. "This matter of an international organization for our craft had been on my heart for a long time," explained the meat cutter Homer Call in 1899, "for I could see how rapidly was the downward course of all butcher workmen. I was confident that if our craft should ever be recognized other than the tools of a great monopoly it must be by united organization." [30]

This view supported not only the desire for a national meat cutters' union, but also amalgamation with the packinghouse workers. "This, I knew, would give strength in time of difficulty," asserted Call. The precise benefits at the retail end were unspecified. There was, Call himself admitted, "a great difference of opinions as to what extent amalgamation should imply. This is a matter of education in studying the avenues of strength." The advantages were, however, abundantly clear to packinghouse men. The refusal of meat cutters to handle the product of "unfair" firms would be a powerful weapon indeed. (A consumer boycott against the Armour Packing Company in 1896–7 was proving a sad disappointment.) When union butchers in Toledo in 1893 boycotted meat from Detroit packinghouses, the result was demonstrably effective in Detroit and, in fact, stimulated talk of a national union with Toledo and Detroit as a nucleus.[31]

Other considerations also dictated amalgamation. Homer Call explained "that fully 60 per cent of the meat cutters are also skilled slaughter house butchers, and in many localities work a part of the day in the slaughter house and the other part in the meat market, making it impossible to draw a jurisdictional line between the two." [32] This view, obsolete though it was, nevertheless carried great weight. There existed a keen sense of community among "all men . . . who use a knife." In packinghouse or retail market, they still worked at the butcher trade, or, as the union's journal later put it, in "branches of the craft."

The four butcher delegates at the Cincinnati AFL convention consequently agreed on the need for a national union with a jurisdiction over both the retail and meat packing branches of the industry. The way would be hard. There were only a few hundred organized butcher workmen, the delegates discovered, and great obstacles to growth. "It was a question whether it was advisable to make the attempt," remarked Homer Call afterward, "but wiser counsels prevailed . . . No great reform was ever accomplished without a commencement." [33]

The four then turned to the business at hand. Under the guidance of the experienced unionist John O'Sullivan, they sketched out a constitution. They next proceeded to elect themselves to the offices of the new organization. George Byer became president, John Hart and W. H. Schwartz vice-presidents, and Homer Call secretary-treasurer. The problem of funds for securing the charter was overcome when Call agreed to put up the necessary ten dollars. President Gompers, on receiving the application, answered on January 2, 1897, "you can organize your national union in the same form as all other national unions affiliated with the AFL." Because the federation was in the midst of moving its offices from Indianapolis to Washington, D.C., a month passed before the arrival of the charter at the Syracuse headquarters (that is, Call's residence).[34]

The Amalgamated Meat Cutters and Butcher Workmen of North America came into formal existence on January 26, 1897. It asserted:

supreme jurisdiction over the United States in which there are at present, or may be hereafter, subordinate Unions located, and . . . the highest authority of the Order within its jurisdiction; and without its sanction no union can exist, or any scale of prices be recognized without the sanction of the Executive Board of the Organization.[35]

The national union thus began with a slate of unpaid officers, a rough constitution and, insofar as the AFL could confer it, the "highest authority" over all present or future unions of butchers. The future alone would tell how far the Amalgamated Meat Cutters and Butcher Workmen would proceed beyond that bare beginning.

The first three years were decisive. Then it was that the national union accumulated its vital resources, overcame the initial dangers to its life, and clarified its internal organization.

The dominant—indeed, the only—figure of the opening years was Homer Call. He alone of the founders kept to the grim work of putting the paper union on its feet. Gompers had good reason to commend him after the first six months.

There are few organizations that have come under my notice that have started out under such adverse circumstances and have done so well in so brief a period. I am confident that this is due mainly to your efforts and unswerving devotion to duty.[36]

The other officers left no mark on the early organization (although Hart later did as the president of the Amalgamated). The first packinghouse representatives dropped from sight after short tenures. President Byer, who had been blacklisted and unable to secure a job, resigned his office on May 4, 1898 to join the police force of Kansas City, Kansas. It proved a fortunate loss. Although under the constitution First Vice-President Hart should have moved up, Secretary Call argued "that the West should furnish the President." The executive board finally decided on Michael Donnelly, a South Omaha sheep butcher who had helped Call several months earlier on the Armour boycott problem. Donnelly, informed by Call of his election during the AFL convention in Kansas City in December 1898, "hesitated, realizing the immense responsibility . . . [but] it was a duty which no honest union man, under the circumstances, could refuse to accept . . . I notified the Executive Board . . . pledging myself to use all honorable and fair means at my command to assist in placing our International organization in the front rank among our sister unions." [37] Dedicating himself to the task, Donnelly became a brilliant organizer of the packinghouse workers. The Amalgamated thus finally acquired in Donnelly and Call an effective executive combination.

The immediate task in 1897 was to create a body of constituent locals. The existing AFL unions were brought in within a few months, beginning with Call's Syracuse organization—Local 1. Others followed at intervals, in some cases with evident reluctance. Gompers had to direct some butcher unions to discontinue per capita payments to the AFL, and Call invoked resolutions requiring affiliation with national unions of the appropriate trade. Vice-President Hart's union came in only in April, followed finally by the Cleveland body in May 13, 1897 "after considerable correspondence, explanatory." [38]

To grow beyond the original twelve was more difficult. The Amalgamated lacked the resources to mount a sustained drive: its receipts for the first ten months were only $1247.96, and for the first three years $5565.30. Growth accordingly rested on volunteer effort. "I was aware that I would be obliged to do my work nearby," reported Secretary Call, "for I was desirous of accomplishing the greatest good with the least expense, starting as we did without one dollar in the treasury." Six of the eighteen new locals up to December 1897, significantly, were located in upstate New York, and at least four others elsewhere were started close to existing Amalgamated locals.

Outside assistance was also important. Secretary Call issued over 1500 letters in the first ten months, large numbers of circulars, and two appeals in the *American Federationist*. The response from other unions was discouraging: "They often would reply, help us." But AFL organizers lent valuable aid, bringing in at least four locals in the first months and a continuing stream thereafter. Federation representatives also steered new organizations through their precarious beginnings. The Nashville organizer, reporting a new local, found only one member who had had any union experience. "We will make it a point to meet with them as you request until such time as they get a good understanding of what organization is." The AFL representatives made a vital contribution in the retail field in the early years.[39]

Meat packing proved more difficult for the federation organizers, inexperienced as they were in dealing with the large, complex, and suspicious labor force in the packinghouses. They complained to the Amalgamated that in St. Louis they had "spent a great deal of time and money in trying to organize the packinghouse butchers, but had completely failed." Even after the formation of a packinghouse union, experienced leadership was vital. During the first encounter with employers, locals in Chicago and St. Joseph were immediately snuffed out, and in Indianapolis, Buffalo, and Cudahy, Wisconsin, they fought for months against the dismissal of members. The packinghouses clearly required men of special capacity and experience from within the Amalgamated.[40]

The appearance of Michael Donnelly in 1899 filled that initial need. He had already built up his own Sheep Butchers' Local 36 of South Omaha. Now, as Amalgamated president, he persuaded the three Omaha locals to donate fifty dollars to cover his expenses to Sioux City, where he "was successful beyond [his] expectations." In East St. Louis he abandoned "the common mode of procedure—that of publishing a call for a meeting of butchers." Instead he quietly visited the homes of influential workmen and built a union nucleus, from which sprang three locals.[41] It was a valuable lesson. Donnelly was hindered during 1899 by the necessity of earning a living at his trade and by the constant calls to settle grievances and disputes. Yet the Amalgamated was slowly gaining membership in the packinghouses as well as in the retail field.

Difficulties beset the first growth of the Amalgamated. There was, of course, a high mortality rate among the new locals. They were disrupted by irresponsible officers, by premature strikes, by indifference, or, as Secretary Call remarked of a Canadian union, they simply "vanished from the face of the earth." The Amalgamated also sustained serious losses among its established unions. Local 9 became involved in a controversy at the G. H. Hammond plant in Indiana immediately after affiliating on March 27, 1897. Secretary Call urged the membership to "submit for the present, rather than be thrown out of work; but . . . all acknowledged union men were discharged, so they have lost their grip." Sheep Butchers' Local 10 of New York City seceded in 1898 in protest against the per capita tax and benefit assessments. Meat Cutters' Local 15 of New York City, promising to be a strong organization, succumbed to attacks by a rival union of Bohemian butchers, "a sect by themselves." Even Local 1 of Syracuse, the cornerstone of the national union, foundered as a result of internal strife. The Socialist faction gained control in 1899, forcing the conservatives under Homer Call to set up a new Local 50. The Socialists lost their charter by vote of the Amalgamated convention in December 1899 and disbanded the following October. Some of the members then rejoined No. 50, which later reassumed the designation, No. 1. The cost of these and similar setbacks was high to the struggling Amalgamated.[42]

The boycott was another serious problem. The Kansas City unions had earlier declared unfair the Armour Packing Company. The Amalgamated, on being formed, was bound to assume the fight. But the boycott proved to be a great handicap, retarding organizing and other union activities. Nor was the company's business noticeably injured, even after the boycott was extended to Armour & Co. in Chicago (which claimed to be under separate management). With Gompers' active assistance, the boycott was at last lifted in the spring of 1898, the company only agreeing not to discriminate against union men.[43]

This was no sooner accomplished than the Coopers' Union declared Swift & Co. unfair. Again the Amalgamated became embroiled in a fight not of its own making. And, again, the Amalgamated was the chief victim. Not only was its organizing work hindered, but meat cutters' locals were disrupted trying to enforce the boycott, and discontent developed among union men still employed by Swift. The Amalgamated determined to rid itself of "that grim skeleton which has insisted upon a seat at our table for so long," whether or not the AFL agreed, "as we have learned by a bitter experience that all that our organization accomplishes by boycotts is what we do ourselves." The boycott was finally ended in the spring of 1901.[44] The Amalgamated had absorbed a costly lesson: the boycott could be effective in a restricted area against a smaller firm, as was shown in the case of Dold in Buffalo, but the weapon was likely to boomerang when employed against the major packers.

A final difficulty, less instructive in character, came from within the labor movement. Secretary Call had discovered soon after receiving the charter that shop butchers in Boston belonged to the Retail Clerks' International Protective Association. Answering Call's complaint, Secretary-Treasurer Max Morris asserted: "Meat cutters in retail establishments are retail clerks as much as dry goods or grocery clerks . . . You must confine your membership to the slaughter house employees, and those engaged in your line in

wholesale trade." There was, it turned out, a genuine overlapping of jurisdictions. The Retail Clerks claimed "Any person, regardless of sex, employed in any branch of retail trade other than the liquor trade." The Amalgamated Meat Cutters, on the other hand, was to be "comprised of all men
working in slaughter or packing houses or meat market who use a knife."
Ironically, the butchers' union, whose amalgamated scope had been partly
designed to prevent just such troubles, was plunged into a bitter jurisdictional dispute in the first year of its existence.

The issue came before the Committee on Grievances at the AFL convention at Nashville in December 1897. Since both unions had equally strong
legal claims (although that of the Clerks was prior), the decision turned
on other arguments. Secretary Call rested the Amalgamated case on the
skilled character of the meat cutter: he was a craftsman, not an ordinary
clerk. The Retail Clerks emphasized the common interests and inextri-
cability of retail workers. The AFL decision favored the Amalgamated; its
jurisdiction "shall include every wage earner from the man who takes the
bullock on the hoof until it goes into the hands of the consumer." But the
Amalgamated did not benefit greatly. Jurisdiction over the many men who
handled both meat and groceries was not really clarified, nor would it be
despite further efforts. A 1902 agreement, signed by Call, Morris, and
Gompers, limited Amalgamated retail membership to men "who are exclusively employed at meat cutting and meat service." Other complications
arose. The Amalgamated had to forego AFL endorsement of its shop card
because the Retail Clerks insisted that no union sign should be permitted
in combination stores in which the grocery employees were nonunion.
Above all, the dispute hindered Amalgamated organizing work. The Retail
Clerks, despite the AFL ruling, refused to relinquish shop butchers, and in
some localities fought the establishment of meat cutters' unions.[45]

The new international thus ran into problems from several sides. In
August 1899 Gompers, who followed the fortunes of the Amalgamated
closely, found it necessary to counteract Call's deepening pessimism

I regret to find you in that frame of mind. I am sure that success can never
crown your efforts to permanently establish your international union if you lose
hope in the future or confidence in yourself. It is only by dint of hard work that
you can succeed.[46]

Secretary Call did hold to the task, and, despite the obstacles, the Amalgamated was in fact making progress. By mid-1900, there were over 4000
paid-up members, about equally divided between the packinghouses and
meat markets.

The Amalgamated was meanwhile taking institutional shape.[47] The general characteristics of the national union had already clearly emerged by the time of the formation of the Meat Cutters. Advised by John O'Sullivan, the four founders had incorporated the major features of the national union into the original constitution. But this document was left "in a very crude state" for the future to complete. Responding to outside advice and also to the specific situation in the meat industry, the Amalgamated gradually defined its structure and functions as a national union. By 1900, it was in good running order.

The development took two lines: first, the clarification of the governmental structure and, second, the assertion of the national authority over the local unions. From the outset, the supreme body was the convention, having as it did the powers of constitutional change, of legislation, and of election of national officers. The 1897 convention did adopt the initiative and referendum on "important questions." But if the executive board did not support it, an issue could be brought to a popular vote only at the request of one fourth of the locals. The referendum actually was rarely used, and then generally at the initiative of the executive board. The convention, annual at first under the constitution, was not in practice supplanted by the referendum. Provisions for the election, elegibility, and credentials of convention delegates were made in December 1897. The next convention, in 1899, changed local representation based on membership to a single delegate per local. Later, representation was restored to a membership basis, partly reflecting in the periodic changes the varying relative importance of the packinghouse section of the organization.

The national authority between conventions rested with the executive officers, constituting the executive board. The first constitution set down fully the duties of the president, secretary-treasurer, and vice-presidents. Except for the abolition of the trustees, who audited the union's books, few changes occurred in the scope of the national offices in the first years. The president (who was also designated organizer) held the bulk of the executive powers. He was required to consult with the executive board only on those matters specified in the constitution. The secretary-treasurer had the normal duties of that post. But, in fact, the office from the start actually had great influence. This resulted partly from the fact that the president, as organizer, was in the field while the secretary wielded the day-to-day authority of the international, and partly from the fact that, beginning with Call, the office was filled by capable men. The vice-presidents, unspecified in number for the first year, were "to act as executives of the several districts

or divisions of districts in which they may reside and render such other assistance to the President as he may require." They were customarily distributed geographically, usually at points of strength, but there was no
specific constitutional provision for this, nor any division of the country into
districts.

Few men were likely to exercise their offices on a voluntary basis. Homer
Call, one of the rare ones, commented at the first convention that he had
"given his time and money in advancing the cause of the Butcher Workmen, for my heart and soul is [sic] in the interest of this, our craft . . . I
am sure that with all the work before me . . . you would not ask your Secretary-Treasurer to continue this work without remuneration." [48] But the
empty treasury argued otherwise, and he continued to labor faithfully for
two more years without pay. Only at the Chicago convention in December
1899 were the president and secretary-treasurer put on a salary of seventy-five
dollars a month. The next convention voted the vice-presidents three dollars
a day while in the actual service of the organization. Professional leaderhip
thus became possible.

These provisions also reflected the financial progress of the Amalgamated.
The original constitution had established a monthly per capita of fifteen
cents, but had not provided for its collection. The responsibility was put in
the hands of the locals at the first convention; they had to pay the per capita
from local dues. An arrearage of six months meant suspension of the local
union, and revocation of the charter after one year (reduced in 1900 to two
and four months respectively). The general office furnished monthly re-
ceipt stamps to be affixed in the books of the paid-up members. Many locals,
however, habitually paid only part of the proper per capita. The international in 1899 consequently required a monthly statement of membership
on official blanks, and the following year made the offense of withholding
international dues punishable by a fine fixed by the executive board. These
measures partly stopped the leakage. Finances were strengthened in other
ways. The necessity for remitting per capita of locals in difficulty, for instance, lessened with the passage of time. The charter fee was increased
from ten to fifteen dollars in 1899. An assessment of 25 cents a month was
levied for the first half of 1900 to cover the organizing expenses of President Donnelly. Finally, the per capita was raised in 1900 to 25 cents a month,
the additional ten cents going into a Defense Fund. The national finances
thus achieved a sounder basis.

The second line of advance was in the assertion of national authority; this
was, indeed, the keystone of the American organizational structure. Two

matters were involved here. First, the local unions had to acknowledge the primacy of their relationship to the national union. This doctrine, accepted as it was by the entire labor movement, caused the Amalgamated relatively little trouble. But the issue did arise in the first years. Gompers, for example, had occasion to admonish locals for by-passing the Amalgamated office. The AFL refused to hold direct correspondence with local unions, Gompers told Secretary Call, "in order to protect the rights and interests of your national organization to avoid a conflict of authority, and at the same time prevent the useless waste of labor in writing to local offices in the same matter." The Amalgamated, as well as other internationals, also ran into difficulties with the central labor body of Kansas City, Kansas.[49] But the second aspect of national authority was immeasurably more difficult: namely, the extent to which the local unions would surrender their autonomy to the international.

The AFL charter and the first constitution, of course, gave to the Amalgamated theoretical supremacy over the locals. But, in practice, local autonomy was only gradually and modestly narrowed by the international. Strike action was the most important area in which the Amalgamated asserted its authority. The original constitution had required the permission of the executive board before a strike. But, again, there was no provision for procedures or sanctions. The 1897 convention eliminated the requirement if less than ten men were involved, possibly in the hope that this would encourage compliance on larger strikes. But the rule remained largely a dead letter. When a walkout occurred in Buffalo in September 1900, President Donnelly "found the same old course had been pursued—violate the constitution and then call on the Executive Council for advice and assistance."[50] The 1900 convention consequently drafted an elaborate strike procedure: first, a three-fourths vote of approval by the local; second, investigation by the international president; finally, vote of the executive board. In addition to the powers of suspension and expulsion, which were apparently never employed, the sanction of the international was greatly strengthened by creation of the Defense Fund: strike benefits of five dollars a week would be forthcoming only for duly authorized strikes. On the other hand, the international left collective bargaining almost entirely in the hands of the local unions in the early years. The 1900 convention rejected a resolution granting packinghouse men the "privilege" of charging extra for overtime and Sunday work on the ground that this was a local matter.

The Amalgamated also intervened increasingly in the relations between locals and membership. The international had from the start set minimum local dues at fifty cents a month. In 1899 a maximum of ten dollars was placed on the initiation fee (reduced to five dollars in 1900). Local issuance and acceptance of traveling cards became obligatory. The constitution and bylaws for the government of local unions, first adopted by the international in 1897, defined the causes for the suspension and expulsion of members. No provision, however, was made for appeals to the international. The meat cutters' locals also had discretion in admitting boys under eighteen and proprietors of one-man shops.

Allegiance to the international developed gradually. The members, since they were touched at a variety of points by the Amalgamated, came to recognize their citizenship in it. The protection of traveling members, in particular, fostered this view. The international also established the category of Membership at Large for residents of unorganized areas. The 1897 convention set up a death benefit system. An assessment of 25 cents was to be levied, the sum to be turned over to the survivors of the first member to die. Then a second assessment would be made, and so on. The ill-considered plan lacked support among the locals and was quietly dropped in 1898 after the payment of one benefit. A regular death benefit system, financed from dues, was later inaugurated.

Improved communication further reduced localism. Most constituent locals rarely if ever saw an international representative, but they did receive frequent letters from Secretary Call, who was a tireless correspondent. Conventions contributed to the sense of community. Only six members attended the first meeting in Nashville in 1897, and none was held the next year for lack of interest. But in December 1899, twenty-four delegates met in convention at Cleveland. "As we had been working, scattered so widely, and never having met each other," remarked Call afterward, "our work was largely one of faith, but now we have met, have grasped each other's hands and . . . the personal acquaintance has given us confidence in each other." [51] A monthly journal, two numbers of which Call had already put out on his own initiative, was authorized by the 1899 convention to foster contact within the organization.

By 1900, therefore, the Amalgamated Meat Cutters and Butcher Workmen of North America emerged from the stage of uncertainty. The international had established its internal order, had weathered the first setbacks, and had built up its membership. The time had come to assume the larger task of organizing the American meat industry.

CHAPTER 3

MEAT PACKING: THE FIRST
CYCLE OF UNIONIZATION

The great prize was in meat packing. Here thousands of workers were concentrated in a few giant centers, with half a dozen companies engrossing the
bulk of the business. A great labor organization might be created here in one
swift stroke. This achievement came within the grasp of the Amalgamated,
and then it was lost. The unionizing opportunity, wasted once, would never
present itself again in an entirely voluntary framework.

Chicago was the key. In 1900 the Union Stock Yards employed more than
a third of the total labor force in meat packing—over 25,000. Here, too, the
labor standards of the industry were set. Chicago, finally, was the fortress of
antiunionism. A legacy of bitterness and repression had remained after the
strikes of 1886 and 1894. When a small union formed in late 1897, its membership of thirty-five was discharged, and its life extinguished. No open sign
of unionism remained. But the unionization of meat packing was impossible
without Chicago. It had to be taken.

Michael Donnelly appeared at the yards in the late spring of 1900. He
found the men "in dread of losing their positions and perhaps blacklisted
in the event of their joining a union." Some, including future Vice-President
William Stirling, were in fact discharged on suspicion of supporting his
efforts. Donnelly quietly persisted among the skilled cattle butchers. Following his earlier experience, he visited the men privately in their homes to talk
organization. A few veterans of earlier struggles, such as George Schick and
Nicholas Gier, were willing to try again. Schick "was introduced to President

Donnelly by Brother Stirling, and after getting acquainted, President Donnelly asked me what I thought about starting a union, and I told him it would be a tough proposition . . . but . . . I would volunteer as one to go to a meeting if he called one, and that I would try to get more." A small group of intrepid butchers met and on June 9, 1900 received a charter as Cattle Butchers' Local 87. The union operated in secret for the first months, gradually increasing its ranks and gaining confidence. At the end of 1900, Donnelly reported over a hundred members in the first Chicago union.[1]

Recognizing that the situation "requires to be handled very carefully and . . . all the time and money possible," Donnelly moved his headquarters to Chicago from Omaha in January 1901, and the slow work continued. Encouraged by the success of the cattle butchers, other skilled groups met quietly with Donnelly to form unions. Beef Boners' Local 135, recorded its secretary T. J. Condon, started "in my home one warm night in June 1901 . . . The boys of the boning rooms in the different packinghouses had thought more or less of forming a union, but . . . we did not want to hire a hall lest the public would know what we were doing. We finally decided to meet in my house . . . Donnelly . . . was invited by some of the boys to come up, so we finally decided to take the bull by the horns."[2] In this fashion, there were organized by July 1901 Chicago locals of Hog Butchers (116), Sheep Butchers (118), Beef Carriers (132), Beef Boners (135), Beef Casing Workers (139), and Sausage Makers (140).

Once well started, the union drive generated its own momentum. Many packinghouse locals, as soon as they could meet the expense, hired business agents to build up their organization. The correspondent of Hog Casing Workers' Local 158, noting the lack of organization among the sheep casing workers, reported that "our business agent is going right after them, and as he always gets what he goes after, they certainly will be with us in the near future." The agents also were active in forming new unions under the auspices of the Board of Business Agents and the Packing Trades Council, and some of the later unions were direct offshoots of the original locals. When they became sufficiently strong, the packinghouse unions began informally to impose union shops. The Sheep Butchers, one of their officials reported, "are strickly [sic] union men and good workers, and will not work with any man who does not carry a union card." Although management raised objections, the unions generally had their way. George Schick, who became business agent of the Cattle Butchers, reported that the superintendent of the new Fowler plant "assures me that all the new men employed must be union men." Organization was approaching the point where an official could confidently say in December 1901, for example, "that there is not a man in Chicago skinning cattle that is not a member of the union." [3]

President Donnelly had been obliged to stay close to Chicago throughout 1901. The movement there, he explained, "being young and inexperienced, left without a leader for any great length of time, was liable to go to pieces." But by 1902 he could turn his attention elsewhere. In the spring, Donnelly started the first of several tours of the Missouri River points. The increasing income of the international made possible also the expansion of its organizational staff, largely by putting members of the executive board on the payroll. The incumbent packinghouse vice-presidents, James Sheehan of East St. Louis and William Jameson of Kansas City, were replaced by better men— William M. Stirling of Hammond and Stephen Vail of South Omaha. The decline of the unattended organizations in the Missouri River centers was now arrested, and, with a more adequate organizing staff, the Amalgamated initiated a new growth. Membership increased steadily at the other western centers.

The new unions suffered numerous dismissals at first. Hurrying back from South Omaha in May 1901, Donnelly found the sheep butchers "in a frenzy and demanding justice" because of discrimination against them. Donnelly realized that "if our movement was to be saved quick action was necessary." Threatened by a strike, Swift & Co. backed down. Foremen and superintendents usually fought every new organization until they concluded, as one sausage maker remarked, "that we were determined to organize." [4]

The major packers, despite the turbulent past, did not seriously resist organization after it gained headway. The Amalgamated campaign coincided with a time of rising meat prices and of a tight labor market. Not being rigidly antiunion and responsive to prevailing circumstances, the packers followed a moderate course. The leadership of the industry was, moreover, changing. The founding magnates—P. D. Armour, G. F. Swift, and Nelson Morris—were passing from the scene, and control was going into the hands of their sons. The second generation was more amenable to a new approach to labor relations.

For its part, the Amalgamated presented a persuasive case. The Armour and Swift boycotts, harmful to the union cause in every other way, gave President Donnelly and Secretary Call the opportunity to explain directly to the packers the merits of unionism and to show themselves as capable and conservative labor leaders—as indeed they were. Secretary Call thus explained this aim to the settlement worker Mary McDowell:

[We wish] to impress upon the minds of the employers that we are not building an organization to antagonize them, but to educate and elevate the

workers in the Meat Industry to make them better men and women . . . and by
so doing add to the sum total of human usefullness [sic] and happiness.[5]

The union would be honorable in its contracts, reasonable in its demands,
and considerate of the employers. When the packers were attacked in 1902
for high meat prices, the Amalgamated rose to their defense. The union, said
Secretary Call, was "run on strictly business principles and . . . worthy of
the confidence of other business institutions." [6]

Labor relations in the stockyards might also be improved. The industry
in 1899–1900 had been plagued by worker unrest. The Amalgamated argued
"that organization makes better employees and prevents trouble instead of
creating it." Donnelly worked assiduously to maintain discipline. When the
Swift beef butchers in South Omaha walked out in February 1900 to protest
the employment of a nonunion man, Donnelly hurried there and ordered the
strike ended. He apologized to Gustavus Swift, Jr., and hoped for, but did not
demand, the reinstatement of the guilty men who had injured "the standing
our unions had attained locally, by leaving forty or fifty cattle dead and refusing to dress them." Donnelly had reacted similarly to an unauthorized
strike at the Kansas City plant of Schwarzschild and Sulzberger the previous
summer. The packers, for their part, welcomed his efforts toward stability.[7]

The secretary of the South Omaha Packing Trades Council, George
Stevens, shrewdly analyzed the packers' "dilemma." Their treatment of the
men had created discontent and stimulated organization, but past experience
had demonstrated that opposition to unions was an expensive course, causing
chaos in the yards and slow-ups in the plants. "They finally came to the conclusion that the best way out of the dilemma was to meet organized labor
and treat it in a compromising way . . . That was the step that gave organized labor in general its standing in South Omaha." [8]

Officially, the packers became friendly. "Armour & Co. approve their
butchers joining the labor unions," stated the *National Provisioner* in 1902.
Hostility still remained in the lower managerial ranks, and newly formed
unions still ran into some opposition. Members of Stock Handlers' Local 190
in Chicago, for instance, were discharged when it organized. "But after a few
conferences with the management we were able to show that we meant nothing radical or revolutionary," and since then "we have been on most friendly
terms with the management." This was the common experience. By the
spring of 1902 Secretary Call stated confidently that "we have gained the respect of a majority of the large employers of labor in our craft." [9]

Greater resistance existed at the fringes of the industry, as Call observed
in June 1901, among "the small packers, who, in order to compete with the
large firms, desire to reduce the wages of their employees." [10] When an or-

ganization was formed at the Kingan plant in Indianapolis in November 1903, the company posted a notice reminding the men of past fairness and amity.

This company, since its beginning has, through its management, treated directly with its men.

Individual applications regarding either work or pay will always receive careful consideration.

The company would therefore ask its workmen to consider well, each man for himself, the ultimate effect of organized agitation upon his own employment.

Will any portion of our old employees go in for a changed relationship with this company, thereby forfeiting their right to individual recognition of services? [11]

Some companies—Kingan, Morrell of Ottumwa, Iowa, Cudahy Bros. of Wisconsin—successfully stifled the union effort at the start. Elsewhere, organization survived the first fight, only to fail after a period of uneasy truce—at the Buffalo plant of the Jacob Dold Packing Company, for example. Permanent organizations, too strong to be dislodged, were established in some independent plants. Local 227, started in August 1902 in Louisville, guaranteed its success by a hard-won victory over the Louisville Packing Company in early 1904. The slaughterhouses in Baltimore, Evansville, Wheeling, Cincinnati, and New York City also became firmly organized, as did small plants in and near the major stockyards. But the majority of independents held back effective organization, and the many local houses remained untouched except where the men were brought into meat cuttters' locals.

The failure seemed unimportant. The Amalgamated effort was directed toward the few major packers who controlled the bulk of the industry, and in their plants the union was advancing unopposed by 1902.

Progress so far had been chiefly among the skilled knife men. Success here raised the crucial issue: the disposition of the mass of unskilled packinghouse workers. The Amalgamated had been conceived as a craft organization. Its original packinghouse jurisdiction—"all men . . . who use a knife"—covered those considered to be members of the butcher craft. "Butcher Workmen in all branches are skilled workmen," stated Secretary Call. The unskilled, although admissible under the first constitution at the discretion of the locals, were not welcomed into the Amalgamated. During its first years, federal labor unions of unskilled packinghouse workers in Kansas City, South Omaha and Indianapolis remained directly affiliated with the AFL.[12]

President Donnelly soon rejected the exclusive policy. In March 1901, he lectured the membership on "Our Duty as Butchers":

There must be no aristocracy in the labor movement. I have worked at the highest wages paid in the packing plants, but I cannot forget that the man who washed the floor while I worked at the tables is entitled to the same consideration I am. I cannot forget that he is a human being, and that he has a family. It should be our purpose to make the injury of the common laborers the concern of the skilled workman.

The skilled butchers did, in fact, respond during several laborers' strikes—for instance, at South Omaha in 1898.[13] But Donnelly had other than humane reasons for urging the organization of the laborers.

Craft labor had long since become obsolete in the packinghouses. The work was carried on by specialists trained to do a single operation expertly. A few of these, it was true, did delicate jobs. In a typical cattle-killing gang of 230, the 19 floormen, backers, rumpers, and skinners received at least 40 cents an hour—the skilled rate. And these few were often inclined to separate themselves, hoping thereby to conserve the monopoly on their skills. They refused at first to join the South Omaha cattle butchers' union. In St. Joseph, where the highly skilled men did organize, all the others in the beef gangs were excluded; 40 men were eligible in the entire packing center. But this was clearly the wrong path. The highly skilled, too few and too scattered, were made vulnerable to replacement from below by the division of labor. If the lesser men were brought into their organizations, promotion and training could be regulated by the union; and the danger of being displaced during strikes would be much reduced. From the outset, therefore, the Amalgamated favored the admission of the lesser skilled men. These, however, were similarly susceptible to replacement by their inferiors. The killing gang was a ladder on which each man prepared for the next higher step. "Today it is impossible," concluded the journal, "to draw the line where the skilled man leaves off and the unskilled man begins." The division of labor thus "places the skilled workmen largely at the mercy of common labor and makes it necessary to organize all working in the large plants under one head."[14]

Other motives besides the defensive argument were at work. It would have been feasible to organize the less skilled into unions outside the Amalgamated. And the men—mostly unskilled—in the processing and by-products departments were doing work that was entirely apart from the killing and cutting jobs. Yet Donnelly vehemently opposed the inclusion of unskilled packinghouse workers in federal unions directly tied to the AFL. "Better no Union than a 'Cheap John' Federal Labor Union," he insisted, quickly bringing the existing packinghouse federal bodies into the Amalgamated. And Donnelly also proceeded to organize the processing and by-product workers.

These decisions revealed two additional motives for wanting the unskilled in the Amalgamated. The first was ambition. The higher-paid men were not sufficiently numerous—only 20.9 per cent of the labor force earned $2.50 or more a day—to constitute a union that would satisfy an energetic leader such as Mike Donnelly.[15] The second, more important reason was his anxiety for orderly and cautious progress, and this in turn precluded large unorganized or independent elements. Donnelly recalled the difficulty caused by the Anchor Federal Labor Union when the Amalgamated had tried to settle the Armour boycott controversy in 1898. He believed that the unskilled had to be under Amalgamated authority in order to insure good relations with employers. "Wherever the employees are all in the union," argued Donnelly, "there is much less danger of friction between the organization and the company."[16] Many weighty considerations thus favored the unskilled packinghouse workers. Skilled labor opposition, while it did not die, went underground during the period of Amalgamated success.

The bulk of the unskilled were recent immigrants from eastern Europe. "This class of people," wrote Secretary Call, "have been gathered together by the employers for the reason of their ignorance . . . and so long as the employers could keep them in ignorance they could employ them at low wages and . . . the most degrading conditions, hence their bitter opposition to our organization in its early period."[17] The immigrants proved unexpectedly responsive to union efforts. Canners' Local 191 in Chicago did not grow, an Irish-American official explained, because "we were not able to convince our fellow workmen, the majority of whom do not understand English, [of] the benefits of the organization." Once a nucleus had been won over, however, the rest of the immigrants were easily organized. "We elected stewards of the different nationalities," the Canners' official stated, "and they found no trouble in having them join us and we find them the very best Union men we have got." This was the usual experience. Through interpreters and immigrant officials, the packinghouse unions were able to attract and hold the newcomers.

Hostility toward the immigrants was not pronounced in the stockyards. The lack of strong ethnic tensions sprang from the packinghouse situation: the close contact on the killing floors, as compared to steel mill or foundry; the evident dangers of leaving the immigrant workers unorganized; and the movement, already apparent, of the foreigners into the skilled ranks. The Chicago Sheep Butchers' Local 118 considered itself "the 'Banner' local, as we have at every meeting seven languages to speak, and every transaction is translated to all languages, thereby giving every member equal rights." The

Pork Butchers' Local had 600 Irish members, 600 Germans, 300 Poles, and 300 Lithuanians. In 1904, a Pole was elected president. The Slavic workmen quickly took their places in the burgeoning packinghouse locals.[18]

Negroes presented greater difficulties. They had first appeared in the Chicago yards in significant numbers during the summer of 1894. Coming as strikebreakers, they had nearly precipitated a race war.[19] But most did not remain afterward in the packinghouses; only 500 Negroes were in the Chicago plants in 1904. Despite the remembered bitterness, the whites saw the need to organize them, and the Negroes hesitantly joined. There was a real effort to develop a sense of solidarity. When, for example, a colored tail sawyer died, he was treated by the Chicago Cattle Butchers' Union "with the honor and respect that is due to every member . . . regardless of race, color or denomination." Good intentions were almost thwarted when one of the white mourners was counted as a Negro. "For awhile it looked like as if a riot call would have to be sent, but . . . an apology was accepted by Brother Huston, order was restored in the ranks, and the long line of mourners resumed their march to the depot." [20] Despite the social tensions, many Negroes took part in the new labor movement.

The female workers proved the thorniest problem for the Amalgamated. Secretary Call had warned as early as 1899 that "women were crowding into the avenues of our trade and labor . . . and the question must be met." No action was taken, however. The low-paid women did not seem a real threat because they were concentrated in canning and labeling operations. The first impulse for organization came from the girls themselves. After a spontaneous strike at Libby, McNeill and Libby in early 1900, the leaders founded a Maud Gonne Club, after the Irish heroine of home rule. They read a newspaper account of a speech on packinghouse unionism by Mary McDowell, head resident of the University of Chicago Settlement "back of the Yards." The social worker, at their request, approached President Donnelly, who agreed to organize the female workers. Women's Local 183 was chartered March 27, 1902 and after a slow beginning (many of the charter members were discharged) grew to a membership of over a thousand.[21] Similar unions were soon formed in other packing centers.

After female organization was well started, opposition developed within the Amalgamated. Women were increasingly being substituted for higher-paid men, particularly in sausage departments, and affected locals fought the encroachment. In 1902, New York unions started a campaign to force the girls out of the local packinghouses. Canning Room Workers' Local 333 in the Cudahy plant in South Omaha went on strike against the introduction of more female help. During the disastrous Chicago sausage makers' strike in 1903, immigrant women entered the struck departments in large numbers. The issue received a full-dress debate at the 1904 convention. The sausage makers' and canners' locals demanded the exclusion of women from their jurisdictions except on traditionally female work. The convention faced a painful dilemma. Sentiment clearly favored the resolutions; but not union principle. "Why did not the Amalgamated act before organizing the women?" asked a female delegate. "It would be unjust," another delegate admitted, "now to discourage their organization, as many of them have families to support." The convention finally voted to demand equal pay for equal work from the packers—an unlikely possibility—and instructed the executive board to try to confine women to work that was not "brutalizing" and at which they had been employed prior to 1902. This dubious compromise was ignored by the sausage makers' locals. They immediately afterward formulated a scale including the clause: "Abolition of Women Labor in the Sausage Departments." The female problem, alone among those raised by the acceptance of the unskilled, was not resolved successfully.[22]

The inclusive membership policy did not turn the Amalgamated into an industrial union. Donnelly, although admiring the industrially organized Brewery Workers, was willing to respect the jurisdictions of other internationals. Tin can workers' locals were transferred on request to the Sheet Metal Workers' Union. Although organized largely through Amalgamated efforts, steamfitters, coopers, engineers, firemen, carpenters, and others were likewise assigned to their appropriate nationals. The fact was that the Amalgamated, despite its coverage of the unskilled, retained a trade outlook.[23]

This commitment also determined the union structure in the packing centers. The constitution stated: "Not more than one charter for each branch of the craft shall be granted to one city." The result was that a packinghouse local was limited to members in a "branch of the craft": cattle butchers, hog butchers, beef boners, beef carriers, sausage makers, canners, oleo workers, and so on. The same arrangement had existed under the Knights of Labor. Although arising from persistent craft thinking, the jurisdictions were actually departmental; the cattle butchers' unions thus took in everyone in the beef departments from the lowliest penner to the splitter. The only exceptions were the female workers, who were put into a women's local irrespective of their work-places. (The Amalgamated, after one bad experience, rebuffed the efforts of immigrant leaders to form locals on an ethnic basis.) [24] Each local union, in addition, had jurisdiction over all the appropriate departments of the packinghouses in the center. The local unit of industrial unionism—the plant—was adopted only outside the larger centers and was designated a "mixed" local.

The "craft" arrangement inevitably gave rise to difficulties. There were frequent jurisdictional disputes. Did head boners belong to the Cattle Butchers or Casing Workers? pig feet shavers to Hog Butchers? tallow men to Oleo Workers or Casing Workers? The doctrine of exclusive local jurisdiction also was troublesome. When the Hammond plant was moved to the Chicago yards, the executive board ruled against the request of Cattle Butchers' Local 75 to retain its charter because Local 87 had jurisdiction over Chicago.[25] These troubles, however, were not seriously harmful to the Amalgamated.

President Donnelly had the good sense to avoid rigidity on jurisdictional matters. His chief concern was to create strong locals. One delegate at the 1904 convention argued on the grounds of "trade autonomy" that 30 or more sausage makers in a locality should always be permitted to form their own union. Donnelly condemned this narrow view "of skilled workmen in some branches of the craft who desire to organize little locals within themselves." Strength came from numbers.[26] Despite some opposition, Donnelly's idea generally prevailed. Many small packinghouse locals were amalgamated or given broader jurisdictions. Although all the packing centers had about the same range of operations, the number of locals varied in direct relation to the size of the labor force. Division by "branches of the craft," accomplished pragmatically, in many ways suited the situation in the packing centers. The danger of fragmentation was avoided; the men were organized by common interest and occupation; uniform local conditions among plants were quickly achieved.

Unity, President Donnelly soon realized, was needed within each packing center. The first successful Packing Trades Council was organized in Chicago in August 1901. Each local received representation on a membership basis, and dues were fixed at 25 cents per union member every three months (later reduced to one cent per month). Nicholas Gier of Local 87 became the first president. A salaried secretary was appointed, and by 1903 plans were being made for a Labor Temple. A second council was formed in East St. Louis in February 1902, and, after the endorsement of Donnelly's actions by the 1902 convention, others soon appeared in the remaining western centers, as well as in New York City and San Francisco.

The packing trades councils brought order to the stockyards, unifying the locals and keeping them on "the straight and narrow path." Acting as the immediate authority in the centers, they settled jurisdictional disputes, adjusted grievances with the packers, endorsed demands and agreements, and "stopped hasty action on the part of some of the locals." Organizing was largely taken over and completed by the councils. They also represented the packinghouse workers in the city centrals and before the public. The Chicago body was instrumental in getting child labor law amendments passed in Illinois. Even the "most skeptical," observed an East St. Louis official, had to see that the councils were "of the greatest benefit and in the packing centers an absolute necessity."[27]

The packing trades councils also coped with the craft unions outside Amalgamated jurisdiction. They had troubled the Butcher Workmen from the start. The Amalgamated, while lacking control over the auxiliary crafts, necessarily became involved in their fights. When the Stationary Firemen placed a boycott on the Cudahy plant in Kansas City, Secretary Call filed a protest with the AFL because "the Butcher Workmen had never been consulted in the matter . . . and our interests are much greater than that of the Firemen." The packing trades council was Donnelly's answer. Its authority would curb the auxiliary unions. Moves against the packers would require the previous endorsement of the Amalgamated majority. The Chicago Packing Trades Council, limited at first to Amalgamated locals, soon broadened its jurisdiction. Eight of the twenty-seven member unions in July 1903 were craft locals belonging to internationals other than the Butcher Workmen.[28] Despite the multiple jurisdictions, unity seemed to be achieved in the packing centers.

The liabilities of craft organization were overcome in the case of meat packing. The unskilled were organized; the local butcher unions followed effective jurisdictional lines; and the trades councils provided discipline and unity within each center. A pattern suitable to the packinghouse situation followed the initial Amalgamated gains in the industry.

Growing numbers and appropriate structure, essential as they were, did not guarantee the success of packinghouse unionism. The further step—and critical test—of entering fruitful collective bargaining remained.

Initially, a pattern of local negotiation developed. The packinghouse unions bargained separately, acting on their own initiative and depending on their own strength. Once immediate management resistance was overcome and committee recognition attained, the unions pushed the most pressing grievances of the membership. The Pork Butchers and Casing Workers of Chicago, for example, were being docked for time lost during machinery breakdowns. The first achievement was "to get paid from the time we take out our checks until we turn them in again." As a start, many unions demanded the abolition of "church time"—the few minutes worked each day without pay—or the contract system under which ten days' pay was kept back by the company. The Chicago Pork Cellar Men's Union quickly ended bribery by foremen.[29]

More basic issues were gradually brought up. The major grievance of some unions was speed-up. In Chicago, the work load of beef carriers "was certainly slavery . . . This was the first thing we changed . . . Where we had only 37 carriers before we were organized we have now 53, and they do no more loading than the 37 used to do." Cattle and sheep butchers, suffering most from the speed-up, first put into effect informal limitations, "to the effect that no butcher rush another." In September 1902, verbal agreements were reached with the packers, setting the output of sheep gangs at 33 per cent and cattle gangs at almost 30 per cent below the unrestricted rate. The men considered the output regulation "worth more to them than the increased wages." Formal limitations were not imposed in other departments, but a general slow-up of work was soon evident.[30]

Hours were the other important subject of change. Many departments had extremely irregular schedules, requiring frequent late starts, very long hours for a few days, and short time the rest of the week. These practices gradually ended. The Chicago Cattle Butchers, for example, soon demanded a regular start at 7:00 A.M. Then, beginning February 1902, they refused to work after 5:30 P.M. without overtime pay. On November 30, 1902, for the first time in the history of the Union Stock Yards, double time was paid for Sunday work. The packers, anxious to avoid these premium charges, attempted to regulate the supply of stock coming into the yards and adopted the practice of holding excess animals overnight. The demands for regular hours and overtime pay eventually led to a stabilized work day of ten hours and a minimum of overtime work.[31]

For many men, the shop committees and the union button were as important as specific improvements in working conditions. A hog butcher, coming to East St. Louis from a nonunion plant, thought it was "living in Paradise to work in a place where you are looked upon as a man and a human being and not as a slave." [32] Some locals informally achieved seniority in hiring and in promotions, as well as a union shop. Wage increases, granted to a few unions, were not of widespread importance at the outset. This issue usually awaited the disposal of other bargaining matters.

The international, for its part, deliberately remained in the background. President Donnelly's appraisal of the bargaining process was tough-minded: concessions were won by unions strong enough to grasp them. The Amalgamated, as a national union, was as yet too weakly organized to confront the packers. "Our organization . . . must act carefully and conservatively," Secretary Call repeatedly warned. "Radical and hot-headed action is liable to destroy what we have already gained." [33] Some of the local unions, on the other hand, had sufficient strength to extract concessions individually; they should, therefore, take the bargaining initiative. The strategy of local negotiation had the great virtue of flexibility, permitting maximum gains to the membership consistent with the safety of the entire organization.

The national leaders here acted as a restraining influence on the subordinate unions. When an impasse in local bargaining was reached, or when trouble flared up from an unresolved grievance, an official would hastily come to settle the difficulty. If the plant management was adamant, the union would usually be directed to back down. President Donnelly repeatedly lectured the members, "especially those in the packing centers, to remember that no good can come from hasty action. Your grievances are numerous and often hard to bear, but they are also deep seated and of long standing, hence it requires time and the application of a more modern remedy than an ill-advised and hasty strike." The international, preoccupied with building and safeguarding the organization, thus functioned as a stabilizing force in the labor-management sphere. President Donnelly resisted the pressure for national action by weaker unions which did not share in the widespread gains of 1902. The men in Kansas City, St. Joseph, and South Omaha, Donnelly reported in April 1903, "are nearly all clamoring for an increase in wages, but with the exception of a few trades they are not yet sufficiently well organized to expect immediate results." [34] Donnelly was determined to bargain only on a basis of strength.

This strategy permitted only local negotiation during the organizing process, but the thorough organization of any branch would signal industry-wide bargaining for that branch by the international. The executive board had, in fact, hesitantly attempted to do this for the cattle butchers in August 1901. A scale, calling for a 10 per cent increase and other gains, had been adopted and presented to the packers. All but one refused, and the issue passed. When adopting the scale, the executive board had in fact agreed not to call a strike in the event of failure. It was prepared to bide its time. [35]

At the fourth international convention in August 1902, Donnelly reported that the cattle butchers "are now well organized." Their scale, therefore, should not only be adopted, as would be other scales at the convention, but should "be demanded" from the packers. The timing of the move was left to the executive board. Another year passed while the cattle butchers' unions by local negotiation brought their scales up to the Chicago level. September 8, 1903 was the "red letter day" for the Amalgamated. The entire executive board met in Chicago with a committee of company superintendents to consider uniform cattle and sheep butchers' scales for all the western packing centers. After long and difficult negotiations, an agreement, dated October 1, 1903, was reached through mutual concessions. The packers granted modest wage raises of about the same range as the previous year—from 10 to 15 per cent. The output limitation for cattle gangs remained unchanged, but important modifications were conceded by the union in the sheep departments. For the first time, an industrywide agreement, albeit covering only two branches and the major packers, was achieved in the meat industry.[36] It was the high-water mark in the early history of the Amalgamated Meat Cutters and Butcher Workmen.

The organization of the western packing industry had progressed far in a few years. Chicago was thoroughly organized, the other centers nearly so. The structural arrangements, although they conformed to the AFL jurisdictional requirements, provided the inclusiveness and unity vital for labor organization in a mass-production industry. Finally, a viable bargaining relationship was developing with the packers. The union's achievement was impressive. Its swift success, however, raised other problems more difficult and dangerous than those of the initial unionizing stages.

"While we have a large organization," observed Secretary Call sagaciously, "we are not a Union yet until we learn to reduce this large body of Butcher Workmen to a system that will enable us to work together harmoniously." This was the nub of the union's weakness: the packinghouse workers, once they entered the organization, proved extremely hard to manage. Secretary Call was compelled to be absent from his office "so much in order to preserve harmony in our ranks." New unions, even the early locals of skilled butchers, had discipline problems. But the difficulty was greatly magnified when the mass of unskilled men became organized in 1902 and 1903. Chiefly recent immigrants, they were not responsive to the arguments for restraint. They expected quick returns for their dues. "There are thousands of raw recruits," complained President Donnelly, "whose first thought is that the organization is for the sole purpose of giving them immediately a higher standard of wages and conditions."[37] Scattered through the ranks, moreover, were malcontents and "impatient spirits"—often local officials—who fanned the dissatisfaction.

Under any circumstances, it would have been difficult to control the untutored and excited mass of packinghouse men. But a special discontent was stirred by Donnelly's cautious negotiating policy—indeed, particularly by its tactic of piecemeal progress. The benefits came too slowly and unevenly. While skilled men in cattle-killing departments averaged hourly increases of from 15 to 20 per cent from 1899 to 1903, laborers in a number of surveyed plants advanced only 6.3 per cent.[38] Grievance procedures and improving conditions did benefit the unskilled men, and some of their unions were, by local negotiation, beginning to raise wage levels as well. In May 1903, Chicago locals of chiefly unskilled labor increased the minimum rate from $17\frac{1}{2}$ to $18\frac{1}{2}$ cents an hour. At other centers organization was not yet strong enough among the unskilled to achieve wage advances, except at Omaha and Sioux City, where the rate reached 19 cents an hour. The Amalgamated leaders urged patience. "We must give the Packers time to adjust their business to the new conditions which we desire, and we feel confident that . . . we will be met at least half way and many improved conditions granted us."

But the packinghouse workers would not wait for that happy day. Gains that had been satisfactory earlier were no longer so in 1903. The newcomers, one Chicago unionist noted, "have so far failed to see any improvement or where they have not [sic] benefited." Unrest spread through the packing centers. Grievances and disputes kept the national officials continually on the road. The packers' concessions on minor issues, Donnelly found, only aggravated matters. "A new man in the union, like a new union, becomes too much elated, too enthusiastic over a point gained." The situation was becoming unmanageable. Donnelly was reported in July 1903 to be "having his hands full in steering our 30,000 members in the Windy City clear of the rocks."[39]

Strikes were the great danger, and Donnelly did his utmost to prevent trouble. His general rule was to authorize strikes only for defensive purposes—for example, to prevent the dismissal of cattle butchers with seniority in the Swift plant in Chicago. But he opposed strikes, particularly by weaker locals, to gain improvements. His restraining efforts were only partly successful. The Chicago sausage makers, for example, struck without authorization for a general wage increase in October 1903. Sympathetic strikes were also becoming frequent. The men in the Hammond plant in Indiana walked out in May 1903 "against the wishes of the International" in support of the Stationary Firemen. There were also innumerable small stoppages, often unexpected even by the local officials. Chicago, commented the *National Provisioner,* "is always in the face of strikes . . . in a state of siege all the time."[40]

Lack of discipline, besides endangering relations with the packers, undermined the internal union machinery. The packing trades councils, Donnelly reported, were "being thwarted by the unwise and dangerous acts recently engaged in by some locals—that of striking without . . . sanction." The councils in fact were sometimes captured by radical elements. At the Morris plant in St. Joseph in September 1903, a strike was called to protest alleged discrimination against cellarmen "in compliance to an ultimation [sic] written and read by a committee from the Packing Trades Council." The strike ended after four days at the order of President Donnelly.

Part of the trouble was the membership of the allied craft locals in the packing trades councils. They were not affiliated with the Amalgamated and hence not subject to the authority of the international. The price of unity in the stockyards seemed too high. "This industrial system," Donnelly told the 1904 convention, "means sympathetic strikes, and if it is not confined to our own chartered locals will become dangerous." The convention thereupon voted to limit the councils to Amalgamated unions, rejecting the alternative of enrolling the auxiliary workers and thus turning the Amalgamated into a bona fide industrial union. In the next month, June, the allied trade unions withdrew and formed their own trades councils.[41]

This action failed to solve the basic problem because the Amalgamated was not able to control even its own local unions. They could not be restrained, Donnelly admitted, from "arrogating to themselves the right to strike. Even after being informed of their indiscreet, unwise and uncalled-for action, they . . . continued to violate the constitution." In further defiance, they sometimes appealed directly to other local unions for aid. Thus, the authority of the international was gradually eroded.

By early 1904, the situation was alarming. Large numbers were dropping out in discouragement; only 36,000 of the 56,000 members in May 1904 were in good standing. Disgruntled locals were threatening to secede, and the popular discontent was being directed at President Donnelly and the executive board. Donnelly's recommendations during the Chicago sausage makers' strike "were not alone ignored, but we have been publicly criticized for having sold out the union." His efforts to maintain good relations with the packers, Donnelly admitted in May 1904, had led to "much unwarranted criticism." There were men, he warned, "who seek to create a division between local unions and the International." [42]

One alternative was to cut away the unmanageable parts. The idea was behind the elimination of the allied trades locals from the packing councils. Similar reasoning led Donnelly to request the AFL to issue a separate international charter for the unions of soap workers affiliated with the Butcher Workmen. Earlier an avid expansionist, he now believed that "a steady, healthy growth is much more to be desired . . . than a rapid growth." [43] But Donnelly could not at this point repudiate the doctrine on which he had built the organization. The reasons for an inclusive packinghouse organization remained valid. And, in any case, the unskilled groups were already organized and a power in the international. There was no turning back.

The crisis came to a head at the fifth Amalgamated convention. The delegates gathering in Cincinnati were in an aggressive mood. Before the first session on May 9, 1904, representatives had met separately by branches of the packing industry. Scales of wages and conditions had then been drawn up as the basis for industrywide negotiations. Most of the national officers were apprehensive. President Donnelly himself advocated only fixing a standard June starting date for all contracts. Certainly the right time to press for wage advances was not during the depression of 1904. Donnelly advised against "endorsing . . . wage scales which demand horizontal increases . . . We should protect what has been gained and . . . act very carefully." [44] The convention rejected his sober view, and the scales were presented and adopted.

A vital concession, however, was made to the national leadership. After a lengthy debate, the convention decided to give the executive board discretionary power to act upon the scales "as in the best judgment they shall be most likely to bring success to the organization." This was what had been done two years before when the 1902 convention had adopted general scales. But now the packinghouse membership would not be denied. "After the convention adjourned," Secretary Call later recorded, "the cry went up for an increase of wages all along the line, notwithstanding . . . a general stagnation of business." [45] President Donnelly bowed to the pressure. The packinghouse members of the executive board met in Kansas City on May 29, 1904 and formulated the convention scales into a general program for presentation to the packers.

It was the fatal decision. Having yielded once, Donnelly would find himself thereafter incapable of overriding the judgment of the ranks. At every critical juncture his hand would be forced by popular pressure. The Amalgamated started on a course, as it would seem in retrospect, that led inevitably to disaster.

Negotiations began in Chicago in the second week in June with the representatives of the "Big Seven"—Armour, Swift, Morris, National Packing, Schwarzschild and Sulzberger, Cudahy, and Libby, McNeill and Libby—

who were bargaining as a unit. The Amalgamated was asking for wage advances in two ways. The "equalization" of rates on the Chicago basis meant substantial increases at the other western centers, particularly Fort Worth. In addition, the general level would be moved upward, with a minimum of 20 cents an hour for unskilled labor. The demands were, in Secretary Call's opinion, "absurd, unjust and untenable." They also seemed that way to the packers. Their representatives stated that, in view of depressed business conditions, wage increases were out of the question. They were prepared to renew without change the agreements covering skilled occupations. The wages of unskilled men, Edward Tilden of Libby, McNeill and Libby explained afterward, "naturally are regulated by supply and demand and ought not to be regulated, arbitrarily, by a joint trade agreement." Since the going price for common labor in 1904 was under 17 cents an hour in Chicago, why should the packers continue at 18½ cents—not to say 20 cents? The packers, in fact, had a precedent for the omission of common laborers from agreements, since they had not been included in the scales for cattle and sheep butchers.[46]

The discussions thus opened with labor and management very far apart. The distance was somewhat narrowed when the union dropped its demand for general increases for the unskilled men, and the packers, setting aside their oppositions to bargaining on unskilled rates, offered 16½ cents as a minimum for common labor. The issue thereafter turned on the unskilled scale. On June 24, President Donnelly made a counter-offer that would have made the Chicago rate of 18½ cents the standard for the industry (except in the higher labor markets in Omaha and Sioux City). The packers rejected this proposal. At a final conference on July 1, attended by the company heads, the industry offered a minimum of 17½ cents, but only for killing, cutting, casing, and beef loading departments.[47] There the negotiations reached an impasse.

A referendum of packinghouse locals, ordered June 29, had already rejected the packers' offer of 16½ cents. Donnelly interpreted this as a strike vote. When no further concessions were forthcoming after the July 1 meeting, he ordered preparations for a general strike. The perilous strategy of forcing last-minute concessions had worked for Donnelly before; perhaps it would again. At 2:00 P.M. on July 11, J. Ogden Armour, the head of the packers' committee, was informed that the strike would begin the next day at noon. The sudden move apparently succeeded. The packers promptly met and drew up a reply: they were willing to submit all their differences with the union to arbitration.

Donnelly should have been gratified. He had repeatedly lectured the local unions on the virtues of arbitration. Now, unfortunately, he could not follow his own advice. The arbitration offer "comes too late," he complained. At the expiration of many local agreements in May, the packers had begun to cut wages, and the strike orders had already gone out. Nothing short of a definite wage proposal would satisfy the men. Donnelly could not capitalize on his successful maneuver because he could not control the membership. "It was then impossible for me to stave off the strike," he later admitted.[48] The Butcher Workmen thus surrendered an inestimable advantage: the unknown quantity of their concerted strength. The packers had not been anxious for a showdown because there were too many imponderables in the untried ground of an industrywide strike. Now, although reluctantly, the packers would have the opportunity to measure the union's power in a national conflict.

The strike made an impressive start. The response on Tuesday, June 12, was nearly unanimous in all the western centers except Kansas City, where there had been a severe flood several days earlier. The New York City plants of the western packers closed the following day. Over 50,000 men were out. The strike was orderly. All work was cleaned up before the men left the plants, and the stock handlers were ordered to remain to care for the animals. Union signs in five languages warned the strikers "to obey the union's rules to molest no person or property, and abide strictly by the laws of the country."[49] Although there were some minor disturbances, the general peacefulness of the strike was widely commented upon, particularly in view of the record of earlier violence. The union for the first time demonstrated its organizational strength.

The greater advantages, however, were on the side of the packers. They had, from all appearances, not anticipated the strike. But when it came, they applied their full resources to the struggle; plans were shortly put into motion to start operations. The business depression made available many unemployed men in the packing centers, and strikebreakers, largely Negroes, were shipped in from other points. Skilled men were more difficult to find, but the division of labor required relatively few experienced men for the key jobs. Foremen, superintendents, and branch managers were pressed into service. Some men also were recruited from small country plants—"scab hatcheries," union men called them. This was admittedly a costly procedure, but the packers were prepared to pay the price of inefficient operation and wasted by-products. Within a few days plants were being started up, first in Fort Worth and Kansas City, and more slowly elsewhere. It was soon evident that the packers could eventually operate at substantial levels without the union men. Nor did the anticipated meat famine materialize. In addition to three weeks' supply of meat in the packers' coolers, there was an expansion of output from the independents and small plants, many of them inactive for years.[50] The packers were thus relieved of serious pressure from a hungry public. They quickly became confident of their ultimate ability to defeat the union, and they acted accordingly.

Negotiations to end the strike on the basis of arbitration began within two days. In addition to the original questions, the packers injected a new issue. They would not dismiss any men who had come "to help us operate our plant in time of troubles." Strikers would be reemployed individually according to priority of application and available openings. This struck at the very base of collective activity. Gompers, who participated in the early talks, felt that "of course the men could not agree to so preposterous a proposition." It "implies that we are defeated," complained Donnelly. The packers' price for a settlement was the humiliation of the Butcher Workmen—a price too high for the union. On Saturday, July 16, Donnelly conceded the other outstanding points: the union would accept arbitration without stipulations and return at the rates prevailing on June 12 rather than when contracts had expired in May. But all the men would have to be reinstated within three days or, in a further concession, seven days. The confident packers refused to recede.[51]

The talks broke off with the union in a painful dilemma. The packers' terms were too onerous to accept so early in the fight. On the other hand, as E. A. Cudahy pointed out, "The longer the men stay out the slenderer their chances of being reinstated become. Men are coming in from the outside and will be taken care of." [52] Each day would make the reinstatement procedure, if it prevailed, more damaging to the union.

The Butcher Workmen had one further recourse. The maintenance and mechanical trades, thus far still at work, now intervened. Letters were sent to the packers (and also to the Amalgamated, in an effort to make their move seem like mediation) demanding a conference on July 20 "to adjust the present trouble and to prevent its spreading to our organizations. The unrest that now obtains among our members makes this request imperative." Under the threat of a sympathy strike, the packers met a union committee of Amalgamated and allied trade representatives in a stormy session at the Swift offices. Finally, an agreement was reached. The vital issue of reinstatement was covered by these clauses:

The packing companies signing this agreement to retain all the employees now at work, who wish to remain, and will re-employ all employees now out as fast as possible, without discrimination.

. . . .

Any former employee not re-employed within 45 days from the date work is resumed to have the privilege of submitting his or her case to arbitration, on question of discrimination, decision of arbitrators to govern.[53]

This was hardly a victory for the union. The strikers were still to be reemployed individually and merely to the extent of existing openings. The packers conceded only a dubious procedure to protect against "discrimination" in reinstatement.

The union leaders nevertheless professed satisfaction, and Gompers sent his congratulations on "a great moral victory." "We did the best we could to relieve a bad situation," said AFL Vice-President Thomas I. Kidd more soberly.[54] The outlook for the strikers had so darkened within a week or so that the leadership was willing to settle for a slight face-saving device. Yet the union's defeat was insubstantial. The packers were still willing to go to arbitration. As for the retention of nonunion men, Donnelly considered this "a minor point with us . . . We can attend to that." His comment, although meant for public consumption, was essentially true. An exodus of strikebreakers in fact began from the makeshift dormitories as soon as news of the settlement spread. Chicago police estimated that 1500 departed from the Union Stock Yards in the interim day before the strikers were scheduled to return to work.[55] The settlement, painful as it was, served the vital purpose of putting a halt to the unpromising course on which the Amalgamated had blundered. The crisis, however, was not yet past. There remained the crucial question of the response of the packinghouse workers. Could they be held to the agreement of July 20?

Large numbers collected outside the packinghouses in the Union Stock Yards before 7:00 A.M. on Friday, July 22. Confusion quickly followed. Some men were picked out of the crowds, others were passed over, and there was grumbling of "discrimination." Afterward, some company officials were accused of cursing and abusing the men. Serious trouble might have been averted but for the cattle butchers. All those who applied at Swift and S & S were taken on, but at the Armour house many cattle butchers, including Nicholas Gier and other prominent strike leaders, were pointedly ignored. The others then refused to work. Gier, who was president of the Packing Trades Council, mounted a barrel near the Morris plant and made an angry speech declaring that the strike was on again. Men were dispatched to the various plants in the yards, and the Chicago packinghouse workers were once more on strike.

There was a rush for union headquarters. President Donnelly, who arrived at 7:30 A.M., heard excited charges of wilful violation of the agreement. Passions were thoroughly roused. It was evident also that the renewed strike was becoming widespread, and that opposition was crystallizing against the terms of the settlement itself. His leadership already on shaky ground, Donnelly quickly concluded that there was no choice but to sanction the strike in Chicago and to telegraph the other centers to resume the strike. "Had there been more discipline in the locals of the large centers," Secretary Call later concluded, "no second strike would have been called." [56] "Radical" local opinion forced the renewal of the fight, as it had earlier forced its beginning.

Discussions failed to patch up the trouble. The packers offered to place a "principal" of each concern at each plant to guarantee fairness, but Donnelly considered the agreement voided. A new proposition was needed to "restore confidence." The packers received an ultimatum from the Committee of Allied Trades on July 23:

That *all* employees be hired back within *ten days*. Any person not re-employed in ten days, his or her case will be submitted to the arbitration board.

That *all* killing, cutting and casing department men be reinstated to their former positions within *forty eight* hours after the resumption of work.

We regret to say that if the foregoing propositions are not acceptable to you the allied trades will cease to work on Monday morning.[57]

Thus, under pressure from below, Donnelly raised the price for peace and invoked the threat of a sympathetic strike of the allied trades (which he had earlier in the strike vigorously opposed). The blustering strategy, as it had earlier, failed disastrously. The packers rejected the ultimatum, and the test of strength resumed.

The allied trades struck on Monday, July 25. The stockyard teamsters, despite the opposition of their Brotherhood, came out the following day. The walkout of auxiliary workers was apparently as effective as the July 12 strike of butcher workmen. But, again, the packers found adequate replacements among the large numbers of unemployed. The New York butchers, who had opposed the second strike, came out on August 10, but by then production at the western centers was reaching substantial levels. The packers were claiming later in August that operations were close to normal in all the centers except Chicago.[58] It soon became clear that the unions would have to make their peace with the industry.

But a settlement was no longer possible. When the Illinois Board of Arbitration attempted to bring the two sides together on July 27, the packers stated that because the unions had "failed to live up to" their word, "we do not care to make further agreements with them." During the first week of the second strike, their position had been somewhat equivocal; they had seemed willing to reenter the agreement of July 20. But that, too, changed after they saw that they could replace the auxiliary workers.[59] Thereafter, they refused steadfastly to recognize the unions.

The decision stemmed partly from the evident failure of the experiment with labor organization. The unions seemed to have intensified, rather than to have mitigated, the instability of labor relations. Sudden stoppages and sympathetic strikes had increased. As the price for signing the cattle and sheep butchers' scale in 1903, the packers reportedly had extracted a promise from the union to maintain the peace. President Donnelly admittedly had done his best, but his authority had never been sufficiently effective, and now it seemed entirely gone. The repudiation of the July 20 agreement, followed by the sympathetic strike, amply demonstrated the irresponsibility of the unions. "What is the use of wasting time, money, and patience any longer," asked one official, "when . . . any agreement into which any of these unions might enter is not worth the paper on which it is written?" [60]

Operations became more troublesome under union conditions. Shop committees raised seemingly innumerable questions of seniority, dismissals, work loads, and so on. Even union men acknowledged that the grievances were "often ridiculous and unjust." The resultant difficulties were probably inevitable in the first stage of unionism, but the packers did not take the long view. The president of one concern unburdened himself to a reporter:

Three years ago my output for June was 30 per cent larger than it was this year and (as he consulted a mass of tabulations) my payroll was ten per cent smaller. The men have handed me a long list of the amounts of work which my several types of employees may do each day. And at that I tell you there are fewer things they will do than they won't. They won't allow me to hire men except under certain regulations. And they won't allow me to promote men except under certain regulations. No man can serve two masters. I am master here. I am willing to pay market rates for my labor, and then I expect labor to do what I tell it to do as long as it is on my property.[61]

Labor costs rose measurably during the union period. One packer's books revealed an increase of about 16 per cent—10.76 cents per head of cattle—in the cost of slaughtering from 1902 through part of 1904. It is impossible, of course, to determine what part of this rise was ascribable to the unions, and what part the result of prosperous times. However, costs continued to rise in the latter part of 1903 and early in 1904 when recession had set in. The union role seems more certain here. The slaughtering cost per head was 67.97 cents in 1902, 75.54 cents for 1903, and 78.73 cents for the first part of 1904. Had labor costs remained constant, profits would have been 10 per cent higher.[62]

The packers, however, were not seeking a showdown in the summer of 1904. They would, of course, have been glad to be rid of labor organization. But a decisive conflict was neither expected nor desired. No real economic necessity was involved. Despite rising labor costs, the profitability of the packing companies was unimpaired because they were benefiting from a relatively greater price decline for livestock than for meat and by-products. The packers, moreover, could not have predicted the consequences of their own and the union's maneuvers during the negotiations. Nor were they eager to risk an expensive fight whose outcome was entirely uncertain. Donnelly's basic blunder was to force the packers to measure the union's unknown strength in a general strike. When the last dimension of union power—the united action of both butcher workmen and allied trades—was tested and found incapable of shutting down the industry, the packers seized the unforeseen opportunity. They declared an end to collective bargaining in the industry.

The strike continued through the month of August. Violence, largely absent in the first phase, erupted in most centers at the start of the second. Negro strikebreakers were attacked and beaten, and rioting broke out at times. Injunctions were issued at several points, numbers of strikers were arrested, and in Sioux City several companies of militia were imported. The union was, nevertheless, able to maintain reasonable order.[63] Strike funds were a more serious problem. The Amalgamated was financially weak; its treasury in May 1904 held little more than $25,000. When he turned to the AFL for a loan, Donnelly was rebuffed.[64] With contributions from other unions and outside sympathizers, however, the Amalgamated was able to set up commissaries to assist the needy. The summer season, too, favored the men. The strikers demonstrated notable tenacity—the more remarkable because the skilled men had no personal stake in the original issues. Although there were desertions, the ranks in Chicago, and to a lesser extent elsewhere, remained intact into September.

But several weeks earlier it had become evident that the unions could not win the strike. A number of attempts to reopen negotiations were made by third parties: Chicago city officials, an agent of the U.S. Bureau of Commerce and Labor, representatives of livestock raisers and retail butchers' associations. None of these overtures was favorably received by the packers.[65] President Donnelly on August 29 asked for an interview without prior conditions. (He had long since dropped his earlier demands.) The packers refused to see him. A few days later, however, they receded slightly from this hard line. Securing an interview with the reluctant J. Ogden Armour, Mary McDowell, Jane Addams, and Dr. Cornelia DeBey pleaded for an agreement to end the strike in order to avoid demoralization and bloodshed in Packingtown. Armour bowed to the eloquence and prominence of the three ladies. At a meeting in the Armour offices on Saturday night, September 3, a strike settlement was reached:

> Union to call off strike.
> Packers to take the men back as fast as needed.
> Rate of wages of skilled men to remain as before the strike.
> The above to cover all points affected by the strike.[66]

The agreement could not disguise the extent of the union's defeat. Not only were the men not granted a single concession, but they were returning to work without a contract and without promise of the continuance of collective bargaining.

It was too bitter to be easily accepted. The referendum of strikers on September 6 rejected the settlement by a large margin, and the next day every member of the Allied Trades Council voted to continue the fight until better terms could be secured. Donnelly knew that the cause was lost. His last desperate measures after the rebuff of August 29 to extend the strike to stock handlers, railroad workers, and independent plants had not met with any appreciable success. On September 5, the New York unions returned to work in a body. In the moment of defeat, President Donnelly finally defied the radical element. With the consent of the executive board and the local support of Vice-President C. E. Schmidt, he officially called off the strike on September 8. The next day, the Allied Trades Council reluctantly did the same. After eight turbulent weeks, the industry was again at peace.

CHAPTER 4

QUIESCENCE, 1904–1917

The union's stake in the great packinghouse strike became measurable only in defeat. The issue, Amalgamated leaders had assumed, comprised the terms of employment in the industry. So it would have been in the event of a union victory. But failure exacted a steeper price. The Amalgamated paid, not in concessions on wages or hours, but with its very existence in the western centers. Beyond that, the lost strike set in motion a general decline elsewhere in the organization and in the national authority. The mighty Amalgamated was reduced within a few years to a remnant organization, and only after a decade did it begin to rise again.

At the end of the strike, President Donnelly had urged "every member to wear his [union] button when returning to work." The right of union membership was specifically acknowledged by the packers. Moreover, while failing to get a contract, Donnelly insisted that the Amalgamated had achieved "a bona fide settlement of the trouble." "Although we have not won all we went out for, we go back as an organization . . . The Butchers will build up a better and stronger organization from the experience of the past." [1] In fact, no insuperable obstacles seemed to oppose the packinghouse unions. Yet, built up laboriously, they easily crumbled away.

The immediate impact of defeat was deeply disruptive. Many union men were not reemployed at once. Although strikebreakers left the plants in large numbers, many competent workers among them remained. In Chicago, 6000 scabs were reported still at work a month after the strike. The packers further reduced the number of available jobs by a new policy of providing full-time work to employees. Rioting broke out in Chicago among disappointed men, and in Sioux City the strike was declared on again by angry union leaders. Once peak operations were resumed, the bulk of the strikers, particularly the skilled men, regained their jobs. But a vital core of activists was shut out of the industry in the critical period after the strike. Construing their agreement not to discriminate, the packers drew a distinction between members and agitators. "A certain proportion who made trouble will never get back," asserted one plant superintendent. The reemployment process had sorely damaged the union cause.[2]

An overwhelming sense of defeat covered the packing centers. Only a few shared the defiance of an Omaha worker who urged the men to "rally to the standard of organized labor . . . although we were recently beaten almost to complete submission." The union journal berated the packinghouse workers for their timidity. "About 50 per cent . . . are lying down whining like a whipped school boy and imagining that all is lost." It was a startling transformation. "When you first started many of you were compelled to meet in secret . . . Therefore, we cannot see why we should be wearing sack-cloth and ashes . . . The organization can be back to its old prosperous condition if the membership only will that it shall be."[3]

More than fear was undermining the union cause. Internal differences, muted during the period of success, now burst into the open. President Donnelly was fiercely attacked for his conduct of the strike. He had to deny rumors that he was building a mansion on Michigan Avenue with his proceeds for selling out the union. Many members quit in embitterment, including the Chicago Sausage Makers as a body. The Chicago Packing Trades Council, without the sanction of the executive board, submitted to all affiliated locals a resolution for a referendum to put the defense fund in the hands of local unions and also to reduce the per capita by five cents. This unauthorized move, in Secretary Call's opinion, "did more to disrupt and discourage the organization than the strike itself."[4] Besides the repudiation of the Amalgamated leadership, divisions appeared among the groups in the packing centers. The skilled sheep and cattle butchers talked of splitting off from the main body of butcher workmen—following, ironically, the logic of the packers concerning the organization of the unskilled.

Thus the immediate aftermath of the great strike was a quick decline of packinghouse organization. The locals of unskilled men disappeared, and the membership of the others shrank radically. The executive board, meeting in St. Louis on September 26, 1904, accepted the inevitability of this reaction. A "waiting policy" was adopted until confidence revived and recrimination subsided; then the Amalgamated would "go on in a conservative manner and push the work of organizing the Local Unions the same as in the past."[5]

But the past could not be recaptured. Reorganizing work, quickening in mid-1905, had little impact. New locals did not form, and the existing organizations remained small and ineffective. Donnelly's puzzlement grew in the face of continuing failure. Why could he not repeat his earlier success?

The grievances which had animated the first organizing drive were largely absent in the post-strike period. Wisely, the packers forebore from sweeping away the improved conditions gained in the union period (although many of the abuses would reappear in time). The new policy of full-time schedules, in addition, was important to its beneficiaries—those, that is, who were not eliminated by the change. Finally, as they had indicated, the packers did not cut the wages of the skilled men after the strike. Unskilled rates, which were reduced, resumed the 1904 levels in 1907. President Donnelly admitted, in his quarterly report ending July 31, 1905, that "the conditions now existing in the meat industry are much better than they were prior to the organization of the workers." [6]

More enlightened policies gradually evolved. In 1907 Swift & Co. established an Employees' Benefit Association which by 1909 had 14,000 members. Other packers, not including Armour, followed this example. A stock purchase plan, permitting employees to buy stock on time while receiving dividends, was inaugurated by Swift in 1906; over 2000 employees used the plan when the company floated a $10,000,000 issue in 1909. Morris and Armour in 1909 and 1911 introduced pension plans, and in 1912 Swift put a weekly guarantee of forty hours in its butcher departments into effect. The unique plan, designed to counter the irregular employment in the industry, was widely adopted (except by Armour). By 1917, fifty-one of eighty-three plants investigated by the Department of Labor provided some form of weekly guarantee. [7] These efforts, although less impressive than the welfare programs in some other industries, nevertheless represented significant gains for packinghouse workers.

The advances particularly benefited the strategic group of skilled men. Secretary Call, indeed, believed there was "an agreement between the employers and a few of the so-called better workmen who are willing to enslave the balance of their fellow workers to get good conditions for themselves." The cattle and sheep butchers, he explained to the settlement worker Mary McDowell in 1908, criticized him for wanting "to organize all who are working in the industry under one head but I yet believe that it is the only practical plan . . . The packers are working hand in hand with the sheep and cattle butchers to destroy the International Organization." [8] Good treatment plus a renewed sense of exclusiveness made the skilled men impervious to appeals in the post-strike period.

The unions, for their part, were demonstrably incapable of providing benefits. The packers considered void even the cattle and sheep butchers' contracts, whose expiration date was October 1904, and of course refused to renew them at that time. It was imperative, Donnelly well knew, to restore
relations with the packers. An attempt to seek agreement for some unions in
May 1905 was rebuffed by the companies. Little satisfaction was to be had
even on grievances.[9] In its weakened state, the Amalgamated had little to
attract the packinghouse workers. The dangers of union membership, on the
other hand, were palpable. Despite the promise of no discrimination, the
reality seemed otherwise. Active men found themselves dismissed for un-
accountable reasons. The experience of Dennis Lane, later international secretary-treasurer, was typical. After the strike, he was reemployed at the Swift
plant in Chicago. When a serious grievance developed in his gang, he and
two other former shop stewards applied to the superintendent. Soon afterward, Lane was discharged and was unable to find work in Chicago. Another veteran, recalling the post-strike situation, had found "everything
lovely" so long as there was no agitation. "But if they heard anything regarding such activities, and they seemed to have a way of finding it out, the man
found guilty had about as much chance to retain his job as a snowball in
Hades." [10]

The need for clarification became urgent. The AFL Executive Council, at
Donnelly's request, in November 1906 appointed a committee to sound out
the packers. They reasserted their policy of nondiscrimination and even intimated a willingness to deal with a butchers' union, so Donnelly reported,
"conducted on a sound business basis." But no change became apparent in
practice. A second committee approached the companies the next year, "because the packers seem to be, and the butcher workmen believe they are,
antagonistic toward the Union." [11] That conclusion, hitherto uncertain, was
becoming inescapable.

Finally, the strike experience had left a seemingly permanent impress
on the workers. Letters from packinghouse locals complained of hostility
among the men: "the packers are not treating us as mean as some men (?)
who were formerly members of the union." [12] A general apathy toward re-
organization was apparent. It was also significant that, when unrest developed among the lower-paid men in the boom of 1907, the union was unable
to capitalize on the opportunity. The failure of 1904 placed a continuing
stigma on unionism in the western packing centers.

Even the surviving locals succumbed during the depression of 1908. There
remained in the Chicago yards a single union, small and ineffectual, retaining

an Amalgamated charter. The packers, for their part, bargained in Chicago only with the strategic Teamsters' Local 710 and with the Hair Spinners' Union. By 1910, labor in the western centers was almost entirely unorganized. What seemed worse, there existed greater obstacles than in 1900 in the path of unionization. The men were unresponsive, and the packers for the first time followed a fixed policy against dealings with trade unions—if not openly against union membership by their employees. The 1904 debacle had permanently altered the labor situation in meat packing.

After the great strike, President Donnelly had made the reassuring observation: "Remember that A.M.C. & B.W. will live no matter what the men in the big packing centers may do." [13] The local meat trade provided an independent source of union strength that would sustain the Amalgamated. But Donnelly was only narrowly correct. The organization would live, but it would not prosper after the loss of the packing centers, for the 1904 defeat set in motion a widening circle of disruptive forces. By 1910, only a small part—certainly less than 5 per cent—of the total number of meat workers outside the western centers belonged to the Amalgamated.

Organization of the local trade, less spectacular in nature, had accompanied the packinghouse growth. "Not a town or hamlet of any importance but what we have reached in some way and the seeds of organization sown," Secretary Call had said in 1902. As it grew, the Amalgamated also exerted an increasing attraction on the independents and benevolent associations. Butchers' Union No. 5 of Greater New York, for example, joined the Amalgamated after a defeat in July 1901:

In this strike the butcher boys found it not good to be alone and pay all the expenses . . . As we were not affiliated with the A.M.C. & B.W. of N.A., the other organizations did not uphold the butchers as they should.

Other independent unions in New York affiliated through the efforts of an Amalgamated organizer, H. L. Eichelberger. One of the founders of the Benchmen's Association of Retail Butchers, William C. Wellman, became a leading Amalgamated official in the New York area. In California, the important Local Union 115 had until 1901 functioned independently as the Journeymen Butchers' Protective Association of San Francisco.[14]

Progress was, however, very incomplete. At the apex of the movement in New York in May 1904, Eichelberger estimated that only a third of the meat workers belonged to the twenty-two organizations in the metropolitan area. New York, moreover, had the services of the only full-time organizer in the east. Several district organizers worked intermittently among meat cutters elsewhere. The executive board agreed in 1902 to split the salary with Local 115 for an organizer to build up the organization in San Francisco. But the resources of the Amalgamated were expended primarily in the western packing centers. Delegates of retail locals, led by Vice-President Hart, repeatedly urged "that more attention be given the Organization of Meat Cutters." [15] The great strike cut short plans in this direction. Nevertheless, substantial progress had been made in the local trade by 1904. The journal in December 1902 listed 96 retail locals and 35 "mixed" locals; in December 1903 the total was 182.

But the gains in the local trade melted away after the failure in the stockyards. Retail locals had been easily started: there was little employer opposition; business was prosperous; a concrete grievance existed over hours; and the Amalgamated had an awesome reputation. The critical stage involved putting the new organizations on a permanent basis. Here the decision to concentrate on the packing centers took an early toll in the local trade. After the 1904 defeat, the Amalgamated lacked the funds to maintain even the small force hitherto available to the meat cutters' unions. The international was soon "obliged to stand off our Locals when it was believed they would survive. In some cases of our smaller Locals they were sacrificed, but to have attempted the rescue meant a greater injury to the general movement." [16] The retail locals thus had to fend for themselves. Particularly in smaller localities, the odds were heavily against survival. The attrition rate among retail unions, high even in the glorious period of the Amalgamated, accelerated after its great defeat and with depressed conditions in the trade. The number, according to lists in the journal, declined from 130 in December 1903 to 67 in September 1905.

A cadre of local unions did survive the initial difficulties to become effective organizations. They existed in upstate New York—the cradle of the international—in medium-sized cities such as Seattle, Wheeling, Duluth, Louisville and Youngstown, and a few large cities such as St. Louis, Baltimore, and San Francisco. But many of these viable butchers' unions also experienced increasing difficulties, in part, because of the inevitable frictions of the early stages of collective bargaining. Sentiment spread among employers, as one prominent San Francisco dealer put it, that "to make our business a success we must run it ourselves, and not have it controlled for us by others." [17] This view was given force and direction by the rise of the open-shop movement.

The packinghouse defeat eliminated the last restraint on employers. On October 1, 1904 the butcher shops of San Francisco replaced their union cards with "Open Shop" cards. Earlier, a strike would certainly have followed, but now Local 115 refused to accept the challenge. The issue did not come to a head despite wage reductions and gratuitous insults. The policy of appeasement probably saved Local 115, but it emerged sorely weakened. By the end of 1906, membership had fallen from 1200 to 290. The San Francisco story was repeated at many other strong points in the aftermath of the 1904 strike. In September, for instance, the important Sausage Makers' Locals 174 and 211 of Manhattan and Brooklyn struck because their employers refused to renew their agreements. The fight was lost, and the men returned under open-shop conditions. Other unions in the local trade declined more gradually in the face of belligerent employers and the depression of 1908. Strong locals under these conditions reached a low point and, in some instances, went out of existence. A few, particularly those in upstate New York, continued to fare well, but by 1910 organization was moribund in many of the earlier Amalgamated centers and nonexistent where the Butcher Workmen had been weak.

There was another aspect to the decline of the Amalgamated in the local trade after the 1904 strike. The defeat stimulated secessionist sentiment that cut off some of the remaining effective affiliates.

The problem of unity had existed from the start. The butcher workmen, employed as they were in a complex industry, constituted a highly diverse labor force. There was, first, the great division between retail and processing workers. The meat cutters, a fairly uniform group, had a narrow sense of identity. They were, complained Secretary Call, "governed too much by local interests. They seem to think that so long as their local unions or locality are well organized that is all that is necessary." The slaughterhouse workers were afflicted by another form of particularism. Divided by occupation and union structure, they considered themselves cattle butchers, sheep butchers, sausage makers, and so on. Some packinghouse locals had a closer sense of identity with their counterparts in other centers than with other unions in their own stockyards, causing the cattle butchers' unions, for instance, to hold a separate convention in April 1903. There was a further split between the western centers and the minor slaughtering points. The wage scales adopted at the 1904 convention, despite a clause protecting higher rates, were limited to the western centers because the slaughterhouse unions from New York, Cincinnati, and elsewhere insisted on "local autonomy." Even in the successful period, the organization was rent by "internal bickering and dissention [sic]." [18]

The disruptive sentiment was also directed against the international. Amalgamated officers frequently came under attack. The locals, Secretary Call felt, mistakenly considered the international "as something aside from their organization, as if it was a body which existed independent of the local organizations." Disaffection, beyond expressing notions of autonomy, resulted from the need of the international to represent diverse interests. There were continual complaints of the uneven expenditure of per capita and of disproportionate gains by some branches. The leadership repeatedly had to remind critics of the wisdom of attending "to one thing at a time and not direct our efforts in too many directions at the same time." The trouble was, Call observed later, that "the Butcher Workmen of the country are a sensitive and suspicious class." [19]

The eastern sheep butchers demonstrated the divisive tendencies of the early period. This highly skilled group in 1902 feared the introduction of the western "ring" system of sheep dressing. The innovation threatened to "make laborers of all of us that call butchering a trade." The unorganized Buffalo men and the independent New York union consequently joined the Amalgamated to gain its support in the fight against the ring system. Although the international had never fought technological change, the executive board agreed to sanction a strike in Buffalo against the New England Dressed Meat and Wool Company in November 1902 after receiving assurances that the Swift subsidiary would back down before a show of union strength. When this did not occur, a crisis developed in the Amalgamated. The Buffalo local, supported by the New York union, demanded a general strike and, on being refused, attacked the international officials for "discriminating against the east for the benefit of the west." Gompers was warned that unless some action was taken, an application would be made to charter a separate eastern organization of butcher workmen. "After much pressure had been brought to bear," the executive board permitted the Boston sheep butchers to strike in sympathy on May 4, 1903. Finally, having expended over $20,000 on the Buffalo conflict—despite the actual ineligibility of the Buffalo local for strike benefits—the executive board called the strike off on February 4, 1904. The Buffalo men ignored the order and continued the fight on their own. The New York local, whose "fine Italian handiwork" Secretary Call thought was behind the strike, remained tenuously with the Amalgamated until the great strike.[20]

The discordant impulse, while troublesome, was not dangerous during the ascendant period of the Amalgamated. Easier relationships no doubt would have developed in time, but, again, the stockyard defeat intervened and activated the centrifugal forces. Secretary Call in 1906 catalogued the various plans "to divide the organization [:] the so-called skilled men are desirous of forming one body; the sausage makers another; a movement is now on foot in Boston to organize the grocery clerks and meat cutters into another; all to be international bodies." [21] This activity generally had the effect of undermining organizations.

In New York City, however, the outcome was a strong dual body. The metropolitan unions, although without grievances or wage demands, had responded in 1904 to the first order to strike the plants of the western packers. They had opposed the second strike call. International officers managed to get a half-hearted renewal of the New York strike after two weeks. As the outlook darkened in the west, a New York secessionist movement developed under the leadership of John Kennedy, chairman of the local strike committee, and J. J. Craven of the Sheep Butchers. A cold-minded and ambitious man, Kennedy charged the western leadership with incompetence (the agreement ending the first strike was "the weakest, most stupid and unfairest settlement ever signed by a representative of organized labor") and with exploiting the eastern unions. The New York locals, he said, had to withdraw from the Amalgamated to save themselves. A new body was formed—the Brotherhood of Butcher Workmen—and an agreement was reached ending the New York strike on September 5, 1904, a week before the termination in the west.[22]

The Brotherhood refused to return to the fold afterward; indeed, it attempted to draw in the discontented Chicago unions. The independent, although limited to the New York area, quickly became a stable organization, for a time probably stronger than the Amalgamated. Kennedy faced only one painful drawback: illegitimacy within the labor movement. Through the AFL, the Amalgamated was able to force the withdrawal of the secessionists from the city central and state bodies in New York. Repeatedly, a reconciliation was sought in the following years, both by the AFL and the Central Federated Union of New York. In April 1908 an agreement was wrecked by Kennedy's insistence on a new name and charter for the Amalgamated because the butcher workmen were "dissatisfied with its past history." The AFL sustained the Amalgamated refusal, and the two unions continued apart.[23]

The final effect of the packinghouse strike was on the international authority. Ineradicably, the Amalgamated leadership was stigmatized by the defeat of 1904. After repeated explanations and justifications, the officers had to plead, as Secretary Call did at the conclusion of the 1906 convention: "You must understand that we are only human and liable to err. Mistakes will happen; if they do you must not become indignant, but give us a chance to rectify them." But the memory of defeat would not be exorcized. Even ten years later, it was necessary to defend Secretary Call by reciting "at length the correct history of the 1904 strike, showing conclusively that the representatives of the organization at that time made the very best possible of an unfortunate situation."

The principal victim of these recriminations was President Michael Donnelly, the creator of packinghouse unionism. Now he was subjected to "a campaign of vituperation, falsehood and abuse." He had enough support to be reelected in 1906, but the fire had gone out of him. His discouragement became evident in his preoccupation with a labor party and the fear of an invasion of Japanese workers in the stockyards. He suffered also from ill health—probably alcoholism—aggravated by two brutal beatings in retaliation for his reform activities in the Chicago Federation of Labor. Finally, in May 1907 Donnelly resigned, as Call explained, with assurances that his departure would make possible a reconciliation with the dissidents and "would add thousands of members within the next three months." [24]

Donnelly's later career is a sad tale. He became a cigar salesman in Kansas City and, failing in that, was reduced by December 1908, according to the *National Provisioner,* to "wheeling crushed stone . . . for $1.75 a day. Such is the ingratitude of republics—and others." Then he dropped from sight. On two occasions, Donnelly's invalid wife appealed to the Amalgamated for assistance. Donnelly himself reappeared briefly in early 1916. Secretary Call received a letter from him in Texas, where he was serving as a camp cook. His friends got him an appointment as an AFL organizer to be stationed in Chicago. "But before that could be accomplished," Call reported, "his old malady developed and he disappeared and we have been unable to locate him since, much to our regret." [25]

The international's resources meanwhile disappeared. The packinghouse strike left the Amalgamated several thousand dollars in debt. A declining membership thereafter prevented the creation of a sound financial base. The receipts for the two years ending May 1, 1906 totaled $57,865 on an average dues-paying membership of 8037. Thereafter, income shrank further. On May 1, 1910 the Amalgamated had a membership of little more than 2000 and $850 on hand. [26] The consequence was the shriveling of the international's functions. Only the secretary-treasurer remained on salary, and by 1910 only one international organizer was in the field. The journal was discontinued in 1908. Funds were lacking even to print the convention proceedings of 1910 for distribution to the locals.

The connection of the locals with the international became very tenuous. The constitution had been amended in 1906 to hold conventions at four-, rather than two-, year intervals, but permitting a referendum after two years for an earlier convention. When Secretary Call suggested such a referendum in April 1908, only half the locals responded: one was in favor, nine opposed, and the rest reported unwillingness to send a delegate to a convention.

The executive board met in discouragement on May 25, 1908 in Syracuse. Only five members were present, two of the vice-presidencies being vacant. The board considered holding a referendum to decide whether to surrender the charter and affiliate the surviving locals directly with the AFL. This unhappy course was rejected, as was the proposal of John Kennedy's Brotherhood of Butcher Workmen to start a new international.[27] The remnant organization, it was nevertheless clear, was in the last stages of existence.

The international union survived this time of adversity. Partly, it was the unwillingness to part with the proprietary right, legitimized by the AFL charter, over the organization of American butcher workmen. Despite its vicissitudes elsewhere, moreover, the international continued to have a solid base in the unions of upstate New York, where both the active leadership and general office were located. The tenacity of Secretary Call also was important; for a time, as in the beginning, the international office was in his Syracuse home. An improvement occurred in the leadership. Donnelly's successor as president, E. W. Potter, was a Utica meat cutter who left no mark on the history of the Amalgamated. He was followed in 1910 by John F. Hart, also of Utica and one of the founders of the international. Hart was a fortunate choice at that juncture. By foresight or luck, he had resigned from the executive board just before the 1904 strike, and was thus invested with a reputation for infallibility, as well as with neutrality in the internal strife of the subsequent period. He was also suited by temperament to conserve the small resources of the Amalgamated. His policy recognized that "there were two things certain—first, an aggressive movement would have been fatal; second, money was necessary, and I decided that a patient waiting game was necessary." [28]

Retrenchment succeeded because it coincided with the resurgence of organization in the local meat trade. During the prosperous period from 1910 to 1914, sixty-three new charters were issued, and seventeen reissued. Perhaps more important, the established unions were increasing in size and strength. Membership rose to approximately 6500 in 1914. The larger income permitted more international activity. Four full-time organizers were on the payroll in 1914, as well as President Hart, and in January 1915 there appeared the first issue of a new monthly journal, the *Butcher Workman*.

An accommodation of autonomous elements accompanied this renewal of the organization. A special article (XIII) established a District of Greater New York with a council holding much greater powers than had been permitted to the earlier packing trades councils. The District Council assumed the right to receive the ten cent portion of the international per capita designated for defense purposes, and also could assess the locals and require their membership. The District Council jealously guarded its autonomy, curtly refusing, for example, Secretary Call's request in June 1913 for information on its financial affairs.[29] The council achieved its main purpose of building up the New York movement within the Amalgamated, although Kennedy's Brotherhood continued to flourish. By 1914, the district comprised ten locals with a membership of over 2000. California developed an organization with more limited autonomy. The secessionist Local 115 of San Francisco reaffiliated in 1912, and that year a state body was organized—the California Federation of Butcher Workmen. The federation exercised considerable authority within the state, but, unlike the New York organization, it did not take over the defense financing role from the international. At the urging of Pacific delegates, however, the 1914 convention empowered the executive board to permit state and district bodies to withhold the defense portion of the per capita tax. The international strengthened its connection with the two organizations by giving them representation on the executive board. Both M. R. Grunhof, president of the California Federation, and John Roeschlaub of New York City were duly elected to vice-presidencies in 1914. The Amalgamated was gradually coming to grips with the divisive tendency of the butcher unions, as well as with the hard problem of servicing the unions in the local trade. Part of the answer was geographical bodies with substantial powers over their member locals but also loyalty to the Amalgamated.

A rapprochement was meanwhile developing with the Brotherhood of Butcher Workmen. In early 1916, President John Kennedy of the independent New York union had told Gompers, as the latter explained to the AFL Executive Council, "that their interests could not be advanced if they remained within their present circle . . . Either there ought to be some amalgamation between the two organizations, or else it would be necessary for his organization to proceed with an organizing campaign in other districts." Anxious to effect a reconciliation, Gompers arranged a conference in New York City on March 6, 1916. There was strong favoring pressure within the Amalgamated, particularly from the District Council of New York which, as Vice-President Joseph Menhart told Gompers, "feel[s] that there is no place for two organizations in this city." No immediate results were forthcoming, but clearly amalgamation was imminent.[30]

Internal difficulties, however, were far from over. The efforts of President Hart to consolidate his position touched off serious discord. Hart, in attempting to develop a following in California, created a bitter enemy in President Grunhof of the California Federation.[31] Conflict also developed between Hart and Call. Secretary Call was reelected in 1910 with the explicit understanding that he limit himself to running the general office, "leaving the entire management of the organization to the President, where according to the Constitution it properly belongs." Hart, remaining dissatisfied, worked quietly against Call's reelection in 1914 and, failing that, largely by-passed the secretary's office. Call had also aroused criticism because of his outside activities (in 1914 he was elected New York state treasurer and was also for a time president of the New York State Federation of Labor) and because of his support of an unpopular organizer, Rudolph Modest, in New York City. Secretary Call's deteriorating position was also involved with the improving relations with the Brotherhood of Butcher Workmen. Kennedy had asserted that, because of his part in the 1904 defeat, "Secretary Call ought to be eliminated from the office." Gompers had replied that Call "really was not responsible for what occurred," and Kennedy had "deferred" to his judgment. Nevertheless, the secretary knew that he could not survive the admission of the Brotherhood. Finally, there were differences over policy in the period of rising union prospects after 1914. Call, having grown old in the service and hesitant because of earlier failure, advocated a cautious approach against "the opinion of others [that] we should be more aggressive." [32]

The internal confusion suddenly crystallized in early 1917 on the issue of a special convention. Recognizing the adverse actions likely to be taken there, Call spoke out against the convention. In the ensuing controversy, the New York Council—the initiator of the convention movement—, Dennis Lane of Chicago and his supporters, and the Hart faction found themselves aligned as the reform party. Soon after the favorable outcome of the convention referendum, Vice-President Menhart began discussions with Brotherhood representatives. President Hart agreed that delegates of the eleven Brotherhood locals could be seated at the convention "with full rights and power." With this augmented support at the special convention in Fort Wayne on July 24,

1917, Hart swept his critics, including Grunhof of California, from the executive board. Secretary Call recognized the hopelessness of his position and retired without an open fight.[33] His successor was Dennis Lane, a Chicago cattle butcher long active in union affairs and for several years an international organizer and vice-president. In the long run, his election would prove the most significant act of the 1917 convention; Amalgamated history for the next twenty-five years would be bound up with Lane's career. The immediate consequences of the confrontation were to put an end, first, to dual unionism in the industry and, second, to the consuming internal conflict within the Amalgamated. The union was, to the extent of its slender resources, prepared for the great task ahead.

The packinghouse workers of the west had gradually receded in the considerations of the Amalgamated leadership. The retail branch, President Hart observed in 1914, "is the obstacle that prevents the signing of our death warrants." The temptation was strong, particularly for a man like Hart who had no ties with the packing centers, to concentrate on the local meat trade. Secretary Call, for his part, was resentful of the actions of the packinghouse workers, especially the skilled men, after the 1904 strike. The Amalgamated, he wrote in 1908, would "want to see the men of Chicago manifest some interest in the matter [of organization] because of the fact that there are broad fields in which to work where we can get results without going into a section that is rent by internal strife." [34] But, even in its darkest days, the Amalgamated did not entirely write off the packinghouse workers. John T. Joyce, the Chicago cattle butcher, was called to the executive board meeting in 1908 as packinghouse spokesman when the possibility of disbanding was considered. Thereafter, the packinghouse interest was continually represented on the executive board, first by C. F. Smith of the Chicago Casing Workers' Local 158, and after his death in 1913 by Dennis Lane of Cattle Butchers' Local 87. Occasionally, an eastern official would tour the western centers and even make a tentative organizing effort. But no sustained work was undertaken.

The Amalgamated shifted the burden to the AFL. Secretary Call, prodded by the Chicago settlement house worker Mary McDowell in 1912, discussed the packinghouse problem with President Hart. They decided to approach Gompers, "laying all the data before him and if possible persuade him to take up the question of organizing the Stock Yards." Call remarked afterward that he had urged Gompers "to plan his own campaign, manage it in his own manner, and organize anyone in the Stock Yards into any craft or calling he saw fit, paying no attention to the Amalgamated . . . only to strive to gather them under the banner of the A.F. of L., and he could rest assured that the International organization would file no protest." When an AFL campaign began in October 1912 under the direction of John Fitzpatrick of the Chicago Federation of Labor, the Amalgamated took a negligible role in the abortive attempt. The Amalgamated accepted only a residual obligation in the packinghouse workers at this time.[35]

The outbreak of war in Europe radically altered the labor situation in meat packing. By 1915, the demand for meat was up sharply, and, simultaneously, the labor shortage became severe. Heavily dependent on a floating labor supply for seasonal expansion, the packers were particularly hard hit by the break in immigration from eastern Europe. Labor unrest spread through the packing centers. Besides many minor skirmishes, serious strikes occurred in Sioux City in March 1916, East St. Louis in July 1916, in the Armour plant in Chicago, in Winnipeg, Cincinnati and elsewhere. Business, a St. Louis packer observed, "cannot be conducted in an orderly manner . . . in this age of unrest." [36] The AFL, acting on the request of the Amalgamated, became active again in the packing industry in August 1916. By the end of the year, at least five of its roster of 32 paid organizers were assigned to the stockyards drive.[37]

The Amalgamated response was tied in with the strife inside the organization. President Hart told Gompers in February 1916 of "considerable friction" with Secretary Call "upon an active campaign for the extension of the organization." Call represented the conservative position: the unionization of "the smaller localities first, building up an organization sufficiently strong to finance a movement to get into the larger centers with sufficient organizers to remain permanently in one place, to hold the results obtained." Besides the question of priorities, a packinghouse drive involved other controversial issues. A main argument for the inclusion of the Brotherhood was, Gompers explained, that "thereafter a united campaign of organization might be inaugurated." Likewise, Hart's reasons for the special convention in 1917 emphasized "reforms . . . that would make it possible to organize the unorganized packing centers of the country." In all these issues, the opinion prevailed, as Dennis Lane said, that "now is the real time for the men and women in the packing industry to get organized." It was the "golden opportunity." [38] (Ironically, Homer Call, who had steadfastly championed packinghouse organization, particularly of the unskilled, lost out as an alleged anti-stockyards man, and his chief antagonist was John Hart, whose career both earlier and later was notable for its indifference to the packinghouse interest.)

The Amalgamated plunged into the packinghouse campaign. Starting only with Lane, its organizing force in the meat centers expanded to four by early 1917. Quickly outrunning its income, the international in December 1916 levied an assessment of a dollar a man, payable in four monthly installments. Despite grumbling and the voluntary nature of the assessment (there being no constitutional provision for it), $2148 was collected, and the organizers remained in the field. Their limited number imposed a strategy of concentration on the points of manifest unrest. When, for example, East St. Louis workers struck in late July 1916, Lane hurried to the scene, called meetings, chartered locals, and reached an agreement with the packers reinstating the strikers and establishing grievance committees. In Sioux City and elsewhere, organizers also built unions in the aftermath of spontaneous strikes.

The permanent results of this work were small. After a year, of the thousand men organized in five locals in 1916, one local with eleven members remained in East St. Louis. Similarly, Local 577 of Cleveland grew from eighteen members to 700, "and then . . . went down and out quicker than it grew." By July 1917, organization was slightly advanced only in South Omaha, Sioux City, and Chicago.[39] Yet the concerted effort of 1916, despite its lack of tangible success, expressed the recommitment to the packinghouse workers. The Amalgamated would not fail to grasp the opportunities for organization presented by wartime America.

CHAPTER 5

THE SECOND CYCLE
IN MEAT PACKING

The American entry into World War I created the second union opportunity
in meat packing. Workingmen became, as never before, receptive to trade
unionism. It was immensely significant also that military exigencies forced
the Federal government to assume an active role in labor relations. The un-
answered question was whether the Amalgamated could surmount the employer opposition and internal instability which had cut off the first organizational growth in the packinghouses.

The Amalgamated and the AFL had reached the limit of their efforts in
packing before April 1917. The organizational upsurge, when it occurred in
the autumn of 1917, was the result of forces other than union agitation. The
foundations of packinghouse organization were laid in the course of the unrest evident in the western centers in the first months of war. The events of
South Omaha were typical. A strike began, not in the small local of skilled
men, but among the unskilled where "no effort" to organize had "been
made . . . up to this time." On Saturday, September 1, fifty truckers quit
work at the Armour plant, followed by increasing numbers in the next few
days. By Wednesday, the local paper reported, "excitement was at fever heat
. . . over the strike situation." When over 4000 employees of the Cudahy,
Morris, and Swift plants joined the Armour strikers the next day, the conflict assumed major proportions. The movement, thus far spontaneous and
leaderless, now acquired professional direction from local unionists and
then outside officials, including Amalgamated Vice-President Fred Schmidt
of St. Louis. The plant managers first refused to meet strikers' committees,

but they backed down after the arrival of a Federal conciliator. The strike was settled on September 12 on terms of a 2½ cent an hour increase, reinstatement of all strikers, no discrimination for union membership, the establishment of grievance committees, and centralization of hiring under one man at each plant. "With this great gain," reported organizer John Blaha in November, "the workers returned to work and started right in to build up their organizations to full strength." [1] The essential task of Amalgamated and AFL organizers was to harness the energy released by wartime conditions in the packing centers.

Chicago presented a different situation. The Union Stock Yards remained quiet through the summer of 1917. Organization was initiated here, not spontaneously, but through the efforts of the local labor movement. The Chicago Federation of Labor in July 1917 adopted a resolution, offered by William Z. Foster of the Railway Carmen and supported by Dennis Lane, for a joint campaign in the stockyards. John Fitzpatrick, the president of the Chicago Federation, immediately created the Stockyards Labor Council made up of the local unions with jurisdiction in the packing industry. Fitzpatrick became temporary chairman, and Foster became secretary. The organizing corps was drawn largely from officials and volunteers of local organizations, including dedicated ladies from the Chicago Women's Trade Union League. Some outside help was also forthcoming from the Amalgamated, the Brotherhood of Railway Carmen, the Illinois District of the United Mine Workers, and the AFL. [2]

The drive began on Sunday, September 9, 1917, with a mass meeting. The Chicago *Tribune* reported 1400 present, but when the call for signing up came, Dennis Lane later recorded, there was "a dull silence for a moment— then many of those in attendance began to slip away." Sure of the sympathy of the men, Lane tried an old circus technique. At the meeting the following Sunday, he planted a number of meat cutters from Chicago Local 546 in the crowd. They stepped forward and put down their dollars (later refunded) and were followed by several hundred packinghouse workers. Membership increased slowly during the next weeks. Only in November did there begin a great influx of recruits. Foster ascribed the sudden success to the attempt to enter negotiations with the packers at that time: "The effect upon the discontented mass was electrical. At last they saw the action they wanted." Amalgamated Vice-President T. A. McCreash reported in December that "they could not take them in fast enough in Chicago." [3]

By the end of the year, organization was well advanced in the packing centers. The President's Mediation Commission put the extent of organization at from 25 to 50 per cent. The Amalgamated received per capita from 28,229 members in January 1918, an increase of about 20,000 from the packing industry since July 1917.[4] The varying estimates indicated that the unions were still in a minority position at the end of 1917, but they were substantial enough, under existing emergency conditions, to take the initiative against the packers.

The labor situation alarmed the industry. Its initial response was to raise wages. A series of "voluntary" increases pushed up the unskilled rate in Chicago from $17\frac{1}{2}$ cents an hour in 1915 to $27\frac{1}{2}$ cents by September 1917. "These advances," explained a Swift official, would result "in the increased contentment among our employees." Toward that same end, packers intensified their welfare programs. Pension funds were established by Wilson, Swift and T. M. Sinclair; a life insurance policy was begun by the Cleveland Provision Company; and J. Ogden Armour contributed $50,000 for a plant gymnasium. Swift & Co. increased its guaranteed work week from 40 to 45 hours in January 1916. Employment bureaus were started to improve hiring procedures and reduce labor turnover. Finally, several major packers united to create a "coordinating center" for their welfare activities in the Chicago stockyards. It was necessary to admit, M. D. Harding of Armour & Co. told the American Meat Packers' Association in October 1917, that "we have paid less attention to the workers than we have to any other end of our business." [5]

This belated recognition could not arrest labor discontent. Once the United States entered the war, moreover, the packers' anti-union weapons became ineffective. The industry was straining to meet the requirements of the Food Administration; the month's run of cattle in October 1917—415,456 head—was the largest in the history of the Chicago stockyards. Work stoppages became inadmissible under these circumstances. Since strikes would result from resistance to union activities, the standard tactics had to be discarded. In fact, the packers specifically agreed not to discriminate in the settlement of the Denver, Omaha, and Kansas City strikes. Labor leaders acknowledged the absence of open opposition, although charging the meat companies with the use of indirect methods, such as planting spies and supporting bogus rival unions.[6] In any case, the employers could not hamper the wartime drive to any significant extent.

This circumstance was not crucial to the packers. The essential feature of the "open shop" was not the prevention of union membership; publicly, indeed, open-shop advocates proclaimed the right to join unions. The sticking point was the refusal to "recognize" unions. Collective bargaining was the vital issue.

Labor leaders half-expected the packers to yield when confronted with a combination of union strength and guarantees of good behavior. Gompers and John Kennedy had been encouraged by the testimony of J. Ogden Armour before the Commission on Industrial Relations in 1915. The magnate had asserted—clearly referring to his earlier experiences with the Butcher Workmen—that "the unions have not worked well at the stockyards, because as soon as they have . . . power they have not known how to proceed." But he also knew "unions that are very well run, and they are a good thing for the employer as well as the employee." [7] The rigidity of the industry's open-shop position became certain only when the unions attempted to breach it.

On November 11, 1917, an Omaha meeting of Amalgamated packinghouse representatives formulated a set of demands: union recognition; the preferential shop; time and a half for overtime; wage increases of a dollar a day for time workers, 10 per cent for pieceworkers; equal pay for men and women. A separate meeting of representatives of other internationals two days later endorsed the Amalgamated program and adopted similar demands for the allied packinghouse trades. A committee consisting of Lane, Fitzpatrick, and Foster approached the "Big Five" packers—Armour, Swift, Cudahy, Morris, and Wilson. The packers received the committee, as Lane reported, in "a civil manner," but acknowledged no labor difficulties or need to deal with representatives of organized labor.[8]

The rebuke precipitated a crisis in the union ranks. The cries for action were intensified by the dismissal of a number of unionists from the Libby, McNeill and Libby plant in Chicago. On Thanksgiving eve the packinghouse workers at the major centers voted by a large majority to grant the leadership the authority to call a strike as a last resort. A Federal conciliator had the union men reinstated, although at a lower pay scale, but he could not persuade the packers either to meet union representatives or to consider the demands drawn up at Omaha.

Three avenues were then open to the union leaders. The passive course of awaiting further developments would almost certainly have reversed the influx of new members. A second possibility was to put the strike order into effect. (Later, Foster and other Chicago radicals would mark the failure to do this as the fatal error of the unions.) But considerations both of patriotism and caution were binding. The packers, Lane feared, were trying to force a premature strike. The purpose of the strike vote, he insisted, had been only to strengthen the unions' negotiating hand. The final avenue, combining safety and action, was to use the wartime powers of the Federal government against the packers. (Samuel Gompers, fully alive to the new opportunities in this direction, had already been urging the Amalgamated to insist on a voice in the Food Administration.)[9] The cooperating organizations adopted this third approach. In mid-December, a delegation headed by John Fitzpatrick set out for Washington.

Secretary of War Newton D. Baker, impressed by the warnings of the committee, wired the President's Mediation Commission, then in Minneapolis, to proceed to Chicago to "prevent the very serious situation . . . if a strike . . . took place . . . and spread to others there and elsewhere as is feared . . . The men are being held back with difficulty."[10] Arriving on December 21, the commission gradually worked out an adjustment along the lines of its settlement of the Arizona copper and Pacific Coast disputes. Both the Big Five and the unions signed agreements with the Mediation Commission early on Christmas morning prohibiting strikes or lockouts in the eleven major packing centers for the war duration. Disputes were to be settled by negotiation and, that failing, by the binding decision of a United States administrator. This appointment went to John E. Williams, fuel administrator of Illinois.[11] It was understood that hearings would begin as soon as Williams took office on January 2, 1918.

The arrangement immediately broke down over the issue of union recognition. This question, unreal as it seemed, was of extreme importance to the packers. Their overriding fear was that, by accepting a wartime solution to their labor problems, they would be drawn into a permanent union relationship. The Mediation Commission, it was true, was not supposed to alter existing patterns in any industry, except to protect the right to organize. The packers' agreement with the commission permitted their employees to select representatives who were "to act solely and only for such employee or employees, and for no other person, committee or association whatsoever." It was certain, however, that the labor representatives would be union officials, and the packers feared that, by dealing with them, they would be crossing the line of recognition. To forestall this, the packers insisted that the administrator hear the two sides separately. The precedent was that the packers and labor leaders had met separately with the Mediation Commission, and had signed agreements with it, not with each other. The issue, however, now created a fierce procedural wrangle. Amid mutual charges of bad faith, the efforts to begin the arbitration proceedings were broken off on January 16, 1918. The final blow came soon after with the resignation of Williams on the plea of ill health.

The unions once again resorted to Washington. On January 18, their representatives had a two-hour conference with President Wilson. They warned him, Gompers afterward told reporters, that a strike was unavoidable "unless justice is done . . . We asked the President to take some steps to relieve the situation, and as a last resort to take over the plants." [12] The outcome was the summoning to Washington of J. Ogden Armour, Louis F. Swift, Nelson Morris, Thomas E. Wilson, and Edward Cudahy. They were under heavy pressure. Government seizure, moreover, seemed a real possibility. The Food Administration already subjected the packing companies to considerable regulation, the Senate had before it a resolution for government control, and the FTC was carrying on a hostile investigation. The *National Provisioner* saw "a widespread labor union plan to force the Government operation of many industries after the fashion of the seizure of the railroads . . . This was the object of the labor delegation's strategy, rather than the settlement of any particular packing house question." [13]

The packers had to give ground. Reluctantly, they attended a joint meeting on January 25 under the chairmanship of Secretary of Labor W. B. Wilson. Having finally disposed of the preliminaries of meeting one another, the two sides quickly reached an accord. The packers were reassured on the wartime status of the unions. There was to be no preferential shop, nor standing grievance committees. A public announcement was made that the settlement did not imply union recognition. "All we ask," said Fitzpatrick afterward, "is that some justice be done to those who work in the packing industry." [14] In direct negotiation, the packers largely satisfied the union demands on employment and shop conditions. The other issues, involving wages and hours, were left to arbitration. The selection of a new administrator, after the packers and unions failed to agree on a candidate, was up to Secretary of Labor Wilson under the terms of the December agreements. He appointed Federal Judge Samuel B. Alschuler of Illinois—as it turned out, a most fortunate choice.

Federal administration of packinghouse labor relations was thus salvaged. "This was a difficult, delicate and disagreeable job for me," wrote John H. Walker, union member of the President's Mediation Commission immediately afterward. "I think both sides are beginning to see the thing more nearly as it really is now, than they did when we took up the matter with them first, I think possibly they could be reasoned with now; it was impossible to do that then . . . The trouble is settled, and . . . both sides will abide by whatever decision is rendered." [15]

The arbitration hearings began in Chicago on February 11. The unions had secured the free services of Frank P. Walsh, a skillful Kansas City trial lawyer and former chairman of the Commission on Industrial Relations. He proved more than a match for the legal talent employed by the packers. Walsh's strategy was, first, to show the justice of the union demands and, second, to establish the companies' ability to meet the increased costs. Both the low living standards of the workmen and the high wartime earnings of the industry gave him a strong case. In addition, the FTC investigation then in progress "furnished us," Walsh later remarked, "with a veritable mine of useful information" from the private files of the packers. This windfall, Walsh observed, "if it did not win the case for us, at least kept all of the packers from going on the stand except those whom we put there, and thus allowed many an important question to go by default." [16] The erratic J. Ogden Armour and the youthful Nelson Morris alone faced Walsh's brilliant examination. The defense of the packers was handicapped also by the publicity attending the hearings. Enormous interest was aroused over the drama being enacted in the Chicago Federal Building. These were characteristic headlines:

LIFE'S HARDSHIPS TOLD BY
WOMEN OF STOCKYARDS

One Lived In Chicago Six Years, Never
Saw Movie, Park, Nor Lake Michigan—
Say All Are Underpaid—Testify
Struggle For Mere Existence Is Hard [17]

Aware of the nationwide audience for their utterances and already under attack on other counts, the packers could hardly deny the validity of union demands for higher wages and an eight-hour day. Their main defense, in the end, was to plead the disruptive consequences of changing labor standards in wartime. When the hearings closed on March 7, 1918 with Walsh's powerful summation, it was evident that only the size of the award remained in doubt.

Tension mounted in the packing centers as Judge Alschuler, with accustomed judicial care, prepared his decision. On March 26, Fitzpatrick and Lane telegraphed Walsh that "a most critical situation is developing as a result of the seemingly unreasonable delay by the arbitrator . . . The men are saying that we hoodwinked them . . . We will be unable to prevent a walkout if the decision is not announced immediately." [18] At last, on Saturday morning, March 30, Judge Alschuler read his lengthy, closely reasoned decision before a packed courtroom of union officials, company representatives, and stockyard workers. The award granted the basic eight-hour day with ten hours' pay; time and a quarter (the union demand was time and a half) for the first two hours of overtime, then time and a half, retroactive to January 14 as earlier agreed; increases of 4½ cents an hour for workers earning 30 cents or less an hour, 4 cents to those between 30 and 40 cents an hour, and 3½ cents to those earning over 40 cents an hour (the union demand was a uniform dollar-a-day increase); no change in guaranteed time, except to reduce the weekly guarantee at Swift plants from 45 to 40 hours to conform to the general pattern. As Judge Alschuler concluded, Dennis Lane reported, "a happy realization of long sought relief followed the tense moments . . . fraught with anxiety . . . A decision had been made that in every minute detail is considerate and fair to all concerned." Jubilation filled the packing centers.[19]

The Alschuler award had the immediate effect of stimulating the organization of packinghouse workers. The unions capitalized on the enthusiasm, claiming as they did the credit for the award. Secretary Lane, returning from a tour of the western centers in April, reported that mass meetings were drawing in thousands of new members. The organizing stimulus spread beyond the Big Five plants of the eleven large centers covered by the agreements. Unions sprang up in the independent plants and minor packing centers across the country in order to share the benefits of the arbitration award. In general, the Big Five and major independents extended the Alschuler award voluntarily beyond its original bounds. Smaller firms, such as those in the Chicago Packers' and Sausage Manufacturers' Association and the St. Louis Local Packers' and Provision Association, did fight the application of the terms of the decision. Only in a few cases—for instance, in Seattle—were they successful. But the Alschuler award, whether or not it was adopted voluntarily, gave a specific and direct reason for packinghouse workers outside the major plants to organize.

The emergence of a wartime labor policy by the Federal government further encouraged organization beyond the eleven large centers. The National War Labor Board, beginning in the spring of 1918, guaranteed the right to organize and to bargaining through shop committees. Butcher workmen outside the Alschuler administration now had recourse to a governmental body which was backed by the emergency war authority of the president. The NWLB intervened in a number of packinghouse disputes, most notably in Pittsburgh, and the publicity attending the NWLB's activity informed the scattered packinghouse workers of their wartime rights.[20]

The packing industry thus experienced a rapid union growth during the war. By November 1918, the Amalgamated was paying per capita to the AFL for 62,857 members—over twice the number in January 1918, and tenfold that of January 1916. The gains did not seem ephemeral. "We are doing well here in the Yards," William Z. Foster wrote Frank Walsh in July 1918. "The organizations maintain themselves very good, in spite of the croakers who said they would fall to pieces as soon as the excitement died out. I think the foundations of unionism have been laid in the packing industry for a long time to come." That seemed a valid judgment as the war drew to a close.[21]

Beyond numbers and internal stability, however, there remained the vital problem of recognition. The packers had, of course, fully demonstrated their hard line during the initial period of Federal administration. The question was whether their resistance could be broken down in the course of the war.

The Amalgamated did achieve a kind of *de facto* recognition. Under the Alschuler administration, employees were to be represented by spokesmen of their own choosing, outsiders included. Although the packers did not have to sign contracts or recognize unions, they did have to deal with such representatives. This was the opening for the Amalgamated. When Judge Alschuler received a complaint from one packing company, he automatically informed Secretary Lane, and then hastily added to the company, "not, of course, in recognition of the union, but because of the customary fact that he or some of his associates in the union have generally represented such employees as were members of the union." For, in reality, Amalgamated officers were the only spokesmen of the packinghouse workers. And, under the workings of the Alschuler administration, they met repeatedly with company officials in negotiations, arbitration hearings, and grievance cases. Continuing contact soon proceeded with evident amity and reasonableness. "It was nothing unusual," observed the *Butcher Workman*, "for the packers to consult union officials with the object of obtaining their aid in solving labor problems." Acting at the request of a Morris superintendent, for example, Lane persuaded the cattle butchers to speed up beyond the standard rate of 16½ head per hour, with proportionately more pay, in order to achieve economies in other departments. Under the Alschuler administration, the unions assumed an important role both for the employees and management.[22]

This arrangement did not satisfy the unionists. They were "thoroughly aware of the urgent need of a sound and permanent agreement of the workers with the packers as to wages and working conditions." [23] The war was almost over when the National Conference Committee, representing the Amalgamated and the other internationals with packinghouse jurisdiction, approached the packers with a set of forty-two demands as the basis for a contract. However, the packers refused to enter a contractual relationship, and the National Conference Committee reluctantly acquiesced. Fourteen of the issues were eventually resolved by direct negotiation, and others were dropped or ironed out informally by Judge Alschuler. The major points, however, were considered in hearings in December 1918 and January 1919. The decision of February 15, 1919 only raised the minimum male hourly rate to $42\frac{1}{2}$ cents and gave time and a half after eight hours, instead of ten. The union leaders could not hide their disappointment over the award. [24]

The conclusion of the second round of arbitration reopened the crucial question of future union-management relations. The war was over; other emergency labor machinery was being dismantled; and, with a peace treaty imminent, the Alschuler administration would soon end. Armour Superintendent J. E. O'Hern, chairman of the packers' committee, called in a group of union leaders to sound out their intentions. They unhesitatingly expressed the desire "to arrange, if possible, an agreement which would take the place of the present agreements at their expiration"—that is, union recognition and a contract. After several weeks of conferences, the packers entirely rejected the proposition. They offered instead, and they so informed Secretary of Labor Wilson on April 12, 1919, to continue the existing arrangement for one year after the signing of a peace treaty "in order to avoid labor controversies and to promote the general welfare during the troublous period of reconstruction." [25]

A difficult decision confronted the unions. To reject the packers' offer meant that "a strike would be almost inevitable." A showdown was unwelcome, for the packinghouse organizations, although large, were still incomplete. Moreover, according to Lane, the paid-up membership fell sharply after the disappointing second Alschuler award. The unions again sought to apply government pressure on the packers, but that tactic, effective in wartime, was no longer so in 1919. The National Conference Committee pressed for concessions: in the method of initiating grievances; in the extension of the agreement to cover all Big Five plants; in an expanded union role in administering the agreement; and in new machinery to eliminate delays. However, barring a few minor points, the packers stood fast; indeed, at one time they threatened to withdraw their original offer. A fierce debate raged among the union leaders. The Stockyards Labor Council adamantly opposed the continuance of the arbitration arrangement. Caution, however, prevailed. The internationals' representatives in early June finally "decided that the practical and sensible thing to do under the circumstances was to renew our agreement [until] . . . one year after the ending of the War." [26]

The emergency phase thus ended inconclusively. The unions still had to achieve the collective bargaining function; without it they could not hope to survive permanently. However, they had earned a year or so (as it seemed then) under the shelter of the Federal administration. "Our duty and responsibility," the National Conference Committee wrote the local unions, "will be . . . to enter into an active campaign of agitation. . . . If we make good in securing one hundred per cent organization during the coming year, we will have no trouble in securing recognition of our organization and union shop agreements." [27] The decision to postpone a showdown rested on two premises: first, that organization could be perfected in the postwar period, not to say, maintained intact; and, second, that completed organization would necessarily lead the packers to grant recognition and a contract. As it happened, no opportunity developed to test the second premise.

The vital organizational problem in the postwar period involved the mass of Negro workers. Substantial numbers had been employed in the packinghouses before the war, ranging in 1909 from 3 per cent in Chicago to 16.1 per cent in Kansas City. Many Negroes had advanced into the semi-skilled and skilled ranks. Their daily earnings in Chicago in 1909 averaged $2.07, compared to $1.79 for Polish and Lithuanian workers, and $2.34 for native-born whites. The prewar Negroes, forming a permanent part of the packinghouse force, responded to the wartime organizing drive in the same manner as the whites. By early 1918, labor leaders estimated, 90 per cent of the northern Negroes in the Chicago yards carried union cards.[28]

Another larger group of Negroes entered the industry during the war. Northern economic opportunity had started a massive migration of southern sharecroppers and laborers. Large numbers found work in the packinghouses. In a few instances, they came as strikebreakers; but most filled openings created by the war. The Chicago yards, which experienced the largest influx, had approximately 10,000 Negro workers—over 20 per cent of the labor force—by the end of 1918.

The newcomers proved exceedingly difficult to organize. Their background made them unresponsive to the union appeal, or, if they were swept into the union, made them hard to hold. Moreover, they lacked the sense of

grievance that impelled the older hands in the packinghouses. Union leaders soon despaired over organizing "these hopeless, oppressed, and discouraged negroes." There were other obstacles. Competing Negro unions, such
as the American Unity Labor Union, confused and divided the southerners.
The community leaders of the Chicago "Black Belt"—the two Negro alder-
men, the social workers, and clergymen—urged loyalty to employers who
had given the newcomers their economic opportunity. Labor leaders charged
that the packers "subsidized negro politicians and negro preachers and sent
them out among the colored men and women to induce them not to join
the unions." [29]

Grounds assuredly existed for the contention that unions impeded the economic advancement of Negroes in the stockyards. Some of the cooperating
unions excluded Negroes either explicitly, as in the cases of the Machinists
and Railway Carmen, or informally, as with the Electrical Workers and
Plumbers. The Butcher Workmen had always been open, but here organizational problems also existed. In Chicago the union was accused of keeping
Negroes in a minority by distributing them among the local unions, and,
when a segregated local was consequently established, there were charges of
"Jim Crow." The eventual solution was to organize the laborers on a neighborhood basis. Local 651 had its headquarters in the second ward—the Black
Belt—and thus became primarily a colored union. However, Negroes also
belonged to other Amalgamated unions in Chicago, and integration was the
usual pattern elsewhere.

Yet there could not be wholehearted Negro support for the union cause.
Walter F. White, writing in the NAACP's *Crisis* of the Chicago labor situation, observed that because of past experience "the Negro workman is not at
all sure as to the sincerity of the unions . . . He feels that he has been given
promises too long already . . . What he wants are binding statements and
guarantees that cannot be broken at will." [30]

The unions had failed to organize the southern Negroes. William Z.
Foster admitted that "we could not win their support. It could not be done.
They were constitutionally opposed to unions, and all our forces could not
break down that opposition." According to a union estimate, at no time
were more than 15 per cent of the Negroes in the Union Stock Yards organized.[31] This group, one fifth of the Chicago labor force, presented the
critical problem after the refusal of the packers to recognize the unions in
the spring of 1919. The Negroes had to be organized in Chicago and elsewhere.

The Chicago Stockyards Labor Council mounted a concerted drive im-

mediately after the unions agreed to continue under the Alschuler administration. The early summer of 1919 saw extensive unionizing activity in the
Union Stock Yards, but there was little response. Organizers' reports complained of "great opposition encountered" and of "the arguments we are
compelled to face by men who are trying to oppose the union." The Stockyards Labor Council played its final card when it announced on July 20
that it would present new wage demands with a minimum rate of 70 cents
an hour.[32]

Whatever hope remained was smashed by the Chicago race riots which
began on Sunday, July 27. The South Side, including the stockyards district and the Black Belt, was the scene of most of the bloodshed of the following week. The dangers to the organizing drive were obvious. "Right
now it is going to be decided," announced the Chicago Federation of Labor,
"whether the colored workers are going to continue to come into the labor
movement." "Night and day," Fitzpatrick said, "the officials of the Stockyards Labor Council toiled and fought to hold in check the forces of organized labor . . . to prevent workers from assaulting other workers." Some
packinghouse workers certainly participated in the rioting, but the unions,
by asserting their support for law and order, remained unstained when the
rioting subsided toward the end of the week. Then, early Saturday morning,
unknown incendiaries fired a group of houses in the Lithuanian section
"back of the Yards." The destruction enflamed the immigrant community
against the Negroes and focused the racial conflict, hitherto generalized, on
the stockyards. Since the hostility of their membership was fully apparent,
the unions could no longer hope to gain favor in the Black Belt. "The
breach," admitted Fitzpatrick the day after the fire, is "so broad that it is
almost impossible now to cement or bridge it over." [33]

The emergency seemed, however, to present an alternative avenue to the
organization of the Negroes. When the riots had broken out, the packinghouse workers from the colored quarter, lying a mile to the east of the
stockyards, had dared not pass through the white neighborhoods in order
to reach their jobs. They had still not returned to work when the fire had
occurred. Now the Stockyards Labor Council injected the union issue into
the situation. The source of trouble in the yards, it argued, was not racial
hatred—no conflict existed between white and Negro unionists—but the
natural anger of union against nonunion men. This fact made it intolerable
to have the Negroes brought into the yards under military guard, as had
been planned for Monday. "Union men will not work with machine guns
pointed at them," labor council officials told the packers and public authori-

ties after the fire. Only one solution would prevent bloodshed. "Bring your workers back to a union shop . . . as fellow union men and women and we will try to hold the situation in hand for you." Having failed to convert the Negroes to unionism, the council now was using the explosive situation to force organization on them.

The weakness of the stratagem became apparent as soon as the union offer was "contemptuously" rejected by the packers. Having raised the union issue, the labor council could not put it aside, nor did it have the means of enforcing its demand. When the packers recalled the Negroes on Thursday, the council called a strike in a last desperate move. The avowed grounds were, first, to avoid the outbreak of violence and, second, to protest the presence of soldiers. But the underlying issue was clear. "The men have a perfect right to strike," asserted one leader. "They have a right to know whether the Negroes returning to work are members in good standing of the unions." The justification for the strike quickly disappeared. No violence developed, or even threatened; and the soldiers were withdrawn immediately, followed by the special police and deputies. When the strike ended on Monday, August 11, the organization of the colored packinghouse workers was a completely hopeless cause in Chicago.[34]

More modest union drives had meanwhile been proceeding at other packing centers. Here the same inhibiting factors were at work, although not the final experience of the Chicago riot. Racial violence, however, had occurred in East St. Louis in July 1917 and in Omaha at the end of September 1919. Nowhere could the unions claim substantial success in organizing the Negro migrants.

The failure meant that thorough organization could not be achieved. But the unions proved incapable of maintaining their existing strength. Packinghouse unionism was shortly plunged into an internal conflict that sapped its strength in Chicago, the vital center of the industry.

The Amalgamated had not departed from the structural lines of the earlier period of organization. The district councils in the large centers, in conformity with the decision of the 1904 convention, were limited to Amalgamated affiliates and, since the 1917 convention, covered a radius of 25 miles, so that they included nonpackinghouse locals. The councils had the same functions as the earlier packing trades councils of handling local affairs and the additional right, after international approval, of keeping ten cents of the monthly per capita for their own defense fund. The district councils grew into substantial institutions. The South Omaha body, for example, acquired its own newspaper, band, athletic club, and, in 1920, labor temple. This vigor created, especially in the later period, an autonomous spirit in a few of the district councils, but they never presented any real problem of control for the Amalgamated.

Chicago was different. Here the central packinghouse body was the Stockyards Labor Council. It had begun in 1917 as an organization of Chicago local unions to unionize the stockyards. Having achieved this, the council established itself as the immediate authority over the stockyards, affiliating locals both of butcher workmen and of the allied trades. The Stockyards Labor Council, operating under the auspices of the Chicago Federation of Labor, was not subordinate to the Amalgamated and was consequently in implicit conflict with its authority over the packinghouse locals in Chicago.

In theory, the paramount attachment of local unions was to their international. But the primary loyalty of the Chicago unions was to the council. This resulted, first, from the fact that the Stockyards Labor Council had created them, and, second, from the immigrant character of the membership. The great foreign concentration in Chicago had been organized by a Stockyards Labor Council adherent and former Woodworkers' business agent, John Kikulski. Establishing his influence over the burgeoning immigrant movement, the magnetic Kikulski thwarted the Amalgamated effort to divide his recruits into departmental locals. Instead, the immigrants were organized into five huge laborers' unions over which Kikulski exerted complete control. Dennis Lane complained that he could "speak but one language—the American language—and [had] no chance whatever to be heard." The immigrant workmen were "more apt to be controlled by the great foreign language orator who had been molding their minds for a year and a half." [35] The Chicago laborers' unions were the solid core of the independent strength of the Stockyards Labor Council.

Nor was there common ground on matters of policy. William Z. Foster had been the first secretary of the Stockyards Labor Council. When he resigned in July 1918 to participate in the steel organizing campaign, his replacement was J. W. Johnstone, who also had a syndicalist background. Although no longer an avowed radical, Johnstone still retained something of the outlook. Cautious policies provoked his ire. The success of the Stockyards Labor Council, moreover, depended on its appeal to the untutored stockyards workers. Kikulski, Johnstone, and others had capitalized on the wartime involvement and rising expectations of the men in the yards. The disappointing post-armistice period made it particularly necessary to sustain the excitement of the packinghouse workers with dramatic words and deeds. This approach was bound to clash with the careful Amalgamated leadership. Friction, apparent earlier, became heated over the question of securing recognition from the packers. The Chicago leadership—excluding John Fitzpatrick—favored a showdown in the spring of 1919 and condemned Lane's cautious desire to avoid "any ill-timed struggle which ultimately brings disaster." The Chicago locals voted down the packers' proposal. When the National Conference Committee accepted, the Stockyards Labor Council refused to endorse "such a miserable agreement." [36]

An open break could not long be avoided. The strike called by the Stockyards Labor Council following the Chicago race riots was without Amalgamated sanction and was in violation of the arbitration agreement. Even worse, the council moved independently to present demands for arbitration by Judge Alschuler. A week after hearings began on August 12, 1919, an Amalgamated conference formulated its own demands. Judge Alschuler heard the rival labor groups separately, but he handed down a single award. The Amalgamated leadership determined to cut down "these unscrupulous pretenders." "The local jurisdiction of the labor council they represent," thundered Dennis Lane, "was stretched to such an extent that their noses were stuck into the national affairs of our movement . . . the only recognized international body authorized to represent the organized workers in the packing industry." [37]

Months before, the Amalgamated had laid plans for "shaping up its organization in Chicago, as it had done in other packing centers." The plan for a Chicago district council, approved by the executive board in March 1919, had been shelved as a result of opposition in the stockyards and in the Chicago Federation of Labor, but Lane now saw the need for decisive action. On July 27, 1919 he called the Chicago locals together to form District Council 9. Eight stockyards locals, holding the bulk of the packinghouse workers, refused to join and were subsequently suspended from the international in September 1919.[38] A prolonged struggle ensued, fiercely contested during its course and unprofitable in its outcome.

The labor council, solidly established as it was in the stockyards, was isolated. "The great weakness of our Chicago Stockyards Labor Council," William Z. Foster observed afterward, was that "the actual control of the international unions remained in the hands of reactionary AFL officials and, alas, we had no organized militant minority to link us up with the rank and file workers in the other packing centers." Although an attempt was made after the onset of the struggle, the council could not expand beyond its local base. Nor could it retain its ties with organized labor in Chicago. The AFL on December 22, 1919 instructed the Chicago Federation of Labor to unseat the member locals belonging to the Stockyards Labor Council. The Chicago body favored the SYLC—its creation—and was at that time at odds with Gompers over the issue of a labor party. But on January 8, 1920 the stockyards unions submitted their resignations, as Johnstone explained, so that the Chicago Federation "may not be placed in the same position with the American Federation of Labor that we are with the Amalgamated Meat Cutters and Butcher Workmen." [39]

Thus far, efforts at reconciliation had failed. A last attempt, spurred by Gompers' expulsion order, collapsed over the insistence by District Council 9 on unitary voting representation for all affiliated locals. Johnstone was prepared to bring his group into Council 9 and to concede his subordination to the Amalgamated, but not to surrender domination over the Chicago movement. On February 15, 1920 an apparent solution was found. The two councils would be replaced by a new one with an Amalgamated charter and jurisdiction over the entire Chicago metropolitan area. A convention on February 22, presided over by Secretary Lane, chose, as president, Martin Murphy—the SYLC president—and, as secretary-treasurer, Johnstone. But the triumph of the insurgent group was short-lived. Lane postponed issuing the charter for the united body. Then, on March 16, District Council 9 suddenly convened and elected the immigrant leader John Kikulski president. Kikulski's defection carried the five immigrant locals out of the Johnstone-Murphy camp and reduced it to a mere splinter group. By this devious stratagem, obscure in its details, the Amalgamated finally dealt a fatal blow to the opposition within the Chicago movement.[40] Kikulski himself did not have much opportunity to enjoy the fruits of his new alliance. On the evening of May 17, 1920 he was shot by unknown assailants; a few days later, he died.

The Chicago strife came to an end. The Stockyards Labor Council, lacking its immigrant base, gradually faded out of existence. But the Amalgamated victory had a high price. The energies of both sides were wasted for many months. The internal conflict, charged as it was with vilification, confusion, and raiding, undermined the vitality of the stockyards organization. Membership was reduced and effective local leadership lost. Finally, the movement emerged in a tarnished state. The struggle had been fought out in full view and, indeed, had directly involved the packinghouse workers. They could hardly preserve intact the faith in unionism vital for the trials that lay ahead.

The Amalgamated paid heavily to maintain the primacy of its authority. It was therefore ironical that, at that very time, the international administration should have been entering another period of weakness and disarray.

The treasury had achieved a fairly substantial footing by the first months after the armistice. Thereafter, expenditures had to be increased. The withdrawal of the AFL organizers in early 1919—Lane estimated that the federation had borne a quarter of the expense of organizing the packing industry —required an expanded Amalgamated staff. Twenty-five organizers were on the payroll in early 1920, plus the members of the executive board. The monthly per capita income of $25,000 was not quite sufficient to support that group, as well as operating expenses and the legal costs of the periodic arbitrations; but the treasury of $146,000 on August 1, 1919, while not large in relation to the dues-paying membership of about 75,000, would nevertheless permit the international to meet its needs for an extended period.

The surplus could not, however, survive any attempt to support a major strike. Among other preparatory reforms, the 1917 convention had amended the constitution to make discretionary, rather than obligatory, the payment of strike benefits. It remained to be seen how the executive board would use this power. In October 1919 a strike of over 15,000 men, the bulk newly organized, broke out in the eastern packing houses. Despite Secretary Lane's objections, the executive board decided to "support the strike to the limit of our Treasury." Almost $100,000 was "dumped into this situation," and when the strike ended in failure, the Amalgamated, as Lane had foreseen, had neither its money nor the membership. The result was that on January 1, 1920 the international had a cash balance of $9567, plus $45,000 in Liberty Bonds, and by May 1920—the last financial statement printed—under $35,000 in total. At the end of August, it was necessary to dismiss eleven organizers.[41]

Meeting in St. Louis in July 1920, the Amalgamated convention attempted to strengthen the union's finances. Per capita was raised from 35 to 50 cents a month. Also, district councils could no longer withhold the ten cents designated for the international defense fund. These reforms did not appreciably improve the financial situation, for a marked decline in membership soon began as a result of spreading unemployment. The Amalgamated was twice in the next year forced to resort to assessments on the membership. It was entering the critical year of 1921 with almost no reserve in its treasury and an income inadequate for its pressing needs.

The St. Louis convention also had overhauled the governmental machinery of the international. The executive board was reduced to three officers: a general president, vice-president, and secretary-treasurer. "Thus," explained Dennis Lane, "decision and subsequent action and matters of consequence will not suffer harmful delay because of an unwieldy Executive Board whose members are scattered all over the country." Instead of vice-presidents, there were to be nine "District Presidents" who were "to act as Executives of the District to which they are assigned." [42] The district presidents were designated "counselors" to the executive board. In practice, however, they continued to have voting rights when the executive board met in full session. There were consequently two executive bodies: an "acting" board of the three executive officers which met every week or two to dispose of regular business, and a "general" board which met, as earlier, once or twice a year to consider important questions and lay down policy.

This reform could not counteract ineffectuality among the chief executive officers. Secretary Dennis Lane was, of course, a tower of strength. He was a tough, dedicated trade unionist, lacking somewhat in the arts of diplomacy and persuasion, but impressive as a man and as a practical labor leader. He dominated the international office. The new general vice-president was Patrick E. Gorman of Louisville, Kentucky. Endowed with an Irish charm that masked a keen, vigorous mind, Gorman possessed enormous promise. He had already demonstrated his abilities in the Louisville movement,[43] but he was young—still in his twenties—and not widely known in the Amalgamated. Gorman was a powerful asset from the start, but he could not substitute for strong leadership in the presidency. Here the Amalgamated was sadly lacking.

John F. Hart had suited the quiet period before World War I. But, as Secretary Lane remarked, "when our International Union grew by the influx of the Packing House workers the job . . . got too big for John Hart." A meat cutter by trade, he kept his distance from the workers of the packing centers, especially the immigrants. Claiming ill health, he spent a great deal of time at his Utica home or in California. Secretary Lane had to assume much of the president's work. In January 1919 Lane, greatly overburdened, refused any longer to do two jobs. Under Hart's subsequent lax direction, the morale of the field staff, so Lane claimed, was soon "at a very low ebb." But, if he shunned the responsibilities of the presidency, Hart relished its perquisites. The St. Louis convention, through the efforts of his friends, awarded Hart $3600 as reimbursement for the period in 1910–12 when he had not been on full salary. That same summer of 1920, while remaining inactive, Hart began to draw heavily from the treasury: $2892.52 in three months. Lane, discovering this, sharply criticized the president for "setting a bad example, and one that will be rebuked by the rank and file—particularly at a time when the finances of our organization are low . . . I am going to be frank in stating that there will be no more [recurrence] of such drawing . . . while I am Secretary-Treasurer." [44]

The deteriorating packinghouse situation probably forced Lane's hand. On the third day of a general executive board meeting, April 27, 1921, Lane suddenly suggested that the president retire from office on a pension of $200 a month. Protesting, Hart "declared his intention to carry on and hold Office, even in his present physical condition." Lane then made it clear that the executive board would have to choose between him and the president. Hart, seeing the game was up, reluctantly agreed to retire on May 1, accepting a pension of $60 a week, plus an additional $500 if he had to undergo an operation.[45]

But Hart's removal raised problems of a successor. The constitution was ambiguous on the method of filling a vacant presidency.[46] Differences already existed on both personal and policy matters between Lane and the young Gorman. Gorman, moreover, was a Hart supporter. Needless to say, Secretary Lane opposed the contention that Gorman automatically succeed to the presidency. A majority upheld the view that the executive board should elect a successor. Gorman was dissuaded from resigning and continued as general vice-president. But the bad blood thus created did not diminish for several years.

The executive board then elected an outsider, Cornelius J. Hayes of East St. Louis. A participant in the packinghouse movement before the 1904 defeat, since 1911 he had been a government meat inspector and an active local politician who had run for Congress in 1920. Hayes had been appointed an organizer in December 1920, perhaps in preparation for the approaching coup. His choice would circumvent the rivalries and ambitions within the executive board, and, since he had no following inside the Amalgamated, Hayes would not threaten Lane's primacy. However, the union needed leadership which was experienced in its complex problems and known and trusted by the membership. Hayes did not fulfill these requirements. The Amalgamated was thus under the direction of men who were at odds with each other or new to their heavy responsibilities.

The organization was also being subjected to increasing internal stresses. These were partly the outcome of Hart's departure. The California Federation of Butcher Workmen, meeting on June 11, 1921, condemned Lane and urged withholding per capita until he resigned. All the local unions received copies of these resolutions (Hart had foresightedly taken with him a list of addresses). Lane had to defend his actions in the pages of the *Butcher Work-*

man, opening to the membership a sorry view of the conduct of the administration. He was able to arrest the secession movement in California, but, in order to do so, the secretary had to absent himself from Chicago in September 1921 at the climax of the crisis with the packers.[47] Disunity was an expensive luxury.

Dissension appeared elsewhere for other reasons. Two district councils in New York City had existed since the reaffiliation of John Kennedy's Brotherhood of Butcher Workmen in 1917: the old No. 1 under International Vice-President Joseph Menhart and No. 2 under Vice-President Kennedy. Differences had quickly developed between the two bodies. The climax came as a result of the eastern packinghouse strike of October-December, 1919. Menhart had been instrumental in organizing the eastern houses. He had called the Springfield, Massachusetts, conference to formulate demands, and he led the strike. Kennedy opposed the walkout, and District Council 2 gave it no support. Discredited by the strike's failure, Menhart soon faded out of the movement, permitting Kennedy to become dominant in New York. District Council 2 absorbed Council 1 in December 1920. But the result was further strife and some local secessions.[48]

Dissidence developed against the international in the independent houses and lesser packing centers. These were not for the most part under the Alschuler administration, although they generally followed the awards. The international was criticized for failing to have the arbitration system extended to the outlying points, and, beginning in late 1920, for failing to take general action when wage cuts began to hit these places. The sharpest expression of this sentiment came from District President Philip J. Guest. Without authorization, he called for a national strike in support of a strike he was leading in Wheeling, West Virginia, in early 1921. His rash act revealed the general frustration that was both breaking the organizations unprotected by the Alschuler administration and turning them against the Amalgamated leadership.[49]

Two years had passed under the Alschuler administration since the armistice—time, Lane had said, that would have to be used to make packinghouse organization strong enough to force recognition from the industry. Instead, the Amalgamated was impoverished, led by divided and untried officers, and weakened by internal dissension. Membership was declining. The enrollment in November 1919, according to AFL per capita records, was 67,877; the average from November 1920 to April 1921 was 39,868.[50] And depression was setting in. If strength might gain recognition, weakness could only lead to disaster. As 1921 opened, it merely remained for the meat packers to extricate themselves from the Alschuler administration and force the issue to a conclusion.

The Big Five had no cause to regret their offer to continue under the arbitration agreements. They had industrial peace, the *National Provisioner* noted in late 1919, at a time when the country was engulfed by labor troubles. Occasional strikes, it was true, had occurred, particularly during the summer of 1919, but these had been short-lived and invariably opposed by the Amalgamated. "It should be a matter of general gratitude," Thomas E. Wilson told the Institute of American Meat Packers in September 1920, "that the industry . . . is well on its way through the period of readjustment without any serious labor difficulties. The readiness with which employer and employee have submitted to arbitration . . . [has] done much to improve the relations existing between them." [51] Nor was the price of peace excessive. Average hourly earnings in meat packing increased 76 per cent from 1917 to 1920, compared to over 82 per cent for all the manufacturing industries surveyed in Paul Douglas' study. The difference was most marked in the 1919–20 period when meat packing hourly earnings rose only 3.2 cents, the manufacturing group 13.4 cents.[52]

The onset of depression in the autumn of 1920 put the situation in a new light. The industry was entering a disastrous period. By-product prices had been declining since January 1920, and meat prices had been dropping since a sharp break during the summer. The packers suffered heavy losses on the devaluation of large inventories. Armour's deficit was $31,709,817 at the end of 1921; the other major packers, although better situated, also lost money. In such adversity, arbitration became a hindrance. Suddenly reopening a recently concluded round of hearings, the packers in November 1920 demanded concessions because of the deteriorating economic situation. Judge Alschuler refused to grant the packers' request for a basic day of ten hours, but he also turned down the union's demand for a uniform increase of a dollar a day, awarding only a 5 per cent bonus on earnings up to 25 dollars a week for the period July 5 through December 5, 1920. The decision was considered to be something of a victory for the packers, but they were finding any restriction on their freedom of action to be insupportable. The relative stability of wage rates imposed by the arbitration system, favorable to the industry in the two postwar years, now had the opposite effect: meat-packing rates were still inching upward while they were elsewhere in decline.

On January 14, 1921 the Big Five packers informed Secretary of Labor Wilson of their desire to withdraw from the arbitration arrangement. Conferences followed with the secretary in Washington. On February 21 he was notified by the packers' lawyers "that from now on our clients will regard the President's Mediation Commission agreement, and the renewal thereof, as having been carried out according to their purposes, and will proceed along the lines which they regard as mutually advantageous to themselves and their employees." [53] Two weeks later wage cuts were announced, effective March 14, of eight cents an hour or 12½ per cent on piecework, as well as the limitation of overtime pay to work after ten hours a day or 54 hours a week.

The Butcher Workmen had already encountered, as Secretary Lane put it, the "reactionary tendencies during this melancholy and perilous period of so-called readjustment." A number of independents had cut wages, and others who had formally accepted the Alschuler administration, had announced their withdrawal. Strikes were in effect or threatening at many points in early 1921. To prevent a return to prewar conditions, Lane warned, the locals in independent plants would have "to put their back to the wall and fight for all they were worth." [54] The union leaders, aware of the Big Five packers' preliminary steps, were not surprised by the repudiation of the Alschuler agreement. [55]

The response was a packinghouse conference in South Omaha on March 9, 1921. Despite strong strike sentiment, the delegates accepted the plan of the executive board to appeal to President Harding. In office for only a week and anxious to apply his Inaugural Address view on industrial peace, Harding instructed the new Secretary of Labor J. J. Davis to take up the matter. Within a few days a conference was arranged for industry and labor representatives to begin in Washington on March 21, 1921. Officials of ten internationals with packinghouse jurisdiction met in Chicago on March 16 to endorse the Amalgamated decisions, promise their full support in event of a strike, and select Dennis Lane and Amalgamated lawyer Redmond S. Brennan as the union representatives to the conference. A strike vote, authorized by the Omaha conference meanwhile was concluded, arming Secretary Lane for his trip to Washington. [56]

The central issue there involved the life of the arbitration agreements. This hinged on the technical point that the parties had agreed in 1919 to continue the arrangement until a year after the signing of a peace treaty. (A joint resolution terminating the state of war was signed by the President only on July 2, 1921 and peace treaties approved by the Senate on October 18, 1921.)

Lane and Brennan, of course, insisted that the companies should live up to
the letter of the continuation agreement. But there was no penalty clause,
nor any means of legal enforcement. Moreover, the government participants
—Secretaries J. J. Davis, Herbert Hoover, and Henry C. Wallace—"were of
the opinion that it was never intended that the Agreement . . . should run
until this late date." They suggested a compromise. Business conditions, they
reasoned, justified the wage cuts. If the unions would accept these, the government would urge the continuance of the basic eight-hour day and the
Alschuler administration for six months. Lane had the choice "of either going
with the recommendation of the Government Officials or plunging our membership into a strike. That was a tremendous responsibility!" After consulting with Gompers and other labor leaders, he accepted the government compromise. The packers did so also with apparent reluctance, and both sides
signed documents, dated March 23, 1921, extending the Alschuler administration until September 15, 1921.[57]

Afterward, Secretary Lane summed up the gains exchanged for accepting
wage cuts without arbitration:

> FIRST—We have prevented a strike at a light season of the year in the industry
> and at a time when millions are out of work and when the packers have
> an immense supply of meat in refrigeration.

> SECOND—We saved the eight-hour day with over-time rates.

> THIRD—We restored the Alschuler Agreement with its protection of seniority
> rights, protection against discrimination because of membership in the
> Union and protection against further decrease without arbitration.

(And let me here explain, what may be misunderstood, and that is—that we have
the right to ask for an increase any time during the next six months, if we believe we can convince the Arbitrator that the cost of living or wages has gone up
since we accepted the cut.)

> FOURTH—We have solidified the interests of all of the trades employed in the in-
> dustry, and placed ourselves in a position where we have six months to
> strengthen our Organization—both in numbers and finances.

> FIFTH—We have shown up the packing barons before the General Public as be-
> ing grossly unfair and have strengthened our Organization by getting
> public sentiment squarely behind us.

From the union standpoint, the vital advantage was the final reprieve of six
months to prepare "for the inevitable test of strength and efficiency of our
movement after September 15th." [58] For the last time, the Amalgamated leaders chose a postponement, hoping that more time would bring strength for the
showdown with the packers.

The half-year extension was also an opportunity for the packers. Wartime labor shortages and the union successes had intensified their interest in labor relations. The Institute of American Meat Packers in 1920 established an Industrial Relations Committee to act as a clearing house for the industry, and the major packers created industrial relations departments. Everywhere there were expanded programs of pensions, profit-sharing, insurance, "Americanization," and sports and social activities. In 1920 the progressive Swift & Co. was erecting "employees' utility buildings" to house cafeterias, dressing rooms, and club facilities. All the large companies were involved, as an Armour official said, in "the practical administration of the golden rule." [59]

The packers, enlightened as they were becoming, had hung back from the employee representation then being widely adopted in other industries. Tentative moves had started in this direction. Swift, for instance, had formed conference committees at three small plants in 1918 and 1919, and the talk among the packers about industrial relations savored increasingly of employee participation. Later, the industry would claim its inactivity had resulted from the terms of the original agreement with the Presidents' Mediation Commission, which forbade permanent shop committees.[60] More compelling, no doubt, were the packers' conservatism, a wartime experience different from other nonunion industries, and an unwillingness to unsettle matters during a period of union strength and business prosperity. The approaching end of the Alschuler administration in February 1921, however, made employee representation obviously advantageous. It would deflect public opinion and perhaps also wean the packinghouse workers from unionism. Immediately after announcing on March 8 the forthcoming wage cuts, Swift and Armour brought forward plans for introducing industrial democracy into the stockyards. According to industry spokesmen, the main government argument for the six-month delay was to give the packers time to start employee representation plans.[61] It was, at all events, so used.

Armour was the first to act. On March 15, 1921, elections were held in the Chicago plant for representatives to four divisional committees, which in turn chose a plant conference board. Two months later, having worked out details with the Chicago board, the industrial relations department installed the plan at the nine other major Armour plants. Finally, twenty-three representatives of the ten houses met with management representatives in Chicago on August 3 to establish a general conference board for the whole company and to ratify the plan. All matters pertaining to labor were to be proper subjects for consideration. To each body—divisional committees, plant conference boards, and the general conference board—management appointed members equal in number to the employee representatives. A majority of both groups was required for any decision. If the general conference board reached a stalemate, an arbitrator acceptable to both sides could render a final and binding decision. Only the general superintendent of the company could convene the general conference board, however, and he also had the right to propose settlements at earlier stages.[62]

The other major packers, except for the failing Morris & Co., in May 1921 followed Armour's lead. The plans differed somewhat in form, if not in substance. At the Swift plants, decisions by the assemblies required a two-thirds vote of the combined employee and management representatives, and unanimity in the committees. Failing agreement, both sides were "at liberty to take such action outside the plan as they may think desirable." Wilson & Co. used the Armour unit voting system in its Joint Representative Committees, but omitted arbitration. In event of a stalemate, the constitution left it to the two sides to "determine the best method by which an adjustment may be reached." But the packers did not contemplate disagreements. They assured each other "that the cause which is right and fair is the one which secures a favorable decision." [63]

Employee representation had an apparently successful start. The packers reported high participation in the elections. The installation of plans was generally an occasion for speeches, and, in the case of the Armour plant at St. Joseph, of parades and noon-hour demonstrations. The representative bodies, moreover, immediately began to take up grievances. Armour reported 149 satisfactory adjustments by August. On the other hand, it was the union which appeared before Judge Alschuler in June to oppose the application for a further wage reduction of five cents an hour. The Amalgamated fiercely attacked the "so-called company unions" as "in reality merely packer-controlled makeshifts of, by and for the packers." The union vehemence revealed the fear that employee representation would not be merely a paper operation. Armour & Co. was confident, as September 15 approached, that it would "have a well functioning machine ready to operate when it became necessary to supplant the wartime agreement." [64]

The unions nevertheless performed the motions preliminary to entering into collective bargaining. A packinghouse conference in South Omaha on August 15, 1921 drew up a set of demands, essentially incorporating existing conditions and providing for shop committees and the arbitration of disputes. The executive board received "full power to act." On September 6 a joint conference of the Butcher Workmen and nine other internationals met in

Chicago "to bring about the closest cooperation possible of trades employed in the industry." To the unions the packers responded that they intended to deal directly with their own employees.[65] At the expiration of the Alschuler administration on September 15, the unions were without a means of representing their membership. The crisis was finally at hand.

But the Amalgamated was less prepared to meet it than ever. The union mobilization in the six-month interim failed completely in the face of the depression, of company unions, and the legacy of internal strife and weakness. Indeed, the ranks shrank alarmingly during that period. The membership was near 40,000 at the end of 1920; it was only half that in September 1921, of which perhaps 10,000 came from the western centers. The Amalgamated leadership, despite a favorable strike vote in early October, decided to withhold a strike call and to await a more promising occasion, or at least a more compelling strike issue.

The packers soon obliged. They were now free to make the reductions of which Judge Alschuler had disapproved in July. The general conference board of Armour & Co. convened in Chicago on November 17 to consider the matter. Management representatives, *Armour Magazine* afterward reported, "showed, from the books of the company, indisputable evidence that operating costs must be reduced . . . The cards were on the table." The next day the conference board unanimously adopted the recommendation of its subcommittee. Swift, Cudahy and Wilson plant assemblies performed the same ritual. Then Morris and many lesser firms acted in the old-fashioned manner. The wage cuts, uniform for all companies despite the procedure, were sharply regressive: 7½ cents an hour for workers earning 45 cents or less; 5 cents for workers between 45 and 50 cents; 3 cents for men at 50 cents or more; and 8 per cent for pieceworkers. The packers could reasonably expect, in case of a strike, to hold their strategic, skilled men who, the union admitted, were most responsive to company tactics. Moreover, although the packers had earlier requested the basic ten-hour day, no change was made at this time from the extra hour at regular rates three days a week permitted by Judge Alschuler in the last award. The reductions were to go into effect on November 28.[66]

On November 21, after the first announcement of wage cuts, the unions agreed to strike all plants affected by the reduction "after attempt to negotiate has been made." The visits to the big packers were, of course, entirely fruitless. Still the leaders hesitated. They decided first to assess rank-and-file sentiment at a series of mass meetings on Sunday, November 27. Reports indicated a widespread impatience for action. The delegates, Lane recorded on

November 30, "deplored a strike taking place at this time, yet they appeared to be unanimous of the opinion that it could not be averted." The next day the strike order was issued, effective Monday, December 5, 1921.[67]

Rarely has an extensive strike started under less auspicious circumstances. Winter was at hand. There was widespread unemployment. The Amalgamated had a diminishing packinghouse membership. It was without funds. It could expect little support from the public, the government, or the labor movement. The packers, on the other hand, were united, thoroughly prepared, and in command of enormous resources. Yet the strike call, badly timed as it was in other respects, had one imponderable in its favor: the worker reaction to the record of lost opportunities and to the cynical actions of the employers.

The response on December 5 revealed the intensity of this resentment. The strike began strongly in Kansas City, Omaha, Sioux City, South St. Paul, and Denver. It was weaker in East St. Louis, and ineffectual in St. Joseph. In Chicago—the crucial point—the walkout started slowly but mounted to major proportions within a few days. The strike was not as effective as at lesser centers, but stock receipts and slaughtering reports, as well as mass meetings of 10,000 and more, belied the packers' contention that only 10 per cent of the Chicago men were out. As the unions claimed, the packinghouse workers—many of them nonunion—were showing "a temper and a determination to win the strike the packers did not anticipate . . . a complete surprise to the 'Big Five.' "[68]

The Amalgamated quickly invoked its single strategic advantage. Secretary Lane traveled to New York to urge a sympathy strike in the important Big Five plants in the metropolitan area. District President John Kennedy was reluctant. He had opposed the calling of the western strike, and he undoubtedly foresaw a repetition of the New York experience in 1904. Moreover, an estrangement had developed between Lane and the New York leader since the retirement of John Hart, arising perhaps from the frustration of Kennedy's aspirations to the presidency. But strong strike sentiment existed in New York, involving more than sympathy for the western strikers. The agreement of May 1921 with the New York plants, which had not come under the Alschuler administration, permitted a reopening of the contract after 60 days. Believing reductions to be imminent, the unions requested an extension of the agreement for a year, beginning November 1, 1921, but dropping the reopening clause. The issue was deadlocked when Secretary Lane arrived. His view consequently prevailed. On December 12, the six metropolitan plants of the big packers were suddenly struck.[69]

The packinghouse strike, unpromising as it was, thus developed considerable effectiveness. In the third week, government conciliators still reported the "strikers holding out firmly." [70] It was nevertheless only a matter of time.

One source of union weakness came from within. The industry had been organized, as in 1904, on customary jurisdictional lines. The AFL had rendered assistance to the Butcher Workmen "with the understanding that when the employees of the meat trust are organized, [they] shall be assigned to their respective organizations." For the most part, the Amalgamated honored the agreement. The major packing centers did have Amalgamated locals of "miscellaneous mechanics," some of whom no doubt belonged to other internationals. The Coopers, the Teamsters, and the AFL Building Trades Department at various times complained of jurisdictional infringements. A difficult problem arose in New York City in 1917 with the affiliation of the Brotherhood of Butcher Workmen, one of whose locals was made up of meat drivers. The Teamsters claimed these men, and the issue dragged unresolved into 1921.[71] These problems, while causing some ill will, were not serious, nor did they reflect any intention by the Butcher Workmen to organize the packing houses on an industrial basis.

That had not seemed necessary in meat packing. Organization was not apparently impeded here by the separation of the workers according to craft. The number in mechanical trades in the packing houses was relatively small and separate from the mass of workmen. Moreover, the Amalgamated itself, as earlier, divided its membership into locals on a "craft," not plant, basis. Cooperation between unions of different jurisdictions also proved satisfactory. Locally, common action between Amalgamated and mechanical trades unions was achieved informally. The only exception was Chicago, where the allied trades belonged to the Stockyards Labor Council until its conflict in 1919 with the Amalgamated. At the national level, cooperation was also informal and sporadic. The internationals came together only to prepare for arbitration hearings or during confrontations with the industry. When the packers tried to withdraw from the Alschuler administration in February 1921, for instance, Secretary Lane had to ask Gompers to call a meeting of the internationals "for the purpose of formulating plans and bringing about the closest solidarity between the workers in the industry." [72] This procedure, unwieldy though it was, worked well enough under the Alschuler administration.

But the *ad hoc* voluntary arrangement broke down when the showdown with the packers finally came. As earlier, national representatives of the allied trades met with Amalgamated officials to act in concert through a "joint conference." The meeting of November 21 adopted a plan to achieve unity in each center through "local boards of control." The Amalgamated also sent a resolution to the presidents of the cooperating internationals for united action:

Be It Resolved
    That in the event a strike is forced on the workers in the packing industry, that all Trades involved make a pledge to stand solidly together until a general settlement is reached, and
Be It Further Resolved
    That there be no Organization negotiate a separate agreement with the packers.

But when the decision to strike was taken on November 30, 1921, Secretary Lane recorded, Frank Wood of the Engineers and J. J. Brennan of the Firemen "bolted by refusing to order their men off the job in any plants except those of the big five packers—notwithstanding the fact that [their presidents] had signed and returned to us a resolution pledging their support to the end. (Their end came before they got started.)" [73] Actually, the firemen and engineers did not quit work even in most of the Big Five plants. The Teamsters, who had taken no part in the joint conferences, also refused to participate. The unions, as one Amalgamated leader complained, were "handicapped in making the force of the strike felt." [74] Without the strategic teamsters, firemen and engineers, the packers would have found greater difficulty in continuing operations. But the outcome, in any event, probably would have not been altered by greater solidarity. The experience did affect the thinking of the Amalgamated leaders; their approval of industrial unionism can be dated from this point.

The strike was meanwhile struggling to a conclusion. Police, deputy sheriffs, and, in St. Paul, National Guard units moved into the stockyards to quell the sporadic violence that erupted in the first days of the strike. The forces of law remained on the scene to break up meetings, harass the strikers and, as one journalist wrote of the Chicago police, maintain order "with a flauntingly martial air." [75] The strikers were further impeded by the recent Supreme Court decision on picketing in the *American Steel Foundries* case. Injunctions were handed down at all the packing centers severely restricting the actions of the strikers. An Amalgamated official told the Chicago Federation of Labor that the "greatest difficulty so far has been with police who will not allow the pickets to perform their duty." The Kansas and Colorado compulsory arbitration laws, which prohibited strikes, constituted a further impediment. District President T. A. McCreash and more than twenty others were sentenced to short jail sentences for refusing to call off the strike in Denver. The big packers, all of whom began to recruit strikebreakers, experienced little difficulty in finding replacements. A plentiful supply of jobless men was everywhere available.[76]

The suffering of the strikers and their families grew intense as winter deepened. Relief committees drew on local support, but they could not provide adequate assistance. The attempt to launch a national appeal began only in the third week of the strike. Weeks passed before the AFL gave its endorsement. Then the Amalgamated requested the federation to print and mail the appeal; this, too, required the consent of the executive board. Before action could be taken in late January, the strike was almost dead.[77] Receiving aid that was "disappointingly meager," the strikers were forced by privation to return to work in increasing numbers. "For," as Secretary Lane remarked, "half a loaf gives more nourishment than no loaf at all." [78]

No basis for a negotiated settlement could be found. The unions from the first were prepared to send the men back to work if the packers either rescinded the wage cut or submitted the issue to arbitration. Various efforts by governmental agencies to settle the strike were rebuffed by the packers. They saw nothing to arbitrate or conciliate.[79]

On January 23, three Federal mediators called at the Amalgamated general office and informed Secretary Lane to "no longer hold out any hope for intervention through the Department of Labor." They also suggested, in view of the deteriorating situation, that the unions should take a vote to end the hopeless struggle. Their advice was accepted at a joint conference that afternoon. The balloting on January 26 heavily favored continuing the fight, but the number of voters, according to Lane, was less than a quarter of the original strikers. The New York men had already returned to work on January 14 under open-shop conditions. The Amalgamated executive board therefore ordered an end to the strike on February 1, 1922 "in the name of justice to the loyal strikers and their families." [80]

The second cycle of packinghouse organization thus reached its end. The unions were once again cast out of the packing centers. The lesson was plain. Trade unionism could not win under its own power in meat packing. Only government intervention could redress the balance that favored the packers. Until then, the packinghouse workers would have to take their chances with employee representation, welfare benefits, and the open shop.

CHAPTER 6

THE MEAT CUTTERS: THE PROCESS OF LOCALIZED UNIONIZATION

Another field lay open to the Amalgamated. The retail trade differed from meat packing in important respects. The work involved craft skill, not the production line; the men were dispersed, not concentrated in massive plants in a few centers; control was minutely divided among thousands of shopowners, not confined to a handful of great firms; competition was entirely local in the retail trade, and employment was relatively insensitive to business fluctuation. Here the Amalgamated secured a firm hold. The spectacular cycle of triumph and defeat in the mass-production segment of the industry had no counterpart in the organization of meat cutters. The retail trade was a hard field; progress was slow; but it was also sure.

The slender organization of meat cutters had sustained the Amalgamated in hard times ever since the 1890's. During the world war, there had been a further retail union growth, although of far smaller dimension than in meat packing. While paying per capita to the AFL for roughly 60,000 in 1920, the Amalgamated refused to contribute to the Label Trades Department for more than 10,000, "which fully represents the meat-cutter branch of our Trade."[1] Doubled in size since 1914, this retail group remained the core of amalgamated strength after the collapse of packinghouse unionism in 1921.

Butchers' unions had appeared at one time or other probably in every community of any size in the country. The start of organization was not, however, the key stage. That occurred when the beginning retail organization became a viable union in an operating system of collective bargaining. The transition succeeded with relative infrequency. By the 1920's, union meat cutters were concentrated in a limited segment of the retail trade: in northern California, in the northwest, in Chicago, in St. Louis, in St. Paul, and at a number of lesser points such as Scranton, Evansville, Dubuque, and Westchester County, New York. No more than a tenth of the nation's 120,000 retail meat employees were organized. But the butchers' unions, where they did exist, had a solid foundation and an effective role in their local meat trades.

Retail unionism was, of course, responsive to external conditions. Periods of prosperity and labor shortage stimulated union growth and bargaining gains. Wartime labor measures by the government, although far more important elsewhere, had an effect. A number of butchers' unions were assisted by Federal conciliators during the war. The War Labor Board itself handled the dispute between Local 534 and the Retail Merchants' Association of East St. Louis. The result was a strengthening of existing locals, and the appearance of new unions elsewhere, for instance, in the hitherto barren field of Chicago.[2] Conversely, retail unionism was impeded by open-shop movements, packinghouse defeats, and economic declines. The depression of 1921 set back organization in a number of places. According to a government survey, the retail meat trade was squeezed from 1913 to 1920 by a faster increase in operating expenses than in gross beef profits.[3] Employer associations moved quickly, once a surplus of meat cutters developed, to cut wages and in some instances to break the butchers' unions.

Yet the broad influences were not decisive. Why did some localities have strong unions, and not others? The marked unevenness reflected the decentralized and localized situation of the butchers' unions. Theirs was an essentially independent existence. The international did not provide much of a link. In both periods when substantial resources became available, the Amalgamated chose to invest in the packinghouse workers, because organizing was more efficient in the packing centers than in the dispersed butcher shops. The retail branch, according to a 1922 resolution, felt "general dissatisfaction . . . because of the unfair division of the per capita tax paid into the International body for organization purposes."[4] Outside organizers, both Amalgamated and AFL, did work in the retail trade and provide assistance in times of local trouble, but the main burden of survival and progress lay with the local unions themselves. Here the second factor entered. The retail trade was entirely local in its ownership and in its competition. What happened in one community had little effect somewhere else. No real connection existed between the development of unions in different cities: each locality experienced a separate growth.

Retail unionism was a local phenomenon. Each meat cutters' union had to undergo separately and independently the hazardous course of internal stabilization, of collective-bargaining development, and of employer acceptance. The locality was the meaningful area for these three aspects of retail meat unionization until the end of the 1920's.

A meat cutters' union, having survived its beginning, soon had to solve the critical problems of internal organization. Foremost was the recruitment of local leadership. The skills—nonexistent at the outset—were always difficult to develop. Even worse was the constant depletion of experienced officers. The most intelligent and energetic, they were usually the first to take retiring cards and open their own shops. Local 358 of Janesville, Wisconsin, lost a majority of its original slate of officers in its first year, including the president, who resigned "due to being appointed manager of market." Or they wearied of the unrewarded labors of leadership. A correspondent from Cambridge, Massachusetts, reported in 1902 that "most of our officers refused, at our last general meeting, to keep their old offices, for they were tired of the hard work they had done for our local." Even long-established unions often lacked qualified leadership. The president of Syracuse Local 1, the oldest in the Amalgamated, urgently applied to the national office in 1938 for "someone to help get our Meat Contracts signed. There is no one here who is experienced in this line and unless we get assistance our union will not survive." [5]

It was worse when officers found rewards for their work. Many young locals were disrupted by "loose business methods" or by outright thievery. Organizer Rudolph Modest was unable in 1906 to organize the butchers of Meriden, Connecticut, because "on three previous occasions the Union always busted up through the dishonesty of the financial officers." He heard in Paterson, New Jersey, "the same cry, what's the use to organize the Meat Cutters; somebody will go away with the money anyhow." The international's response was to require the locals to bond their financial officers. The constitutional provision, however, in many instances remained a dead letter. Venal officials continued to be a problem into the 1920's. [6]

Leadership generally reached a turning point with the decision to hire a business agent or to put the secretary on salary. A local correspondent in 1902 explained the reasoning:

It is going to cost us a great deal at first, but in my opinion it is the only way that we can make a success of the union . . . We will have a man in the field all the time, and he is a hustler and a man with eight years experience in labor work.[7]

Local leadership was thus gradually professionalized. By the 1920's, the strong retail locals were led by a corps of careerists, such as J. S. Hofmann of Seattle, M. S. Maxwell of San Francisco, James E. Kelly of Yonkers, New York, M. J. Kelly and James F. Laverty of Chicago, and Louis Wentz and Walter Gieseke of St. Louis. The vital factor was to make trade unionism an attractive career to available men of ability.

The key was finances. It was not accidental that the big city locals, having large potential membership and income, recruited the ablest administrators. To this end, small locals sometimes amalgamated or extended their jurisdictions, or, as in the case of the Westchester County unions, pooled their resources to employ a first-class business agent. Since the early period, this had been among the reasons for district bodies. Higher dues were the other means of supporting paid officials. Secretary Lane uged the 1922 convention to establish a three-dollar-a-month minimum on local dues. His recommendation was opposed because it "would have the effect of keeping the packinghouse workers out of the organization." The convention compromised on a dollar minimum. By 1930, the increasing commitment to the retail trade permitted a further increase to a $2.50 base. The higher local dues would "be of great help in strengthening the weaker ends of our movement," asserted the executive board.[8]

The results of adequate financing could be seen in the case of the powerful Local 546 of Chicago. Raising its dues to $3.50 a month, the union employed "high class fellows . . . who could go out and do business." Secretary-Treasurer M. J. Kelly received $150 a week, and the business agents $100 a week. (The international president and secretary earned only $100 a week in 1922, and the vice-presidents $10 a day.) A delegate from the Portland, Oregon local, stopping in Chicago en route to the AFL convention in 1926, noted "with amazement and admiration the perfect organization of the men employed in the retail meat trade . . . The local's business representatives were recognized as men of affairs and . . . all of them make the rounds in modern cars." On his return to Portland, Local 143 purchased a new Dodge for the business agent "appropriate for one representing an up-to-date, progressive organization."[9] Such bodies in the Amalgamated saw the need for a well-paid, efficient staff, and they benefited accordingly.

The second internal problem involved the stabilization of membership. Although they entered new unions readily enough, meat cutters were not apt union members. For one thing, they worked in small shops side by side with their employers and with the reasonable expectation of becoming proprietors themselves. The initial gains were usually achieved easily and tended to satisfy the recruits. A New Jersey local, for instance, reported that "as the boys have their Sundays, they think what do we want to belong to a union for?" There were frequent complaints in the early years, as President Donnelly reported of St. Joseph, that butchers "allowed the union to lapse after the Sunday closing and other concessions had been granted." [10] The transformation of recruits into permanent unionists posed a critical problem in the organization of the retail trade.

One solution was to provide insurance benefits. Secretary Call in 1899 had pleaded for the adoption of a program that would give the members "something tangible for their money . . . There are some that feel it is the only way an organization can be held together." The 1904 convention finally voted to establish a death benefit effective January 1, 1905. For payments of five cents a month, each member in continuous good standing for six months was insured for fifty dollars, and for a hundred dollars after a year. The death benefit by itself was not thought sufficiently attractive to "induce the men working in the craft to join the Union or keep themselves in good standing after joining." Further efforts to broaden the coverage were unsuccessful.[11] Still, the insurance was better than nothing. Many of the stronger retail locals, in addition, established their own beneficial systems, generally with more comprehensive coverage—one of the reasons, in fact, for the resistance to a national program.

As their strength grew, meat cutters' unions began to make membership a condition of employment. In 1906 Local 12 of Duluth, a well-established union, reported a dozen nonunion meat cutters: "We will take such action that they will either come in or get out of town." [12] Pressure on employers could bring in almost all eligible men within the jurisdiction of an effective union. The union shop gradually became part of retail union agreements until it was nearly universal by the 1920's for established locals. The contract of the Herrin, Illinois organization provided also for a check-off for members in arrears.[13] Beyond union security provisions, retail locals tried to control access to jobs. A preferential hiring clause—the closed shop rarely—was inserted into many contracts. The effective monopoly of job opportunities led to efforts to limit union membership. Local 534 of East St. Louis used a permit system, allowing outsiders to work temporarily during labor shortages on a permit without being accepted into the union. Strong locals also tended to raise their initiation fees; Chicago Local 546 was charging $113.50 in 1926.[14]

Thus local membership gradually was stabilized, first, by the union shop, and, to a lesser extent, by the control of access to work.

The final step in this direction involved regulating the labor supply. The international had made no provision for apprentices. Since before the world war, however, meat cutters' unions had begun to regulate apprenticeship, usually permitting one boy to a market plus another youth for a given number of journeymen. This tactic of limiting the labor supply, typical of craft unions, had little success because the butcher unions could not control the outside sources. Organization was kept weak in the favored cities of Los Angeles and Denver by the continuing influx of meat cutters. Strong unions were also threatened. Seattle Local 81 had to cope with the pool of low-paid, unorganized meat cutters across the bay in Vancouver. Even the powerful Local 546 of Chicago felt the danger. The union did not need aid in its own affairs from the international, asserted Secretary M. J. Kelly, "but this is what we do want, we want the scabs stopped from coming to our town." [15] Union security could never be absolute while the bulk of the nation's meat cutters were unorganized.

Membership stabilization had another aspect. Meat cutters might be on the union rolls, but they would become committed members only when their needs were being served effectively by the union. This occurred only gradually. For collective bargaining had a slow growth in the retail trade.

The overriding concern of the early unions had been the work schedule. Eighty hours or more was a week's work at the opening of the century. All meat cutters' unions therefore were first of all concerned with early closing and Sunday closing. As a rule, they ran into little resistance from proprietors. The issue, also, was largely political. Many states and localities forbade the operation of meat markets on the Sabbath. The enactment of such legislation often had Amalgamated assistance. The New York law of 1901, for example, resulted chiefly from the agitation of upstate locals, although the New York Retail Butchers' Protective Association belatedly added its support. [16] The Minnesota law several years later was also the product of union lobbying.

The initial gain was substantial, but not productive of a collective bargaining relationship. The hours reform, achieved without real bargaining, did not lead automatically to further negotiation. And the political strategy was at the expense of collective bargaining. One shrewd official remarked that the Amalgamated undermined its own position by gaining the New York Sunday Closing Law. For a great many unions, the first achievement on work schedules was also the last one.

Surviving locals, however, gradually became bargaining agents. The key issue continued to be hours. In some cases, the unions requested a half-holiday during the summer months or a further lowering of the closing hour. Syracuse Local 1 in the spring of 1906 concentrated on achieving a 9:30 P.M. closing on Saturdays.[17] Some contracts also began to prohibit work on legal holidays as well as Sundays. Finally, a shift became perceptible in the method of regulating the work time: some contracts began to stipulate the number of hours of work per day instead of opening and closing hours. In 1920, the international officially endorsed this change on the ground that store schedules were not the unions' concern, but some locals adhered to the older pattern for many years afterward.

Wage scales were rarely subject to negotiation in the early stages of collective bargaining. Most contracts, it was true, prohibited a reduced weekly wage as a result of shorter hours, and, less frequently, provided for additional pay for overtime work. But the negotiation of a standard wage rate was apparently not expected either by union members or the employers. The Appleton, Wisconsin local, for example, reporting on the record of its first year, noted parenthetically, "as far as wages go the men are getting pretty fair pay and some of the men have bettered themselves since the union was organized." If the wage issue was injected into negotiations, a Colorado Springs official observed, his union would certainly "meet with quite a little opposition on Scale of wages, no matter what rate we place them." [18] Generally, compensation remained, as it had before organization, a private matter between the meat cutter and his employer.

A change eventually occurred in well-organized localities. Satisfied by the shortened hours, men began to grumble about money. The union experience, moreover, changed men's thinking about the appropriate ways of determining wages. In mixed locals, meat cutters were stimulated by the example of other meat workers. When the Wheeling packers signed a wage agreement with Local 7 in 1904, the market men in the union also demanded "a scale to work by." [19] The shift became possible with a stable organization and a functioning contractual relationship. Local 1 of Syracuse, which early had both, was working under a wage scale before 1900. The surviving locals by 1910 commonly had a minimum rate—scope remained for private agreement above that rate—ranging up to eighteen dollars a week in Pacific Coast cities.

The great breakthrough in labor standards came during and immediately after World War I. In 1920, hours in union markets ran between 55 and 60 a week, a reduction of at least ten hours from prewar levels. The most common schedule was a nine-hour day during the week and twelve hours on Saturday.

Minimum union wages started at thirty-five dollars a week, forty dollars in Chicago and on the Pacific Coast, and forty-three dollars in Westchester County, New York. In general, rates more than doubled. The depression of 1921 forced concessions by many locals, but earlier standards were resumed the following year. Thereafter, the wage-and-hour pattern remained relatively stable, with small scattered gains. Wednesday half-holidays during the summer became more common in the 1920's, and for the first time some locals won a week's vacation with pay. Employers also began to assume the cost of laundering work clothes and of the maintenance of tools. By the 1920's the established meat cutters' unions were fully engaged in the collective bargaining process.

A workable relationship with the employers was the third element in the development of permanent retail organization. In the early period, meat cutters' unions sprang up under amicable circumstances. Letters from new locals in hitherto unorganized places were nearly unanimous in reporting the friendliness of the proprietors, and the willingness to accept the first set of demands for a shorter store schedule. That fund of good will was quickly dissipated.

Butchers' associations greeted the new unions as allies in the early closing movements. But, if the labor organization survived the first period, employers invariably faced demands for further concessions. When the Elmira, New York local requested another reduction in hours, a prominent butcher complained:

It was only a short time ago that the meat cutters were working nearly every night as late as 10 o'clock, and . . . Sunday mornings also. We had no protests then. I do not consider that this present demand is called for.

The early closing movements, moreover, lost much of their force when business declined. The temptation to remain open then overcame some butchers, and others necessarily followed. But the union blocked the way. In St. Louis the butchers' association in early 1904 decided, since many members were doing it anyway, to permit stores to open for three hours on Sunday mornings. A contract was in effect with the meat cutters' union, however, and a heated controversy ensued.[20] Under these circumstances, the union became an embarrassing hindrance. The proprietors discovered soon enough that theirs were not the interests of the union.

Hostility sprang from a variety of other causes. The ordinary functioning of a local union caused friction. Rochester merchants, for instance, complained bitterly of the "personality and manner" of "walking delegates . . .

suddenly thrust into prominence from positions of obscurity." They "arrogantly stalk into our places of business and treat us with scant courtesy and with insult by peremptorily ordering changes to be made. This interference with our business was unbearable." [21] Inexperienced officials also tended to resort too easily to pressure tactics and to become involved in jurisdictional disputes. In the early period, the symbol of union irresponsibility was the refusal to handle the meat of "unfair" packers—"absolute dictation as to how a man who employs union labor shall conduct his business." The Amalgamated very early learned to eschew this tactic against the major packers, but not against local packing firms. A number of retail fights consequently developed, most notably in the boycotts of the Dold Packing Company of Buffalo in 1900 and of Rochester firms in 1903. Finally, beyond concrete issues or personal experience, butchers' associations often fell under the influence of the open-shop movement of the first decade of the century.

Rising employer resistance meant that a viable relationship normally depended on coercive power. The retail unions primarily used consumer pressure. One local reported in November 1900 that twenty customers had been taken from shops refusing to sign its contract. "This mode of working the bosses is somewhat new to them and is getting them worked up. They had expected a strike." Proprietors quickly broke ranks under the new tactic. The market card was an effective boycotting device. It became "a tower of strength to the Meat Cutters Locals." Withdrawal of the card could constitute strong sanction against an unfair market, particularly in union towns. A boycott thus became possible even in the face of an injunction against picketing or the distribution of leaflets.[22] Strikes served also primarily to exert consumer pressure. Meat cutters' unions could not hope to close down markets by withdrawing their members, since a proprietor himself worked and generally could find nonunion help. But they could keep customers out of the shop.

Strategically, the success of the consumer boycott depended upon disunity among the meat dealers. When a united front appeared in Oakland, California, in 1904, Vice-President C. E. Schmidt saw the danger:

There remained no union market in the city for organized labor to throw its patronage to, so that all were compelled to be served by non-union meat cutters or go without meat. To have such a condition continue would mean the disruption of our union.

Butchers' associations, long existing in many places, reorganized, and others formed, to deal with labor problems. One New York retailer observed in 1903:

We realize that we must meet the force of organized labor. The grocers and butchers must keep up with industrial progress. Unorganized, the time might come when we would find ourselves unable to cope with labor conditions that might arise.

But unity was very difficult to achieve in the competitive retail trade. When the Rochester Retail Meat Dealers' Association decided in October 1902 to sign no agreement with Local 95, the business agent hastily "made the rounds" to secure a number of firms, "and that broke their backbone." [23]

The retailers' associations needed means of enforcing a common front. In Rochester in 1903 each member was required to put up a bond which would be forfeited if he dropped out. A more effective device was an alliance with local meat suppliers who could withhold meat from a renegade butcher or, more subtly, restrict his credit or give him inferior service. The effectiveness of the wholesaler-retailer alliance was demonstrated in the Buffalo strike of 1900, again in Rochester in 1903, and Oakland in 1904. The union's packing-house defeat in 1904 completed the possibilities of supplier coercion. The restraint on the major firms now withdrawn, they could join in the assault on unions in the local trade. The consequences became clear in San Francisco in September 1904. The Western Meat Company, controlled by major mid-western packers, joined with the independent slaughterers to form the Butchers' Exchange. An open-shop drive was immediately launched in the retail trade. The suppliers warned the dealers to keep open-shop cards in their markets or "be dealt with in a prompt and decisive manner." [24]

Labor could counter a solid retailers' front only by itself entering the butcher business. This in fact occurred in few instances. The Amalgamated local in Rochester started thirteen markets during the strike of 1903. In Oakland, a cooperative undertaking was launched. The California Cooperative Meat Company, incorporated on April 5, 1904, became a going concern with a wholesale house and retail markets. But union enterprises did not have general utility. For one thing, they had to start "with every business man's hand against us" and "no friends, only such as we were able to buy, and often at a high figure at that." [25] In both instances, the labor entrepeneurs experienced great difficulty in finding an adequate meat supply. There were other practical difficulties. The California Meat Company suffered from inexperienced management and from disaffected and dishonest employees. Moreover, a heavy investment was involved. After the 1904 strike, the international could not sustain such expenditures, nor could a local union ordinarily expect the wide backing such as that forthcoming from the Oakland labor movement. The drain was great also on the leadership resources of the Amalgamated, because several key men were lost to the business ventures.[26] The California Meat Company, once it became profitable, drew away from the declining butchers' union. Although proclaimed at the time, the two union enterprises would have no significant successors.

On the other hand, the situation which forced the union to enter business was itself an infrequent occurrence. The supplier-retailer alliance, despite its evident effectiveness, was a rare event, since it demanded an unusual identity of interests. In Buffalo and Rochester, the meat suppliers had cause to act with the retailers because the latters' fight against the union was occasioned by a boycott of the suppliers' products. In San Francisco, the packers would benefit by the destruction of retail unionism because they were dealing with the same Local 115. But where suppliers had no interest in retail affairs—the usual case—they were unlikely to extend themselves to force a united front against a meat cutters' union.

The power struggle was too demanding on both sides often to be fought to the logical conclusion. In general, retail unionism collapsed at an earlier stage, or an accommodation was reached gradually or after a decisive test of strength (for instance, the butchers' strike in Chicago in 1919).[27] A stable labor-management relationship existed for all the strong meat cutters' unions by the 1920's.

No important unresolved issues remained after the great advance in standards in 1918 and 1919. Between the 1926 and 1930 conventions, there was not a single strike of importance in the retail trade. Bargaining through employer associations also served to stabilize relations. An Amalgamated official later remarked:

> An individual employer can not be expected to have perspective. His identity is so merged with his business, that he loses all sense of value . . . For this reason, I have found it easier to deal with the employer groups on an industry-wide basis, than with single employers.

Experience furthered harmonious relations. It was common for former union officers to lead employer associations, for instance, at Chicago and St. Louis. "Organization is the only means of education for the workers," President Joe Flaherty of the Chicago Master Butchers' Association told a meeting of Local 546 in 1922. "Both organizations of Master Butchers and Meat Cutters ought to work together for all the meat industry." [28]

Meat cutters' unions proved valuable allies. They helped to achieve local meat inspection laws, enforce Sunday closing ordinances, and fight chain store competition toward the end of the decade. Employers certainly valued the uniform conditions which a strong union could give to the trade. That was largely why, despite the disapproval of the international, some unions continued to regulate store hours. "As far as our organization is concerned, we insist that the Union help to maintain regular opening and closing hours," asserted I. W. Ringer of the Seattle Retail Meat Dealers' Association.[29] Mutual interest cemented labor-management amity.

The one vital ingredient, given the competitive character of the retail meat trade, was thorough organization of the jurisdiction of the local unions. Only then could organized shops avoid penalization for dealing with the union. Secretary M. J. Kelly of Chicago Local 546 acknowledged his union's duty to prevent unfair competition of "meat bootleggers with legitimate union markets." This was one reason why the international constitution, since 1899, permitted local unions to accept as members the proprietors of one-man shops. In a few instances, for example in Sioux City, the employers' association made union recognition a requirement for admission to the association. Even with complete organization, cheating remained a continuing problem, but the locals proved reasonably effective in policing their contracts.[30]

By 1929, the Amalgamated had a scattered body of stable retail unions. Their effectiveness, it was true, varied with local changes. Organization, for example, weakened in St. Louis "due to the lethargic methods of the Local," and new officers had to be elected to reestablish the position of Local 88. The Los Angeles union was gripped in the summer of 1924, Secretary Lane reported, by "a group of malcontents, who alone were responsible for the movement having been torn to pieces in that city." Local 111 of Scranton, which had been very effective in the early 1920's, inexplicably declined until by 1930, with a contract still in effect, the union stopped functioning altogether.[31] But for the most part, the meat cutters' unions, once established, did not relinquish their sources of strength.

Ethnic ties had long helped organization in the local trade. In the early part of the century, the most stable retail union in New York City had been the Bohemian Meat Cutters' Local 273. German unions became a mainstay of the Amalgamated in its darker years. It was noteworthy that, having consistently refused before the great packinghouse strike, the Amalgamated began to print its journal partly in German after 1906. The ethnic factor tended to decline in significance among older unions, but it became increasingly important in one segment of the retail trade in the 1920's.

Approximately 10,000 Kosher butcher shops existed in the United States, half in the New York area and the remainder in other urban centers. The

Jewish population also supported an extensive poultry trade in connection with the meat markets. This field had never been well organized; the early Hebrew butchers' unions had been notably unstable. Since 1909, however, a Kosher union had operated more or less continuously in New York City. After a disastrous strike in 1913, the union recovered under the leadership of Isadore Corn and grew strong during World War I. Kosher meat cutters and chicken killers—known as "shochtem"—also organized during and after the war in Chicago, Detroit, Cleveland, St. Louis, and Los Angeles.

The New York organization, the most important of the Jewish unions, only gradually developed a firm affiliation with the Amalgamated. The United Hebrew Trades had chartered the union and supported it through its early struggles. After joining the Amalgamated briefly during the world war, the New York body again became independent. The efforts of District President John Kennedy eventually brought it back into the Amalgamated in October 1921 as Local 234. After another brief secession in the late 1920's, the New York Kosher unions of meat cutters and shochtem finally became a permanent part of the national movement.[32] Elsewhere the separatist tendency was not evident among the Hebrew unions, who had for the most part the same bases of attachment to the Amalgamated as did other retail unions.

Although "the interest of our organization is paramount to them with the other aspects only secondary," Patrick Gorman observed in 1930, the Kosher unions did constitute "a rather cosmopolitan group with varied political ideas and connections." Several things set them apart from the run of retail unions. The religious element had important consequences, particularly for the shochtem. These chicken slaughters required certification by the local rabbinical authority. Employers could claim, as did the president of the Newark Poultry Dealers' Association, that it was inappropriate for religious functionaries to belong to trade unions. Organization was impossible where rabbis adhered to this view. Once organized, on the other hand, the religious tie made the shochtem unions very strong, since the rabbis controlled the labor supply. Labor leaders, sensitive to the dangers, favored separation of the union and religion spheres. "Whenever a question arose on the Jewish religion, we left it to the Rabbis," stated Joseph Etkin, business agent of the Chicago Kosher unions. "Whenever there was a dispute between a butcher and a Shochet, it was decided by the local union." The division sometimes broke down. During an internal fight in the United Hebrew Trades in Chicago in the winter of 1930, for instance, the entire membership of Local 598 was put under ban, throwing 150 men out of work.[33] On balance, however, the religious connection became a source of strength rather than weakness.

The Kosher poultry trade, combining as it did fragmented ownership, a highly perishable commodity, and a captive market, was fertile ground for the gangster element on the fringe of the immigrant Jewish community. The New York industry, in particular, was notorious for being dominated by racketeers. They extracted, according to the estimate of city authorities, $5,000,000 from the trade annually. Inevitably, they entered the unions. Charles Herbert, the business agent of Schochtem Local 440, had a long police record, as did his brother Arthur ("Tootsie"), who was agent for the chicken drivers' union. "There was not a decent trade unionist," Joseph Belsky of Local 234 afterward remarked, "who would dare to touch the Schochtim Union because of the corruption under which it was dominated." In the summer of 1935 Charles Herbert was voted out of office at a special election after being publicly attacked by the New York Market Commissioner. A fierce intraunion fight followed. The international intervened on the side of the deposed business agent, believing he was being nailed "to a cross because of unfavorable publicity." For a time Herbert regained control, but the waning of his influence by the end of 1936 permitted the installation of another administration. The New York trouble thus passed.[34] The shochtem locals, despite their special problems, came under responsible leadership.

Finally, the Jewish locals were active politically. This had been true also of the earlier German-speaking unions and, after the war, of some of the packinghouse organizations. A few unions, including those in Cleveland, Philadelphia, and in Canada, had at some time been captured by IWW and One-Big-Union elements. But the Amalgamated and its retail unions in the 1920's and later were politically inactive or nonpartisan. The Hebrew unions, on the other hand, followed the same Socialist tradition that actuated the garment workers and other Jewish trades. And, like them, the butcher unions in New York were threatened by Communist attack in the late 1920's. Expelled from Hebrew Butcher Workers Local 234, the leftists fought as a rival local, attached to the Food Workers' Industrial Union, until 1935 when they returned individually to Local 234.[35]

The Hebrew unions, set apart in some ways from the rest of the retail unions, had an importance beyond their relatively small numbers. They provided a toehold in areas otherwise entirely without butcher organization. Prospering Kosher unions existed in Boston, Detroit, and Cleveland long before other Amalgamated unions were established there. Local 234 made the greatest contribution. It fathered the organization of the thousands of non-Kosher meat cutters in New York City in the 1930's and assisted them through their early struggles.[36] One recognition of this was the election of Joseph Belsky, the secretary of Local 234, to the international executive board in 1936. Militant and effective, the Jewish organizations were a significant addition to the ranks of butcher unions.

The vigor of the retail unions was proved in the early years of the Great Depression. Amalgamated membership, far from declining, was on the rise during the initial months. The international paid per capita to the AFL for 11,498 members in January 1929; and for 16,589 members in December 1930. "No organization of labor," stated Patrick Gorman to the 1930 convention, "has faced the fearful economic depression of the past two years with more success." [37]

In some ways, hard times strengthened the retail locals. For one thing, union membership became more prized; it was the prerequisite for employment where job access was controlled by the unions.[38] Membership also was a form of social security. Many locals tried to assist unemployed members. Hebrew Butcher Workers' Local 234, for instance, in 1931 established a plan of work distribution by which each working brother gave up a day or more, depending on his wage, to the unemployed men. Other unions established relief funds; the St. Louis and Chicago locals levied assessments of a dollar a week for this purpose during the winter of 1931–32, and informal assistance was often provided in smaller unions. Economic adversity tightened the hold of the union on the membership.

Then the depression deepened. The initial impact had been less pronounced on the retail meat trade than in other areas. As the jobless ranks grew, however, unions began to send notices to the journal for men to keep away from their communities. The meat cutters' unions were bothered, Ray Wentz of St. Paul Local 114 wrote, by nonunion men who "are constantly butting in and offering their services at free lance rates." [39] The economic pressures on the retailers were simultaneously mounting. Their demands for wage cuts were largely forestalled or compromised. But the danger of strikes under unfavorable circumstances was great. For example, in Sacramento in June 1932 the local union called a strike over a deadlocked wage issue. Only 20 per cent of the men responded. The Meat Dealers' Association thereupon informed the international office that, if its offer of $37 was not immediately accepted, outsiders would be brought in and "all further negotiation will cease with the local union." The threatened open shop, wired the organizer on the scene, made "settlement only way to save organization." The employers' offer was finally accepted, and the crisis passed.[40] Yet, clearly, retail unionism could not entirely escape the consequences of widespread unemployment and employer pressures.

The Amalgamated paid per capita to the AFL for the seven months ending March 31, 1933 for an average membership of 10,731—a drop of a third since December 1930 and of 1000 for the same seven months in 1928–29.[41] Still, the figure meant that retail organization was passing through the darkest time of the depression intact. The industry was very far from being completely unionized, but a solid base of local unions existed in the retail field before the arrival of new organizing opportunities in the 1930's.

The strengthening of the international accompanied the emergence of a group of stable retail unions. The packinghouse strike of 1921 had left the Amalgamated leadership daunted; the treasury empty and the income inadequate; and the authority over the locals uncertain and threatened. All these weaknesses were resolved in the following decade. By 1933, the international was in a position to operate effectively and to assume a larger role.

Cornelius Hayes, who had replaced John Hart in 1921, did not long remain president. On April 1, 1923 he left the office of an organization which seemed to have little future. His resignation did not precipitate another succession crisis. The constitution, clarified at the 1922 convention, automatically raised the first vice-president when the presidency became vacant. Denied in 1921, Patrick E. Gorman now took the office. Secretary Dennis Lane had considered resigning in 1922, but had been dissuaded. The enmity between the seasoned secretary and the young Gorman did not last; there soon developed, as Gorman said in 1926, "perfect harmony between Secretary Lane and myself." [42] The combination, in fact, proved to be a peculiarly happy one. Dividing the burdens of leadership, the two men gave the Amalgamated effective national direction.

Gorman became the outside man. President at twenty-nine, his first years were a kind of apprenticeship. In New York City, for instance, he was frequently with Vice-President Jack Walsh "at 'Big' Eddie's or Joe Sheridan's place on the West Side, where [he had] an opportunity of coming in contact with the fellows in the packing houses, and talking to them regarding their wages and conditions, and in general attempting to arrive at some plan through which [the] organization might be made better." Gorman thus became intimately acquainted with the industry and its problems. Between 1926 and 1930, he visited practically every local union in the

Amalgamated. Gorman specialized in employer relations. He developed an unusual capacity for negotiation and mediation, and his reputation in the trade grew. "My only regret," the secretary of the National Retail Meat Dealers' Association commented privately at a later date, "is that there are not more Gormans and that he cannot split himself up into about 20 pieces so as to make himself available in about 20 situations at one time." [43]

Dennis Lane provided the internal cohesion. The constitution did not establish the primacy of the secretary-treasurership, although Homer Call's long tenure had started a tradition in that direction. Lane's authority evolved primarily from his seniority, his tough character, and his control of the political machinery. His was the final word in the day-to-day operations of the international. He wielded the authority over the local unions in matters of finances, of assistance, and of discipline. Significantly, the rebellious California Federation of Butcher Workmen directed its opposition against the secretary and his policies: "We simply would like to know: Who is the International, is it the membership or is it Mr. Lane?" [44] President Gorman, characteristically, played the role of mediator in the dispute. That the secretary's office was the locus of power became evident at Lane's death in 1942. Gorman resigned the presidency, and with it the public attributes of primacy which clung—and still cling—to the office, in order to become Lane's successor. The new president, Earl W. Jimerson of East St. Louis, remained in the shadow of Patrick Gorman.

The second level of national officers also improved. The system of district presidents was scrapped in 1922 after a two-year trial. The convention reverted to an executive board of president, secretary, and nine vice-presidents. The board was relatively inactive during the next decade because of weak finances and few critical issues. Most of the business requiring the decision of the board, such as votes on strikes, assessments, and supervisions, was carried on by telegraph or mail. Only in the mid-1930's did the practice develop of holding regular semi-annual meetings. Even then, it remained customary to delegate thorny problems to the president and secretary, and to follow their lead on matters of policy.

The more important function of the vice-presidents was, as the constitution stated, "to act as Executives of the several districts, or divisions, to which they are assigned." They upheld the international authority and fostered the movements in their districts (which were not, however, given any formal definition). Their caliber was rising in the 1920's, partly because, like M. J. Kelly, they were being drawn from seasoned leadership at the local level. Increasingly, the men who filled the vice-presidencies after 1922 had the two prime requirements for the job: first, ability as labor leaders and, second, standing within their own territory. Of the nine incumbents in 1933, all but one was destined to hold office until death or retirement—hitherto a rare occurrence.[45] The Amalgamated was manned by effective leaders as it entered the New Deal era.

Inadequate income was a severe handicap after the packinghouse strike. It became necessary to eliminate from the payroll all but three of the organizers and district presidents "because of no finances to carry on work." The organization struggled along on its reduced income—for the first half of 1922, 21 per cent of the amount received for the first half of 1921—unable at times to meet death benefit obligations and without funds for emergencies. The provision in the 1922 constitution permitting assessments of 50 cents per member for three months in a year was used by the executive board in 1924, 1925, and 1926; but the expedient was a poor one, hard to operate, inadequate in its returns, and invariably a cause of resentment in the ranks. When a strike in 1926 in Louisville threatened the existence of the strongest remaining packinghouse local, the international was forced to borrow $21,000 from Chicago Local 546 and to appeal for a weekly donation of 25 cents per member for the strikers. The small staff worked irregularly and often part-time; on three occasions between 1922 and 1926 there were total layoffs for periods of up to eight weeks. The effect on the staff was of course demoralizing. Vice-President Jacob H. Davis failed to attend the executive board meeting of April 15–17, 1925 "because he had just been given employment at his old job, he could not very well lay off now, as it might throw him out of work for the balance of the Summer." [46]

The issue was met squarely at the Louisville convention of 1926. "The time has arrived," the executive board told the delegates, "to relieve the organization of this poverty-stricken condition . . . by the introduction of sensible business finance." After an extensive debate, the convention adopted by a vote of 47 to 8 a per capita increase from fifty cents to a dollar a month. A tense period followed before rank-and-file opinion crystallized. The opposition, said Vice-President Jack Walsh, "was aided and abetted by some of our Business Agents, who were fearful that with the payment of the additional per capita tax, their local union could not meet the Business Agent's salary." The increase occasioned one or two secessions, and remained unpopular elsewhere, notably in California, but was generally accepted.[47] The Amalgamated became solvent. For over twenty-five years after 1926, there were no further advances in the international dues.

Finances were strengthened in another way. The international had always experienced difficulty in collecting full per capita payments from the local unions. Gorman saw local meetings where he was "positive that several hundred more were in attendance than what the International was receiving tax upon." Gradually, however, the constitutional sanctions and the supervisory machinery were improved. In 1930, there was an overhauling of the reporting system and of the local records at the international office. The president was empowered to remove from office any local officer found guilty of holding back international per capita. And vice-presidents and organizers were instructed to audit twice annually the books of local unions in their territory. Finally, after further discussions by the executive board and the rapid increase in new locals, four district auditors were appointed in 1938 to inspect regularly all local accounts. Leakages were thus reduced, if not entirely stopped.[48]

In one area, the international met total frustration. Amalgamated officials, particularly concerning the retail branch, had always placed great stock in the beneficial functions of the organization. So did Secretary Lane after the packinghouse defeat of 1921. The front page of every *Butcher Workman* carried a death benefit report together with a brief message on union protection. In conjunction with the per capita increase in 1926, the maximum death benefit was raised from $200 after two years' membership to $300 after three years. A sick benefit system had long been the next goal. But in the 1920's much attention focused on the problems of older people in industry. Lane conceived a home and pension plan quite different from those used by the Carpenters, Pressmen, or Typographers. Each Amalgamated member would be assessed twenty-five dollars payable in monthly installments of one dollar. The fund would purchase an extensive cattle ranch which would be a self-supporting home for men over sixty-five with twenty years' membership, disabled workers with ten years' membership, and orphans who would receive vocational training. There was also an option of a pension of thirty dollars a month. At the Detroit convention in 1930 the opposition, which proved to be considerable, was overborne and the executive board resolution was adopted, 57 to 19. But criticism, centering on the Pacific Coast, did not subside. After postponements because of internal politics and the depression, the home and pension plan was formally discarded in 1936.[49] Lane's cherished idea, ill suited as it was to the New Deal era, came to nothing.

The preparatory years were significant, finally, in the relations between the Amalgamated and the local unions. Constitutionally, the authority of the international had long been defined. Only two noteworthy changes occurred after 1922. The convention of 1926 gave the international the control over the money and property of seceding locals, clarifying a constitutional ambiguity and strengthening the Amalgamated's hand against dissidents. The president was also empowered to "assume charge of the affairs of any Local Union at any time when trouble of an internal nature threatens the effectiveness of the local," and "to remove any local officer from office upon evidence of official misconduct." This constitutional change of 1930 only legalized the procedure of trusteeship which had already become a practice in the Amalgamated.[50]

The other improvement involved a major reduction of secessionist activity. "We had perhaps as much of it," President Gorman said in 1926, "as any affiliated branch of the American Federation of Labor." [51] By 1933, the bonds of unity had tightened.

New York City had always been a separatist center. Even before the packinghouse strike of 1921, the movement there had been deeply split; some locals had seceded after the East Coast meat strike of 1919 and the departure of Vice-President Joseph Menhart. Most remaining unions left after the 1922 convention, disgruntled by their involuntary part in the packinghouse strike and influenced by John Kennedy. Their leader since the days of the independent Brotherhood, Kennedy did not return to executive board in 1922, and he thereafter was openly against the administration. The secessionist movement proved a failure, because it rested on a crumbling organizational base. Unlike their experience in 1904, the packinghouse unions lost their grip on the big New York houses as a result of involvement in the western strike. The efforts of the AFL and of Vice-President Jack Walsh, Kennedy's successor in the New York area, led to the reaffiliation of the remnant unions within a few years (although independent pockets remained as late as 1935).[52]

The Pacific problem was of greater import, because the California movement was a strong point of the organization. The roots of the trouble were remote and tangled, extending back before World War I. At bottom, the issue in the 1920's was a conflict over authority. The California Federation of Butcher Workmen had assumed many of the functions of the international. When the Amalgamated gave strike permission in 1921 to the San Jose local in face of the disapproval of the California Federation, state Secretary F. M. Sanford rebuked the executive board:

It seems to me that matters in which locals of our State Federation are concerned, the International should refer to us for action and back up our decision, as we should know what is best for . . . this section.

The Amalgamated constitution, Secretary Lane responded, "has never placed the power of a state branch above the powers of the International Union." [53] There were similar tensions over the appointment and direction of organizers assigned to California and over the impositions from Chicago which drained money from the California locals with little apparent return to the state movement.

These differences were intensified by the emerging alliances which put the dominant factions of the international and state branch in opposite camps. The ruling California party, led since 1920 by Milton S. Maxwell, had sided with President Hart. His removal in 1921 and the consequent near-secession in California hardened differences between Lane and Maxwell. Simmering throughout the 1920's, the conflict finally led to an open break as a result of the old-age home plan. A convention of California locals openly defied the international by voting to withhold per capita until the abandonment of Lane's scheme. On January 31, 1931 the international executive board revoked the charters of the three San Francisco locals and also of the state branch until it could be reorganized "under leadership that would be obedient to the laws of the International Union [and] until a loyal and satisfactory sentiment was evident among the membership in California." Milton Maxwell then proceeded to transform the state organization into an independent union—the Western Federation of Butchers. [54]

Many circumstances militated against a permanent breach. It was a grave matter, particularly for retail unions, to be apart from, if not actually at odds with, the central labor bodies. At the 1931 convention of the California State Federation of Labor, delegates from loyal Amalgamated locals introduced resolutions condemning the Western Federation of Butchers as "an outlaw organization, having no standing with or representation in organized labor and . . . unworthy of the support of any union man." Amalgamated pressure meanwhile mounted. Legal action was taken to seize the treasury of San Francisco Local 115. A new state branch was chartered, and the establishment of rival local unions became imminent. "Continued division will in time break down all that has been gained," Maxwell feared. "Our employers are very fair, and we should not strain these relations to a point where the employers will hesitate and ask themselves if they are to be the victims of this division." His concern was intensified by the possibility of a collective bargaining showdown growing out of the grim economic situation. [55]

Maxwell was therefore receptive to a reconciliation. President Gorman, in California since October 1930, employed his considerable talents to effect a compromise, as did a committee from the San Francisco Labor Council appointed by AFL. By the end of 1931, an accommodation was reached by which the international would return to good standing the California membership of the state branch. The complicated terms were incorporated into a contract (and several "Interpretations") and submitted, with Maxwell's support, to the California locals. In January 1932 the secession came to an end. The California body gained many of its aims, and the Amalgamated the reaffiliation of 1620 dues-paying members.[56] The connection was cemented in December 1933 when Milton Maxwell was brought into the international executive board. The Amalgamated finally achieved lasting internal unity.

Paradoxically, the international progressed because it had a limited role in the 1920's and early 1930's. Packinghouse organization had created heavy demands for every vital service—organizing, collective bargaining, funds, and leadership. The Amalgamated had broken down on both occasions when it had gone into the western centers. Retail unionism, on the other hand, imposed no such burdens, for it was a form of organization local in its limits and its essential strength. The international did, of course, service the retail locals. The important fact, however, was that, unlike the mass-production field, such work could be limited to the capacity of the international without disastrous consequences. And the strengthening of the retail movement permitted the increase in per capita and the recruitment of seasoned local leaders from which the international immensely benefited.

The retail situation also contributed to the conclusion of internal dissension. Retail bargaining, since it was local, could not be the cause of fundamental differences with the international. A compromise had not been possible with the Stockyards Labor Council in 1919 because of its commitment to a program which, from the Amalgamated viewpoint, would wreck the entire packinghouse movement. There was no such policy issue in the conflict with the California Federation of Butcher Workmen. Relations with employers in California were a local affair and not an essential concern of the international. The quarrel, lacking as it did vital differences over external policy, was resolvable. The very autonomy of the retail unions thus helped to end the secessionist problems of the Amalgamated.

The international would have to expand its functions in the 1930's. But, as never before, it would then be prepared to assume the heavier task.

CHAPTER 7

NEW FORCES IN THE
RETAIL TRADE

In 1933, the Amalgamated had in its organization at the utmost 10 per cent of the retail meat employees in the country. The existing meat cutters' unions were stable and effective, but unionism seemed unlikely to extend to new areas at any but the slow and accidental rate of past years—if, indeed, the depression did not reverse the trend. Then unforeseen changes accelerated the pace of organization. First, the retail trade was becoming big business during this period; and, second, the power balance between labor and management shifted as a consequence of the New Deal. The possibilities of retail unionism were suddenly transformed.

After an existence of many years, the chain store system entered its expansive phase after World War I. The Great Atlantic and Pacific Tea Company grew from 991 stores in 1914 to 15,418 in 1929. The volume of the B. H. Kroger Grocery and Baking Company in 1931—nearly a quarter billion dollars—represented over a tenfold increase in fifteen years. By 1929, chain stores accounted for 39.2 per cent of the retail grocery business. An increasing degree of concentration accompanied this development. The growth of A & P came almost entirely through opening new stores, while Kroger and Safeway, the next largest firms, expanded partly by the acquisition of smaller chains. The three giants controlled over half the total chain business in the early 1930's; and American Stores, First National, and National Tea accounted for another eighth.[1]

The development had not at first involved meat retailing. Barring a few early exceptions, the chains kept strictly to staple goods.[2] But by the mid-

1920's, they were moving reluctantly into perishable food lines. A & P began to experiment with meat departments in 1924 and by February 1931 had 3969 combination stores. Other companies were doing the same. The 1929 census indicated that 17,249 of the 52,618 chain units dispensed meat; ten years later, over half did, accounting for well over a fourth of total meat sales. A revolution had hit the retail meat business.[3]

The butcher unions of course felt threatened. There had not been, Patrick Gorman observed, "among the retail trade the powerful opposing financial combinations as we had in the packing industry." [4] But A & P, Kroger, and the other great chains, no less than Swift and Armour, could absorb the costs of the fight against organized labor and could, in addition, support the welfare programs that were widespread in American industry in the 1920's. American Stores, for instance, had a mutual aid society offering generous benefits, a company newspaper, hospital, visiting nurse arrangements, and an enlightened personnel policy. The men knew, an executive explained, "that we are watching them sympathetically, ready to give them their chance." [5] In 1925, A & P started its stock purchase and group insurance plans, and other companies were similarly active. The decline of the small proprietor apparently made the organization of the retail trade a more formidable task.

Yet weapons were available to those meat cutters' unions which were strongly entrenched in the local trade. The chains were, for one thing, entering a new and difficult field. They had to recruit competent butchers, and these craftsmen were likely to be within the union in an organized town. Chain store employees, part of a large and impersonal concern as they were, had a kind of inclination to unionism that did not exist in the close, personal atmosphere of the independent butcher shop. Above all, the chain stores were susceptible to boycott tactics. No united front was now possible in the retail trade; the independents, indeed, applauded the boycott of the hated chain stores. For their part, retail unions whose membership was employed in independent markets could support activities against the chains almost indefinitely. The Dayton, Ohio, local maintained a full-scale campaign against Kroger and A & P throughout 1930. It was common practice for well-financed unions to assist unemployed members by paying them to picket unfair stores.[6] The chains, although their resources matched those of the big packers, were more exposed to union pressure.

The strong Amalgamated locals consequently were able to enter the chain stores. Friendly relations developed between Safeway and the California unions in the 1920's. A master agreement, covering the northern part of the state, extended as far south as Bakersfield and Santa Barbara by 1933. The MacMarr Stores had contracts with Amalgamated locals in the northwest, and the relationship continued when the chain was absorbed by Safeway in 1931. The National Tea Company had union contracts at the strong midwestern points, including Chicago and, beginning in 1931, Minneapolis-St. Paul; so did Kroger, a more reluctant concern, in Chicago, Memphis, and some Illinois towns. A & P alone was rigidly open shop. But President Gorman had cause to conclude in his 1930 report on chain store growth that "we have met that issue successfully." [7]

Once gained, the foothold opened an entirely new organizing opportunity. The great obstacle had been the local character of the retail meat trade, since union success at one point was not transmissible. Each locality had to repeat the full process of unionization. From an organizing viewpoint, nothing could be more inefficient. The chain store development now brought centralized management to the retail trade. For the first time, local limits could be transcended in organizing the meat cutters. The quick response to this fact became the core of the Amalgamated strategy. The international leadership seized the initial local contacts to approach the central offices of the chains.

The Kroger campaign illustrated the strategy. Hostile at the outset, the firm had almost been placed on the unfair list at the Amalgamated convention of 1930. The declaration was limited to A & P, however, after President Gorman reported a "more friendly attitude" by Kroger within recent months. Behind this union restraint, Gorman later explained, was the fact that Vice-President M. J. Kelly, while representing Local 546 of Chicago, "had succeeded in reaching the chief executive of the Kroger Company." From that point, there were continuing efforts to broaden the contact. The executive board, meeting in January 1931, instructed President Gorman to "make every effort to bring about negotiations with the Kroger Company through it's [sic] head officials at Cincinnati." Gorman finally managed to start a sustained relationship with the top Kroger executives in the course of contractual negotiations for the St. Louis and Cincinnati locals in early 1934. What Gorman wanted to impress on the company was the reasonableness and responsibility of the Amalgamated. "We are desirous of cultivating a harmonious relationship," he wrote to President Albert Morrell of Kroger, "one of Live and Let Live and to find the common ground that we may reasonably occupy together." [8]

Gorman's correspondence and discussions with chain store executives had a continuing theme: the promise of industrial peace, fair demands, and honored contracts. In exchange, he asked the chain management to permit and so far as possible encourage the unionization of their stores. This approach, persuasively argued by Gorman, resulted in informal understandings, essentially of good intentions, with Kroger and National Tea.

A more advanced stage was being reached with Safeway, which had the highest degree of unionization. Because of its negotiations with the California state branch, Safeway was accustomed to union contact above the local level; and the company was seeking contractual concessions for its low-volume stores. The "National Agreement" signed at the Safeway offices in March 1935 specified its future policy on the organization of its stores.[9] The agreement, Gorman judged, "gives us the opportunity to establish local unions in places where, so far, we have been unable to do so." The result actually proved somewhat disappointing. Local resistance to the concessions given the company, Gorman confessed to President L. S. Warren of Safeway, "makes it rather difficult for us to effectuate a compliance with the agreement as we sincerely intended when we conferred." The agreement consequently had little more effect than the informal understandings with other chains; yet it was a considerable asset for organizing purposes just because it was tangible. President Warren, in fact, grumbled that the union "used the agreement itself at times to misrepresent the true condition to prospective members."[10] The value of the Safeway Agreement was reflected in the repeated efforts by Gorman to persuade Kroger to sign a similar document.[11]

The Amalgamated was seeking three kinds of organizing concessions. There was, first, the question of nonunion territory. The standard policy was, as the Kroger personnel director said of his firm, "that membership in the Union is strictly a matter of personal choice, and that it is not our province either to promote or oppose such membership." Kroger would not desert this position, but the Amalgamated was able to see to the enforcement of the neutrality once the understanding had been reached. Gorman could, and did, report the discriminatory practices of Kroger functionaries to the Cincinnati office. Similarly, he had reinstated three men discharged by the National Tea Company during a union drive in Milwaukee in 1934. The Amalgamated, of course, preferred a more positive policy. Informally, it was sometimes possible to secure the active backing of friendly local executives. There were also occasional suggestions that the main Kroger office "pass the word along" to a district manager "to go along with the meat cutters' union and thereby get your share of the business and profit by a friendly feeling."[12] These efforts, however, were not overly successful in the early period.

Safeway was the apparent exception. According to the National Agreement of 1935, the company would "not only offer no objection to its meat cutter employees becoming" Amalgamated members, "but it will encourage such affiliation." This major concession should have meant the swift organization of Safeway. But the impact of the provision was limited. The union, Gorman complained in 1939, had "attempted to make it effective . . . where we were just getting a start." Nevertheless, he could "cite a dozen places where the Company not only refused to encourage their employees to join the Union but actually offered opposition."[13]

In some ways, union recognition—the second question—was more significant than organizing employees in the retail trade. For one thing, even after joining up, they could not be considered permanent members until they were under contract. More important, a contract itself became the means of organization: unionizing employers, rather than employees, had always been a primary tactic. Ostensibly, chains assumed that the men would be organized before a contract, but that was not a central consideration. How many members meant organization? Since the basic concern of the chains was competition, the critical question was whether or not the rest of the trade was nonunion. The Amalgamated granted that chains should not be penalized. Gorman, for example, wired the Cleveland local not to press Kroger for a contract:

Kroger does not object to an agreement if other concerns are presented one. (Stop) Get these boys right on contract. (Stop) Have them work on other stores for the time being and we will have no trouble with Kroger when we have had some success with their competitors.[14]

The Safeway Agreement was very precise on this point. Contractual negotiations would commence whenever a local union had—excluding Safeway employees—55 per cent of the meat cutters in the locality (defined as the area within a 25 mile radius of the union office). Lacking machinery, the arrangement proved unworkable.[15] Clearly, however, competitive considerations made local organizing work a crucial counterpart of the tactic of unionizing through the employer.

The third issue, of less consequence, involved union membership when there was an agreement. This was more a matter of principle than self-interest for the chains. They did subscribe to the open-shop philosophy of the New Economic Era. Thus Kroger preferred a strike in Milwaukee in December 1935 to forcing three employees to join the union. But concessions in practice did provide union security. The chains usually accepted preferential union employment. The Kroger strike in Milwaukee was settled by giving union members preference in layoff as well as hiring, and there were obscurely worded clauses that in fact meant a union shop. The Kroger contract for Cincinnati stated:

When contacted, if the union is unable to furnish competent help, the employer is at liberty to hire other help. When such other help is hired, then in keeping with the high ideals that motivate the parties to this understanding, it is anticipated that such other help may become members of local #610 but it is understood that this clause is inserted for the purpose of closer cooperation between the parties and membership in the Union is not necessarily compulsory.[16]

The theoretical open-shop objection had little material significance for the Amalgamated.

On balance, the first contact with the chains at the national level provided only modest organizing benefits. The top officials agreed, so far as they could within the decentralized framework of their companies, to prevent management opposition. No doubt some quiet encouragement was also forthcoming, and, where competitors were already organized, union recognition was granted without undue concern about the number of union members among the employees. Otherwise, the difficulties characteristic of the retail trade were not substantially lessened. The Amalgamated still had to organize the butchers locality by locality. It was gradually doing so; between 1930 and 1936, 2500 Kroger employees became union members.

The developing union-company relations held greater promise for the future. Gorman did succeed in establishing confidence in the Amalgamated. Key executives, in addition, were receiving a vital education. Discussing a hostile district manager, L. A. Warren of Safeway explained to Gorman that the official was "probably unfamiliar with the methods of organized labor . . . It was just as new to me when I first met you; since that time I have improved my education quite a good deal on the subject." [17] Chain store officers, like others in management, were not disposed to hand over the advantages of nonunion operation. But they—unlike the packers—would not fight, and they were likely to cooperate, if the balance of power shifted toward organized labor. Many forces were working in this direction during the 1930's.

Foremost was public intervention in the unionization process. The right to organize and bargain collectively now came under government protection. The change began with the inclusion in the National Industrial Recovery Act of Section 7(a). Labor boards took action in a number of cases

of alleged union discrimination in the retail meat trade. The National Labor Relations Board settled a major conflict between A & P and seven Cleveland unions, including Meat Cutters' Local 427, at a conference in Washington
on October 30, 1934. While not signing a contract, A & P agreed to reinstate all strikers, submit future disputes to arbitration, negotiate with
committees, and inform all Cleveland employees in writing of their right to
join a union. A considerable success was thus achieved against the nonunion
bulwark of the retail trade through the agency of the NRA. It seemed at
the time, President Gorman later remarked, "a new Magna Charta for
the workers." [18]

But the Recovery Act quickly failed to fulfill its initial promise. The Amalgamated was excluded from the formulation (except through testimony at
public hearings) and the administration of the code governing the meat
trade. Nor did the NRA improve the bargaining position of the retail locals.
The wage standards of the Retail Food and Grocery Code were so low that
Gorman and Lane had to warn the membership "not to be shocked. . . .
These are merely the minimum wages and would be ridiculous for us to
consider from a trade union viewpoint." [19] Far from supporting rates, the
code standards encouraged meat dealers to demand wage cuts, because union
meat cutters were earning up to three times the code minima. Compliance
quickly broke down. The correspondent of Local 157, for example, reported
that in Schenectady "there is practically no NRA in any line of business.
Every code is violated day in and day out. . . . The butchers are afraid to
join for fear of losing their jobs." [20] Long before the Supreme Court decision on the NRA, the guarantees of Section 7(a) and the wage and hour
regulations had become dead issues in the retail trade.

The Wagner Act gave real force to the intent of Section 7(a). At first,
there was a question of jurisdiction over the retail trade. In practice, the
larger chains came within the Federal labor law. The independent dealers
and small chains were subject to the state labor relations acts which, beginning in 1937, were passed in New York, Pennsylvania, Utah, Wisconsin, and
Rhode Island.[21] Once Supreme Court approval came in 1937, the Wagner Act
started an enormous union upsurge. In the Buffalo area, for instance, four
locals were formed, and 2000 men organized by the end of 1937. Butcher
unions had repeatedly failed there in the past, "but thanks to the recent labor laws enacted the workers have today protection which has recently been
made use of by the Amalgamated Meat Cutters." Similar reports came from
many points; the guarantees of the Wagner Act were unquestionably effective in the meat trade. The Amalgamated paid per capita to the AFL on a

membership of 24,000 for March 1937; 34,700 for July 1937; and 43,522 for December 1937.[22]

Important though it was, the public protection of the right to organize lacked the overriding significance in retail that it had in the mass-production industries. Meat cutters were a dispersed labor force across the country and in each locality. Organizing them remained a formidable task even without opposition, and employers, if they chose, could make the job still harder. The Amalgamated had always relied on pressure on employers as an organizing method. It had all the more reason to do so after centralized management entered the retail trade.

Union powers of coercion were greatly augmented during the 1930's. The use of union shop cards as a method of drawing trade to fair employers and injuring unfair employers had not come under court ban, but the more effective means of enforcing a consumer boycott—picketing, circulars, and unfair lists—often resulted in injunctions, and more regularly when organizing was the purpose. The Norris-LaGuardia Act of 1932, which was copied in seventeen state laws, ended this crippling restriction. The courts' power to issue injunctions was drastically reduced "in any case involving or growing out of any labor dispute . . . regardless of whether or not the disputants stand in the proximate relation of employer and employee." In a case involving the Milwaukee Meat Cutters' Union and F. G. Shinner & Co., the Supreme Court ruled in 1938 that a "labor dispute" existed even though not one employee was a union member or a picket. "Since this decision," Gorman wrote elatedly, "there has been picketing galore in many places." [23] Picketing was further safeguarded by the *Thornhill* decision (1940), which declared state antipicketing laws to infringe on the constitutional right of free speech. Amalgamated locals were free, for the first time, to exert the full economic power of a consumer boycott on employers.

Another effective tactic received increasing use. Butchers' unions entered joint organizing drives with the Teamsters in some places, for instance, in Los Angeles, Philadelphia, Buffalo, and San Diego. The potency of the alliance was demonstrated in the A & P dispute in Cleveland in 1934 when the Teamsters completely halted store operations. Many Amalgamated locals began to rely on this support. Hearing of an order from the Teamsters' national office to stop such cooperation, a distressed local official wrote to Gorman: "If such is the case we may as well withdraw our pickets and call everything off, because that is where we were putting the pressure on by shutting off their meat supply." When two locals were contending before the executive board for jurisdiction over the independent D.G.S. Stores in

Washington, D.C., one representative rested his case on the fact "that because of his close connection with the Teamsters Union that his local union could easily organize these stores." Although bad blood between Dan Tobin of the Teamsters and Secretary Lane prevented an alliance at the national level, many Amalgamated locals were able to secure Teamster support.[24]

This asset was only part of the favorable drift of events in the 1930's. The growth of organized labor tended to strengthen the consumer boycott. The Yonkers, New York, local enlisted the CIO organization at the Chevrolet plant in Tarrytown against two nonunion markets. They agreed to sign a contract when they were threatened with one hundred pickets on Saturday and notices of their unfairness on the bulletin boards of the plant.[25] Although this was an extreme instance, clearly the expansion of trade unionism made more formidable the picket line and boycott. And local public officials and police—in the past a source of trouble—now were often sympathetic to union efforts.

The effect was evident on organizers and local leaders. Demonstrating great militancy, they pushed their campaigns with a vigor and determination—and occasionally a disregard for law and order—hitherto unknown among meat cutters' unions. "We learned one definite lesson," asserted a Cleveland official after a series of successful strikes. "That lesson is that butchers want action. And when we learned it we gave them action." A new confidence developed. A Cincinnati leader pleaded with Gorman in 1939 for permission to take on A & P:

> I am not afraid with the wonderful movement we have in Cincinnati that we can seriously injure their business . . . A good fight against the A. & P. Co. will not harm us [and] will result in a bigger and better local 410 in Cincinnati.[26]

No challenge was too great any longer.

The advance of retail organization became irresistible under the dual impact of the guaranteed right to organize and the coercive ability of the Amalgamated. The power shift was evident in a sterner approach to the chain managements. Their policy of withholding recognition pending the organization of local competitors was no longer acceptable. Gorman coolly told the complaining National Tea Company that friendship with "our organization could be developed more if you could . . . sign agreements . . . without being governed too closely by what other merchants are doing." Meat cutters' unions took root and rapidly expanded in such formerly inhospitable regions as southern California. Sectional chains—Loblaw in upstate New York, First National and Economy Stores in New England, and

American Stores in the Middle Atlantic states—were coming under contract for the first time.

The final test of strength was against A & P. Here the Amalgamated had achieved no contracts before the Wagner Act except in Seattle. Now other locals launched the assault against the massive chain. Often they benefited from a sympathetic meat trade. "All employers in Cincinnati hate them," one official observed. "If we must begin the fight, their competitors, I know will rally round us." By the end of 1936, A & P meat departments were organized in St. Louis, Minneapolis, and Milwaukee. Gorman and Lane reported on October 30, 1936 that, despite "the bitter anti-union policy," A & P was opening "a way for us to organize their people" by leasing their markets to employees. It was doubtful "that the leases are genuine, nevertheless, they have paved the way for us to get the markets lined up and have our shop cards in their stores." Faced by a mounting drive, one A & P official in St. Louis agreed to "pass the work down the line so that it would be easy to enroll all employees. He kept his word and Local 88 of St. Louis has announced now that A & P is fair to organized labor. We are inclined to feel that they will resort to the same method wherever an intensive organizing campaign begins." [27]

But Gorman's intention was not to organize A & P forcibly locality by locality. He preferred his strategy of reaching the chief men. The Amalgamated now had a powerful bargaining position, for its ability to organize A & P stores was evident. The time had come to find an approach to A & P at the top.

The opportunity came from an unexpected direction. The expansion of the chain store system had raised a storm of opposition. Beginning with an act in Missouri in 1923, chain-taxing laws became increasingly common, particularly after their constitutionality was upheld by the Supreme Court in 1931. The climax of the assault began in February 1938 when Wright Patman of Texas introduced a punitive bill in the House of Representatives. Annual taxes would be levied beginning at $50 for the tenth store in a chain and rising to $1000 for each store in excess of a thousand, multiplied by the number of states in which a chain operated. A & P faced a charge of close to half a billion dollars annually! The Patman Bill had substantial support in Congress and across the country, and, at the time, its enactment did not seem inconceivable. [28]

A & P took the lead to fight this eventuality. The company announced a publicity campaign under Carl Byoir, the renowned public relations expert. [29]

The support of organized groups was important, and none more so than labor. Since the futility of A & P's open-shop policy was already apparent, the company decided to make, in effect, an exchange with organized labor. AFL President Green, Gorman related in April 1938, "has had some meetings with the highest officials of the A. and P. He is optimistic enough to believe that within a very short time a thorough understanding can be reached with this company in a national way. Naturally, if this is done, the company will expect the American Federation of Labor to use its influence in defeating certain obnoxious chain store legislation that crop up from time to time." These negotiations culminated in a dinner given by A & P President John A. Hartford during the AFL Executive Council session in Miami in January 1939 for officials of unions having jurisdiction in A & P.[30]

The AFL did its part. A resolution was slipped through the convention in October 1938 without debate that called for a study of the Patman Bill. Outright condemnation could not have been achieved at that point when union hostility to A & P was still strong. However, the wording of the preliminary clauses made the resolution seem to be an endorsement of the chain stores, and it was so employed in the publicity campaign. The Amalgamated Meat Cutters contributed significantly in the later stages of the fight. Gorman himself played a key role in the defeat of the Bricklayers' attempt to have the AFL support the Patman Bill at the 1939 convention at Cincinnati. As it happened, the butchers' union there had a long-standing grievance against A & P and threatened to picket the convention, thus shattering the chief argument of opponents to the Patman Bill that the chain was now fair to organized labor. Gorman managed to avert this embarrassment. It was he also who made the most effective defense of the chains on the convention floor. The convention subsequently voted merely to continue the inquiry into the Patman Bill. The following year, Gorman testified persuasively against the measure in Congressional hearings.[31] Perhaps more than any other interested union, the Amalgamated put A & P in its debt.

Meanwhile, the company was responding in kind. President Gorman conferred in late September 1938 with Carl Byoir, who was handling labor as well as public relations for A & P: "I got along splendidly with him and believe we agreed upon a program that will establish for us a more general, friendly relationship with the Company." Amalgamated locals had been progressing rapidly in the preceding months. Now it was formally understood that A & P approved the organization of its meat cutters and would sign union agreements covering them without delays. In November 1938, an A & P executive from Chicago, Charles A. Schimmat, was appointed labor relations director as, Byoir remarked, "a much needed step toward facilitating negotiations." In his congratulations to Schimmat, with whom he was already on friendly terms, Gorman observed "that this idea originated during my conference with Mr. Byoir." The message ended on a comradely note: "Good luck, Charley, and if you have any problems over the country that you want me to step in with you—do not hesitate to call upon me." [32] The start could hardly have been more auspicious.

In contrast to the other major chains, A & P adopted a highly centralized procedure. The negotiation of initial local agreements had to await the arrival of Schimmat. Obstructionist tactics by hostile managers were thus prevented in some, but not all, places. On the other hand, since Schimmat alone carried the negotiating burden for the entire company, many locals had to endure long waits before securing recognition. The delays occasioned considerable unrest in 1938–39, and some grumbling against Gorman. He himself became uneasy. "There is no question in my mind but that the A. & P. Stores are hedging here and there . . . They are evading entirely the densely populated areas, where their stores are located by the hundred." Gorman thought it would be worthwhile for a hostile editorial from the Bricklayers' magazine to "reach the high officials of the A. & P. Company. It would be bad for these people to feel that organized labor swallowed their program hook, line and sinker merely upon their promises." [33] From the Amalgamated standpoint, it may have been fortunate that the Patman Bill remained a live issue into 1940.

By the summer of 1940, the Amalgamated had contracts with A & P in 63 localities. Many eastern points were still not covered, including New York and Boston, but there were agreements for Washington, Baltimore, Philadelphia, and one pending for Pittsburgh. The midwest and far west were quite thoroughly unionized. According to Gorman, 6000 of the 9200 A & P meat cutters were union men in 1940. [34] Two years of top-level contact had breached the open-shop stronghold of the retail trade.

The centralized management of the chains made efficient the unionizing process among meat cutters. A & P was only the most striking example. Gorman's strategy had been employed earlier with Safeway, Kroger, and National Tea, and he experienced signal success in the late 1930's with the heads of such other chains as Loblaw and First National. The larger part of the retail trade, however, was still in the hands of thousands of small dealers with one or two employees. Yet an attempt was made to apply the centralized technique of organization even among the independents.

Trade associations had a long history in the retail meat field. Always flourishing during periods of stress, they increased in importance after World War I largely in response to the chain store threat. Most local bodies were affiliated with the National Association of Retail Meat Dealers. Some officials who took a leading part in the affairs of the National Association such as George Steindl and A. J. Kaiser of Chicago and I. W. Ringer of Seattle were on friendly terms with the Amalgamated. It therefore became possible to seek a general agreement for the independents of the retail trade.

Discussions began in Chicago in June 1937. Using the Safeway Agreement as a starting point, the two national organizations worked out an understanding. At its convention in early August, the Meat Dealers adopted the agreement, despite opposition from eastern delegates. The association recognized the Amalgamated as "representative of the meat cutters and butcher workmen . . . and does hereby designate and accept the Union as their sole and exclusive bargaining agent." It was also "expressly agreed that the Association will not, directly or indirectly, prevent or hinder the unionization of employees by the union or the establishment of local unions." Finally, both parties would "cooperate to enlarge and widen their respective organizations to include a greater percentage of the persons engaged in the retail meat industry." [35]

Some organizing benefits were forthcoming. The president of one local association, for instance, immediately called a meeting and "suitable negotiations were entered into." The Milwaukee Retail Meat Dealers' Association, which had hitherto refused to recognize the union, now did so, although "we still have a lot of members who are opposed to the whole idea of unionism." The National Agreement also was of value in bringing under contract nonunion shops in localities where collective bargaining was already in effect. [36]

In the end, however, the agreement proved a disappointment. Since the National Association of Meat Dealers was a voluntary organization, its officers—unlike the chain managements—could not compel obedience at the local level. To attempt to do so would have negated one of the main purposes of the National Agreement: namely, to build up the membership of the association. Moreover, the force of the agreement was vitiated by the failure of other provisions. The hostility of some local associations was not dispelled. The National Agreement did provide a measure of endorsement for the Amalgamated, but ultimately the union had to look to its own strength. Gorman concluded in 1945 that "we would have just as many contracts in as many cities where retail meat dealers are organized as we have now even if no agreement existed at all." [37]

Decentralized ownership continued to hinder the Amalgamated in the independent segment of the retail trade. According to a union estimate, a slight majority of the retail membership in 1940 came from the chains.[38] Still, the gains among the independents were steady and considerable. There was now a reversal of the earlier pattern in which chain stores had been organized by locals with a strong base among the independents. By the late 1930's, unionization in newly opened territory was occurring first in the chains, and then spreading to the rest of the local trade. Indirectly, unionism in the independent trade was forwarded by the centralization in the chain segment of the industry.

The advance of retail unionism reordered the internal relations of the Amalgamated. The meat cutters' unions had enjoyed a large measure of local autonomy, particularly in dealing with their employers. Neither the market nor the ownership of the retail trade had extended beyond the jurisdiction of the individual union. Its actions had no consequence elsewhere. The international, therefore, had limited its concern to the protection of the retail locals, that is, to questions of strike action. Only in a few instances did the international attempt to direct the bargaining policies of the meat cutters' unions, for example, to regulate the length of the work day rather than store hours. Some locals persisted in fixing the openings and closings of markets. "We have no objection to the continuation of such a system so long as our local unions are not threatened by strikes and lockouts," Gorman admitted in 1939.[39]

The international role in the retail field began to expand as a consequence of the organizing strategy of the 1930's. Consistently, the Amalgamated aimed at establishing amicable relations with the major chains. Their price was, as Gorman said, proof "that our organization is sound and our contracts are going to be religiously lived up to." [40] That meant, in essence, that the international would have to judge local actions not only for the probable effect on the union, but also for the impact on the employer. To the extent that the international did so, the accustomed autonomy of the retail unions would be narrowed.

Freedom to strike became increasingly circumscribed. Permission usually was withheld from actions which violated contracts or hit the employer unfairly, for instance, pulling a strike shortly before a holiday. When the

Memphis local in 1937 decided to walk out in sympathy with the Retail Clerks against Kroger, Gorman ordered the union to propose arbitration: "If Clerks have a strong case they should not fear arbitration (Stop). If they refuse arbitration you should fulfill the terms of your contract with Kroger." The effect on the company was what Gorman desired. The general manager sent his thanks: "This confirms expressed thoughts as to value of a contract being more than paper to be filed away and forgotten or torn up at will." [41]

In general, too, the international became sparing in permitting strikes over bargaining issues. The policy of restraint manifested itself most sharply in the case of A & P. "After twenty-five years of bitter opposition," Gorman told the executive board in March 1938, "our International has sufficiently sold itself to the A. & P. Tea Company to the extent that they are treating with us and signing our contracts. We must keep that confidence intact and not do anything that would jeopardize the move to thoroughly organize the A. & P. stores over the country." A tight rein had to be imposed on impatient locals. The southern California locals, Vice-President T. J. Lloyd informed Gorman in April 1939, had been waiting for over six months to begin negotiations. "I must play ball, I know, with the International program but I do not feel it is right to hold back our men any longer." Ohio was the tensest area. The business agent of the Cincinnati local, Michael Schuld, told Gorman that after a year of delays "my patience has been exhausted." He agreed to postpone a strike in May 1939 only under the threatened revocation of the local's charter. [42]

It was not ordinarily necessary to go to that extreme. The authority of the international over strikes, although in the past it had often been a formality, was entirely clear in the constitution, and the growing Amalgamated treasury made the local unions reluctant to forego strike benefits. Unauthorized strikes still occurred, particularly after 1939 when war prosperity came and living costs rose. There were also frequent occasions when strike permission was necessary. However, Gorman spoke the truth when he told Safeway that "We detest a stoppage as much as any employer and I don't believe there is an International Union connected with the American Federation of Labor that will go as far as our organization to avert trouble." [43] The implementation of that policy necessarily narrowed the earlier freedom of action of the local unions.

The international also expanded its role in actual retail bargaining. It had long been customary for local unions to submit their demands for approval before presentation to employers. This became a constitutional requirement in the 1940's. The constitution also had, for many years, defined the procedure by which the international could enter stalemated negotiations. If the local sought strike authorization, the president or his deputy could seek an adjustment in cooperation with the union. That failing, the executive board had to consider the pending issues as grounds for authorizing a strike. Disapproval of the union's demands meant that these could not legitimately be the basis for economic action. The international also had indefinite, residual powers by which it could influence local decisions.[44]

The Amalgamated thus could enter the collective bargaining arena of the retail trade. Highly skilled in adjusting differences, Patrick Gorman worked assiduously to ease negotiations. A constant stream of correspondence came to him from retailers, no less than local unions, on stalled bargaining. His role, indeed, tended to assume more the character of mediator than labor partisan. In one instance, in fact, the secretary of the Seattle Meat Dealers' Association suggested that a dispute with Local 81 be submitted to Gorman and Lane for a binding decision. International-chain relations, once these became established, sometimes put Gorman in a more positive role than that of mediator. A concession on apprentices, he revealingly told an Auburn official, would be forthcoming "if your Local Union, frankly speaking, desires to 'play ball' with our International Union and the Loblaw Stores."[45] The Amalgamated thus used its authority over the locals for the greater good of the organization.

Yet, while expanded, the international role remained within rather narrow bounds. The subordinate retail unions—particularly the strong ones—were accustomed to a free hand, and they resented the intrusion in their negotiations. Nor could they ordinarily be directed to act against their wills. The international had only negative powers, and even these carried no veto over concluded agreements. Moreover, the negative powers which were available—over demands and over strikes—had only restricted application. Considerations of union politics and of local strength inhibited the easy use of international controls. The fact was that the expanding Amalgamated part in retail negotiations involved a far larger measure of persuasion and influence than of dictation over locals. All this referred to the moderation of local decisions. As far as the initiative in retail bargaining was concerned, the international was without real power.

The fact was illuminated on several occasions when the Amalgamated did attempt to make bargains in the pursuit of organizing advantages. The Safeway Agreement of 1935, for instance, gave several important concessions to the company. Gorman hoped that the locals would "look at a proposition from a general concept of organization betterment." Some strong locals admittedly could enforce harsher terms on Safeway. "But such achievements in themselves are not an asset to the National Organization . . . The first duty . . . is to organize the people of our craft . . . This agreement . . . gives us the opportunity to establish locals where, so far, we have not been able to do so." [46] Unfortunately, the local unions lacked this broad vision.

The chief issue involved small meat departments where meager sales made it unprofitable to employ a meat cutter at standard rates. In the Safeway Agreement, the international accepted the principle of relief in such cases. Supplementary agreements, making this specific, were signed simultaneously with California and Seattle representatives. The California agreement provided that journeymen in low-volume markets would receive $30 a week—the regular minimum was $40—plus 12½ per cent of sales over $200. But both the California state branch and the Seattle local voted down the arrangement. Notwithstanding Gorman's efforts, the low-volume provision was nowhere put into effect. Company president L. A. Warren acknowledged Gorman's "sincerity and desire to carry out the agreement . . . I only wish it would be possible for you to make an agreement which would be binding on every meat cutter in the United States." The continuing vigor of local autonomy in effect nullified the Safeway Agreement.[47]

The low-volume problem was ultimately solved without credit to the international. An accommodation was gradually made at the local level. Some unions, such as powerful Chicago Local 546, gave relief at an early point. Others finally bent to the pressure of events. In Los Angeles, Safeway closed 170 unprofitable markets by 1938. Then unions saw the wisdom of concessions. Only then, too, was the international able to employ its constitutional authority. The stubborn Pasadena union was informed that strike permission would not be forthcoming to locals refusing "to accede to a reasonable adjustment" on the low-volume issue.[48] The final answer came from the industry itself, because supermarkets were beginning to supplant the neighborhood markets. Local autonomy had prevented the Amalgamated from capitalizing on the problem for organizing purposes.

The agreement with the National Association of Retail Meat Dealers was similarly undermined. Non-employing market owners formed a substantial part of many locals—40 per cent, reportedly, of the large Seattle union in 1937. The National Agreement permitted proprietors who were union members to transfer to meat dealers' associations and prohibited compulsion on nonmembers to join the union. This considerable concession did not go down well with the local unions. The secretary of the Seattle association reported a conversation with the local Amalgamated leader: the agreement was good where the meat cutters "were not organized, but it was not doing anything for him, and he wanted to know if I was going to make a campaign to get his members away from him, that he had spent a lot of money getting them in, and they weren't going to give them up." Actually, the agreement did not require this without local consent; but it did prohibit the recruitment of unaffiliated proprietors. Receiving a complaint from St. Louis, for example, Secretary Lane in November 1937 instructed Local 88 not to picket one-man stores for organizing purposes. Again, in 1940 the St. Louis union was warned that, while it could retain proprietors who were members, others would have to be permitted to join the meat dealers' association. "It is the order of the International Executive Board that you do this," Gorman wrote sternly. The union grudgingly acquiesced. However, except where the international intervened directly, aggressive locals continued to recruit proprietors. On membership matters, no less than on substantive issues such as low-volume stores, they would not subordinate their immediate interests to the larger benefit of Amalgamated.[49]

Strains inevitably resulted within the organization. The refusal to grant strike permission, particularly against A & P, caused grumbling. Gorman did not alter his restrained course, but, as he pointed out to the A & P labor relations director, "in taking these positive positions with our Local Unions I am to a certain extent sticking my neck quite a way out. When I do this, I naturally make an enemy not only of the officials of the Local Union but many of the membership and, of course, this may have its repercussions later." The agreement with the Meat Dealers also stirred complaint among the locals (and some charges that the international had exceeded its authority). Gorman accepted the risks. "I feel that in my relationship with employers I really have tried to play fair," he commented philosophically, "and occasionally, the huge cry goes up from our own members that Gorman is a 'bosses man.' "[50]

The resistance to international policy sprang partly from tensions within the local unions. Safeway President L. A. Warren, who was knowledgeable about the West Coast unions, concluded that the repudiation of the Safeway Agreement occurred because "within nearly every one of these locals there are factions with allegiance to different types of employers and with an interest in candidates attempting to dislodge the existing officers."[51] The emergence of a new group of local leaders compounded the problem. At the decisive organizational moment, there had appeared in a number of large cities men from outside the Amalgamated officialdom who had grasped the opportunity, built up the local movement with little outside help, and in so doing established themselves in the positions of local power. These tough, resourceful men took a proprietary view of their unions, and they were unlikely to be submissive in the face of international authority.[52]

No overt or concerted opposition, however, developed. The Lane-Gorman administration had long been in power and was firmly entrenched; its policies, while irksome to some locals, were patently successful; and the growing resources of the international provided larger means for rewarding friends and punishing enemies. Of greater importance, the international proceeded circumspectly and with due regard for established practices and the sensitivities of the local unions. The intrusion on local autonomy was of a negative sort. The international more vigorously applied its control over strikes, and it took a larger part in mediating disputes and in moderating the position of the local unions. But the unions did not surrender control over the bargaining process or their freedom to accept or reject changes. The substance of local autonomy remained.

The use of centralized retail management for organizing purposes put a strain on accustomed relations between the international and the locals. The tactic also encountered obstacles from an unforeseen quarter. The Wagner Act and its state counterparts prohibited employer encouragement, no less than opposition, regarding union membership and established the principle of majority rule in the choice of bargaining agents. Employers were required to keep entirely outside the organizational process. This fact blunted the sharpest Amalgamated weapon in the retail field: the ability, through conciliatory tactics and coercive power, to come to terms with an employer for the organization of his men.[53] It was a paradoxical situation. The Norris-LaGuardia Act, as interpreted in the *Shinner* case, gave a union complete freedom to picket a shop for organizational purposes. But the labor relations legislation prohibited an employer from responding to the pressure. Gorman propounded a hypothetical situation:

Suppose in a case where no union members are employed picketing begins. Suppose that the business of the concern involved is jeopardized badly, because of the pickets and the employer feels that he would like to treat with that organization to end the matter. He could not do so without violating the Wagner Act. What could an employer do legally in such a case?

The law, on the face of it, negated the Amalgamated power to influence an employer.

A certain amount of leeway in reality remained. The Amalgamated counsel, commenting on Gorman's hypothetical case, observed:

In actual practice, the employer could pass the word out to his employees that he personally is on friendly terms with the organization, and that he has no objection to his employees joining, and believes it would be to their best interest. He could do this in the best of faith, because continued picketing might ruin his business, and cause the employees to lose their jobs.[54]

The Wagner Act prohibited coercion, but it could not render an employer absolutely neutral.

The Amalgamated operated under this assumption. Gorman, for instance, wrote to Joseph Bappert of the Kroger Company:

I certainly would not ask you to violate any provisions of the Wagner Act, but I do believe it would help materially in a very mutual way if the local management in the various units of the Kroger Company would work with us cooperatively . . . If you can consistently encourage your Terre Haute management to legally work with us we can establish a friendly set-up that will be beneficial.

Similarly, after the Toledo local picketed the A & P stores, Gorman urged Schimmat to advise the company representatives, who were hostile, "to work with us in a friendly way" so that "we may quickly reach the point where we can sign a contract for Toledo and thus forget any friction that might have resulted there."[55] The employer could still play an important, if unobtrusive, part in an organizing drive.

Another legal grey area had greater significance. Under the Wagner Act, the prerequisite for designation as a bargaining agency for a given unit was the support of a majority of the employees. NLRB procedures were instituted only if the claim to a majority was challenged. Otherwise, the responsibility for ascertaining the fact rested with the employer. The formal position of the chains, of course, was to enter negotiations only when a union did have a majority. But a critical matter of discretion existed here. The employer could demand proof or, if he were so disposed, merely accept the claim of the union. The Amalgamated encouraged the second course.

Finally, an employer had some scope even when more than half his employees were in the union, because the union's position remained precarious until it had gained a contract. An employer's willingness to act without delay prevented the melting away of a momentary majority. The value of a contract was further enhanced by the usual inclusion of a union-shop clause. (The chains had dropped their earlier opposition to the arrangement.) The quick conclusion of an agreement normally secured the position of the union.

The Amalgamated was not engaged in a cynical attempt to circumvent the intent of the Wagner Act. Gorman wrote at one point that the Wagner

Act "would not be worth the paper it is written upon" if "an employer may sign a contract with a union, irrespective as to whether or not that union has a majority of its employees. What would prevent a labor hating employer . . . to sign a contract with the Company Union?" But Gorman did not see the Wagner Act abstractly. First, he believed that the choice of a union was never made in an atmosphere of complete freedom. Employees assumed the opposition of the company unless it indicated otherwise. The employers' power in this area was driven home when the Amalgamated lost part of its gains with A & P in the south after an irresponsible organizer had antagonized the management there. Second, it was inconceivable that wage earners should reject organization. "I agree with you," Gorman wrote to a Kroger official, "that if a group positively refuses . . . their employer's store should not be picketed. All things being equal, however, these cases are so few and far between that the so-called opposition cannot be given too serious consideration." [56]

The Amalgamated experienced a large measure of success in enlisting employer support. Unquestionably, encouragement was forthcoming in many places. After an agreeable meeting with the union representative from Kansas City, a Safeway official informed Gorman "that we may be able to be of assistance to him in accomplishing his organization work among our employees, and I think that before very long our relationship will be firmly established." While uncommonly open, this statement reflected the policy of many top chain executives. They frequently enabled organizers, as Gorman desired, "to begin the work under the most friendly circumstances possible." [57]

Contracts posed a more delicate problem. Many agreements were probably signed under dubious circumstances, particularly in the rush of organization from 1937 to 1939. A major chain consummated a contract for its eastern Massachusetts stores without actual evidence of a majority—which the Amalgamated did not in fact have. Similarly, A & P saw no need for a meeting of its Youngstown employees for organizing purposes. Instead the local official was requested to give Gorman copies of other Youngstown contracts: "Tell him you will pass these along to me and that we will get action as soon as possible." [58] The employer, however, could act only within certain limits. There had to be some semblance of organization. Gorman could not prevail upon A & P to sign a contract for his home town of Louisville because of the nonunion sentiment of the meat cutters there—which Gorman himself acknowledged. Moreover, the chains became increasingly cautious as the NLRB and the courts gave effective application to the Wagner Act.

The nub of the situation was best expressed by an Amalgamated attorney: "The employers and organizers have enough tact to effectuate a working agreement without fear of entanglement under the Wagner Act." [59] That was so in many, perhaps a majority, of cases, but on frequent occasions the union did become "entangled" in the provisions of the Federal or state labor relations acts.

Difficulties arose in several ways. The competitive situation could make unions undesirable or, more frequently, a doubt would somehow openly be raised about the majority status of the union through lack of "tact" by management or labor representatives. In either case, the company would demand positive evidence of a majority. In Louisville, the local Kroger management was angered by a strike in one of its stores and was also aware that the other chains were unorganized. Kroger therefore wanted proof of a majority before negotiating a contract. Even after inspecting the application cards, company officials demurred on the ground that the men had been "high-pressured into the union." Finally, an informal election demonstrated the union's majority. On the other hand, a similar situation in the American Stores in Johnstown, Pennsylvania, resulted in no contract after the investigation of a Federal conciliator failed to find a majority for the union.[60]

The greatest frustration was experienced in the A & P stores in Cincinnati and Detroit. Negotiations had commenced in both places, but there were delays during which, according to the union, local managers undermined the organizations. A & P then refused to sign agreements without elections. Gorman appealed to A & P President Hartford in November 1940. A Detroit company official, Gorman claimed, had acknowledged in 1938 that the union had a majority and promised an early contract, causing the local leaders "to rest upon their oars until Mr. Schimmat should arrive." In Cincinnati, the A & P labor consultant, Chester M. Wright, allegedly had promised the union a contract if it withdrew charges filed against the company with the NLRB. The A & P president, consulting his legal department, was advised "that unless it clearly appears that the local Union represents a majority of these employees, we have no right to bargain or sign an agreement concerning the terms of their employment." The company had recently been cited by the NLRB for encouraging employees to join an AFL union. In fact, NLRB representatives warned A & P lawyers that a contract in Cincinnati "which disregarded the free choice or wishes" of the employees would violate the Wagner Act.[61] An employer, whatever his inclination, was immobilized once the question of representation was raised.

The Amalgamated was therefore fortunate to lack strong rival unions in the retail meat trade. The United Retail and Wholesale Employees of America, which had CIO jurisdiction over meat cutters, rarely constituted a serious threat. The CIO union did not concentrate on the retail food line; the meat cutters, a skilled group, inclined toward the AFL; and they were usually put into a separate bargaining unit from retail clerks for representational purposes. Only a few important contests resulted from CIO competition, for instance, an NLRB election in October 1939 in the Union Premier Stores of Pennsylvania. The Retail and Wholesale Employees also recruited members in the A & P stores in New York City and in scattered areas such as Dayton, northern Indiana, and Little Rock. In the end, CIO captured relatively few meat cutters and seldom forced the Amalgamated into elections.

Independent unions were somewhat more troublesome. These sprang up in some chains, either spontaneously or, so the Amalgamated charged against A & P, with company support in New York City and Buffalo. An independent union raised a question of representation before the New York State Labor Relations Board for the meat cutters of H. C. Bohack & Co. In this case, the Amalgamated local did not have to undergo an election because of the evident minority status of the independent. A narrower victory was won in the Providence, Rhode Island, division of First National Stores. The Amalgamated local's majority was challenged despite the recent renewal of its contract. Benefiting from an NLRB ruling on appropriate bargaining units, the Amalgamated won by a small margin in Rhode Island. The independent union was certified without opposition in the adjacent parts of Massachusetts and Connecticut. Perhaps the worst setback at the hands of an independent occurred in the A & P stores in Boston.[62]

The Atlantic and Pacific Managers' and Clerks' Independent Protective Union created a more complicated problem in New York City. It affiliated with the Amalgamated as Local 263 in the late spring of 1939, but a majority voted for no union in a consent election conducted soon after by the state board. The local leader blamed the defeat on A & P interference in the election, not "by those in the higher offices of the company but . . . by minor officials." Anti-Semitism also was a factor. The bulk of the A & P employees were Catholic, while the New York butcher unions, even those not engaged in the Kosher trade, were largely Jewish. This was one reason for permitting a separate union for the A & P group. However, Gorman concluded, this act did not sufficiently offset opposition to affiliating with what the A & P workers conceived to be a Jewish organization. The primary difficulty, however, involved the A & P union itself. "To win the confidence of laboring men, you must prove that such confidence is justified," a local Amalgamated official remarked after the election defeat. The A & P union leaders had not done so either in the independent period or in "their fancy maneuvering" to affiliate with the Amalgamated, and now the Meat Cutters, having taken in the independent, assumed the burden of the latter's unpopularity. Despite vigorous organizing work by the New York retail unions, a majority of A & P butchers rejected the Amalgamated, as well as the CIO, in elections in 1941 and 1945.[63] Independent unions could be troublesome even when they did not remain independent.

Labor legislation thus reduced the organizing efficiency in the retail meat trade. Employers, rival unions, even the workers, availed themselves of the new laws to offset the increasing power of the Amalgamated. But progress was only somewhat slowed thereby. It was possible, indeed common, to circumvent the literal intent of the Wagner Act. The independent dealers were largely out of range of the state as well as Federal laws. And the advantages far outweighed the drawbacks of this legislation. The protection of the right to organize benefited the Amalgamated as much as any union, and was a decisive element in the advance against the major chains. In the long run, too, the inconveniences had a healthy effect. The necessity of undergoing elections kept the rank and file at the forefront of the thinking of the Amalgamated officialdom. It might have been otherwise had the increase been chiefly of men "turned over" to the union.

Although its task was more expensive and prolonged, the Amalgamated did generally experience success. The union prevailed eventually even in such hard cases as A & P in New York and Detroit. Only in a few instances did the Amalgamated sustain a permanent defeat among the meat cutters. By 1941, unionization was far advanced in the retail meat trade.

CHAPTER 8

PROLOGUE OF PACKINGHOUSE
UNIONIZATION, 1933–1937

Meat packing remained to be taken. Twice this mass-production field had come within the grasp of the Amalgamated, and twice it had been lost. In the 1930's the conditions at last became right for permanent organization. But the process took an unforeseen turn. The quickening course of packinghouse unionism led off in a new direction.

The Amalgamated packinghouse branch had diminished to insignificance after the disastrous strike of 1921. Besides the loss of the major centers, traditional areas of strength also declined. Following the national strike, the New York packing plants became nonunion after two decades of recognition. Louisville Local 227—the strongest remaining packinghouse organization—was disrupted by a prolonged strike that began in April 1926. Small locals survived only in scattered minor plants across the country.

Amalgamated interest could not be long sustained in the face of the evident inaccessibility of the packinghouse workers. A resolution at the 1923 AFL convention for a national drive led to a meeting on February 14, 1924 with Gompers and Frank Morrison. The plan was diverted into a modest campaign in the New York houses and even there had no success.[1] The subsequent years saw almost no Amalgamated activity in the major stockyards. This suited some of the retail officials who felt that packinghouse expansion always came at the expense of the meat cutters. It seemed just as well during the uneventful 1920's to hold to the securer ground of the retail trade.

Yet there remained a basic commitment to the packinghouse workers. A resolution in 1926 to replace the steer's head with the market card as the Amalgamated emblem was put down by Secretary Lane; he wanted a symbol representative of the entire industry. Lane and Gorman did not forget the spectacular (if abortive) success during World War I when the government had forced "the Packers into a National Agreement with our Organization, and by so doing enabled us to increase our membership by more than 100,000 members." [2] That pinnacle, Amalgamated leaders knew, could be regained only with renewed government intervention. In that event, the Amalgamated would hasten to reenter the meat-packing field.

The opportunity came in the opening stages of the New Deal. The National Industrial Recovery Act required the inclusion of Section 7(a) in every code of fair competition. To avoid delays in the recovery program, the administration promulgated on July 20, 1933 a temporary "blanket code"—the President's Reemployment Agreement—pending the conclusion of codes for individual industries. The meat packers, among others, insisted on modifications in the blanket code. Having achieved this after a week of hard bargaining, representatives of the Institute of American Meat Packers signed the agreement on August 6, making the packing industry "subject to all the terms and conditions required by" Section 7(a). [3] The packinghouse workers thus received "the right to organize and bargain collectively through representatives of their own choosing . . . free from the interference, restraint or coercion of employers" and from the requirement "to join any company union or to refrain from joining . . . a labor organization."

The Amalgamated had already swung into action. Gorman and Lane had attended a Washington meeting on June 6 called by William Green to discuss the pending Recovery Act. Thereafter, Gorman later stated, he and Lane "worked day and night to perfect extensive organizing campaigns." By the end of 1933, the organizing staff was the largest in the history of the organization. An important financial concession was made in December 1933 to the incoming workers: they were absolved from local dues, paying only the international per capita of a dollar a month; the Amalgamated itself would cover the expenses of the packinghouse locals. The arrangement met the competition of the low-dues company unions and independents and, by putting the local officials on the international payroll, permitted a greater degree of national control over the packinghouse organizations. [4]

Section 7(a) evoked an immediate response in the stockyards. Amalgamated representatives spread the word about the new guarantees. A circular used in Omaha, for instance, twice reproduced Section 7(a) and concluded:

Lets Go. We Must Make The Most of The Opportunity This Law Offers. Failure Means Continuance Of the Damnable Condition Which Has Existed For More Than Three Years—And Worse.

Assured of their rights, packinghouse workers flocked into the Amalgamated. In those early days, the international headquarters was "taxed to capacity" filling requests for charters, speakers, and information. Secretary Lane claimed that 50,000 joined up in the last two months of 1933.[5] Active locals existed in all the packing centers, and some plants, particularly among the independents and at the lesser points, became well organized.

The Amalgamated strove to maintain order in the first hectic months. President Gorman, for example, instructed the Omaha organizer to "do all within your power to prevent any walkout of men in any of the plants . . . We should all try our damnedest to prevent any further sporatic [sic] strikes as they will retard rather than advance the progress we intend to make." Trouble threatened everywhere, but the only serious Amalgamated strike occurred among the Chicago stock handlers on November 27, 1933. Gorman and Lane could claim with considerable justification: "The packing industry, because of the strict discipline we have exercised over the membership, has had little labor difficulty."[6]

The plan was to "move forward together" in dealing with the packers. A Chicago conference of packinghouse delegates on December 4, 1933 adopted a Memorandum of Conditions drawn up by the executive board, including demands for an eight-hour day, time-and-a-half for overtime, grievance procedures, seniority, shop committees, arbitration, and restoration of 1929 wage rates with a minimum of 50 cents an hour. Each local union was instructed to present the memorandum at the plant level, while the international would approach the general offices of the Big Four. On January 6, 1934 Gorman and Lane submitted the memorandum as the basis for negotiations. After a lapse of several weeks, the companies uniformly declined to engage in collective bargaining with the Amalgamated. Past experience had no doubt prepared the union for this response. Gorman had earlier forecast that "in many cases we will have to secure collective bargaining step by step in accord with the law."[7] Unfortunately, the inadequacy of the Recovery Act was already becoming evident.

The response to 7(a) had surprised the packers. Their welfare programs in the 1920's had expanded to include free group insurance, vacations with pay, and increased safety work. They had assumed the effectiveness of their labor-relations programs. Vice-President James D. Cooney of Wilson & Co. gave the standard view:

Our Company's fair and considerate treatment of its employees during the past has entirely precluded any just complaint on the part of our employees about any of their working conditions . . . Under these circumstances the job of selling union membership to our employees is a very difficult one. To overcome this natural resistance, the organizers have resorted to coercion, intimidation and misrepresentation and . . . manufactured controversial issues between our employees and this Company.[8]

Among themselves, however, the packers now acknowledged their problems. A Swift & Co. official noted "the instances of open dissatisfaction which we see about us, and perhaps with us." The workers, he admitted, had some legitimate grievances, but these could be rectified and the loyalty of the employees restored. Then the packers would not need to fear outside unions even under the new legislation. For, concluded the Swift man, "the NRA does not require the giving up of previous industrial relations policies except as modifications may have been necessary in hours and wage rates . . . Our destiny is still partially in our own hands." [9]

The company union was the essential device. Employee representation plans were already in effect in thirty or more companies, and others, such as the Rath Packing Company, acted after the passage of the Recovery Act. A real effort was made to give substance to the plans. Wage increases of 10 per cent in December 1933 and 8 per cent in September 1934 were, according to the packers, the outcome of negotiations with employee representatives. Similarly, Wilson & Co. introduced a vacation plan at Cedar Rapids in 1934 through the Joint Representation Committee. The members, a union man wrote, "are called for a meeting not knowing what its [sic] all about and click, click, the magic trick is over. They sit there with their mouths open, then when they come to they start congratulating themselves." The packers, inept as some of their efforts were, seriously sought to sell the representation plans to the packinghouse workers.

The response varied. In some places, men boycotted the elections or participated under pressure from foremen. Elsewhere, as a correspondent reported of his plant in Des Moines, the company union died "a natural death" because of the popularity of the Amalgamated local. The stock handlers' union in Chicago, in fact, grew out of a company union. On the other hand, employee representation was effective in some plants. Swift & Co., always the most enlightened of the big packers in labor matters, had notable success with its conference boards. The St. Joseph organizer reported progress among the Armour men, "but we are having quite a time getting the men from the Swift Plant." He added later that "the Company Union is our worst enemy."[10] The approval of the workers, however, soon became of secondary importance; it was enough to have merely a decent semblance of a company union.

The NRA was proving a paper tiger. The trouble was not so much with the interpretation of Section (7a). It was true that the company union was not prohibited, but the National Labor Board had asserted the right of employees to be represented by agencies of their own choosing and had designated elections as the appropriate procedure. The means, however, were lacking to enforce this policy. When the Amalgamated requested negotiations, the packers responded that collective bargaining was already in effect through employee representation. The union could then appeal for an election. But the regional labor boards could not compel companies to submit to elections. The executive order of February 1, 1934 gave the NLB the power to order elections, but it still could not subpoena either documents or witnesses. The National Labor Relations Board, established on June 29, 1934, did have adequate powers for election purposes. The St. Joseph Stockyards Company and the Cudahy Brothers Company were both directed in 1935 to hold elections.[11] But by then the capacity to enforce compliance had eroded. In any event, the critical period in the packing industry had long since passed.

For the NRA had not been able to protect even the right of union membership. No less than on the recognition issue, the regional labor boards proved inadequate to the task of preventing discrimination. Some reinstatements did result. But there were long delays, dubious decisions, and a growing awareness that no real enforcement powers lay behind either the regional boards or the parent NLB. The impact could be seen in the case of Local 28 in South Omaha. Two committeemen had been discharged by the Dold Packing Company on February 3, 1934. The local representative believed it was a clear violation of 7(a):

Our organizations [sic] progress is at stake if this matter is not cleared up to satisfaction of our membership, the whole South Omaha is aware that these men were discharged for Union activities and feel if they are not justly dealt with. Then no other case that may arrise [sic] can be handled through our organization properly.

The union did lose the case for lack of evidence; the official could not "get more members to sign affidavits as they are afraid of their jobs." And the local Communist paper gleefully reported that as a consequence of the AFL failure "to organize any struggle to get them back, most of the workers on the hog and beef kill threw their union buttons away." [12] The repetition of this experience in many places resulted in the swift shrinkage of the packinghouse membership in early 1934.

This fact determined the Amalgamated strategy. After the packers refused to deal with the union, Gorman and Lane admitted, many locals became "imbued with the strike idea, feeling that this action would bring the packers to a much quicker agreement." But the weakness of the packinghouse unions, no less than past experience and personal inclination, made the Amalgamated leaders anxious "to be absolutely sure that each step we make will not flare back upon us in an injurious way. A cessation of work . . . is unthinkable until we have went [sic] to the very end in our effort to, by reasoning, secure an understanding." [13]

The Amalgamated resorted to the great precedent of government intervention in meat packing: the world war experience. On March 12, 1934 President Gorman sent a formal request to Senator Robert F. Wagner, chairman of the National Labor Board, for an administrator with powers to fix wages, hours, and conditions and to protect the union rights of the packinghouse workers. Several days later, Gorman and Redmond S. Brennan, a lawyer who had represented the Amalgamated during the Alschuler administration, conferred in Washington with several influential men, including Senator Wagner. The documents of the Alschuler administration were circulated "to show the advantages of such a set-up, not alone for the workers in the industry, but for the packers and the general public as well." After a delay caused by the threatening strike in the auto industry, the National Labor Board called a conference with the leading packers for April 27, to be followed by a meeting with Amalgamated representatives. But the packers refused to attend. The labor board, lacking as it was in real authority, could do no more than direct the Amalgamated to handle its grievances, as it had earlier, through the regional boards. [14]

No alternative remained. Gorman and Lane conferred in Chicago on May 4, 1934 with representatives of the locals that had been "constantly clamoring for action." A general strike would have been suicidal in view of the organizational weakness. But on May 8 all packinghouse locals received telegrams "that they were free now to proceed in their own way with the prosecution of their demands." The international would support any local that chose to fight. [15]

The next two months saw notable strikes at the Wilson-Cedar Rapids plant, the entire St. Louis industry, the Iowa Packing Company (Swift controlled), the Wilson and Armour plants and the stockyards at Oklahoma City, a number of New York plants, and, finally, among the Chicago stock handlers in late July. Most of the strikes had a measure of success. The locals that struck were strong and militant. In addition, the time was favorable because of the government program for slaughtering drought cattle during the summer of 1934. And the regional boards, ineffectual in other ways, served a useful mediating function in a number of controversies. A few employers, in fact, acceded peacefully to the show of force, for instance, the Decker Packing Company in Mason City, Iowa, the Cudahy Brothers Company in Milwaukee, Morrell & Co. in Sioux Falls, South Dakota, and Kingan & Co., Indianapolis.

Packinghouse locals, by their own strength of organization, thus finally assumed their proper function of collective bargaining in the late spring and summer of 1934. It was a limited achievement. The packers who were forced to acknowledge that the unions did indeed represent their employees took a very narrow view of "collective bargaining." The standard practice was to listen to the union's demands and then either assent or decline. But the packers did not bargain; they did not make counter-offers; nor did they sign contracts. The outcome of discussions was ordinarily a company announcement to the employees that did not mention the union. In some cases, the company insisted on multiple recognition.[16] Still, the unions were reasonably satisfied with the initial progress they had made on the hard road to collective bargaining.

The greater difficulty was that these agreements covered only a small part of the industry. The bulk of the Amalgamated gains came in the Iowa industry, mainly independent, together with a scattering of minor plants in Chicago, St. Louis, and elsewhere. But the Big Four at Chicago and the river points—the heart of the industry—were unaffected by the drive for recognition. Swift, Armour, Wilson and Cudahy employees were not inclined to strike. At St. Joseph, only forty-four union members attended the meeting to discuss a course of action. They were, the local representative reported, "somewhat skeptical on taking a strike vote. It was feared here that Swift & Co. would not come out. And Armour & Co. could not pull out enough men to hurt them much." Secretary Lane responded that "there must be more interest and militancy displayed by the membership" or "they will not get to first base."[17] Nor did they in the major plants of the industry in mid-1934.

One slim hope remained to capitalize on the NRA. Despite extended talks, the packing industry had not accepted a code of fair competition by the summer of 1934. Ultimately, the Institute of American Meat Packers asserted its opposition to any formal code on the grounds that the industry was adequately regulated in its trade practices through a marketing agreement with the Department of Agriculture under the Packers and Stockyards Act and in its labor practices by the temporary President's Reemployment Agreement in effect since August 1933. The packers wanted to avoid any unforeseen consequences of a code. For one thing, the union could attain a certain formal standing with the packers by participating, first, in the formulation of new labor provisions and, thereafter, perhaps in the administration of the code. In addition, the guarantees of Section 7(a) could receive greater enforcement if—as in bituminous coal, petroleum, and elsewhere—a separate labor board was established for the industry. "The steps necessary to obtain a formal code and imposing such a code at this time would have a very disturbing influence on the industry," wrote the president of the Institute in August 1934. A code "would benefit no one and would tend to disturb the satisfactory situation that exists." [18]

No way of coercing an unwilling industry existed. Then, on July 13, 1934, AFL President Green called Gorman's attention to a new administrative order permitting the NRA to impose a code for any uncodified industry in which labor conditions "constitute an abuse inimical to the public interest and contrary to the policy of said act." Green was certain that the Amalgamated could get "a code for the Meat Packing Industry . . . under the quoted section." Reacting, the international on August 7, 1934 requested all packinghouse locals to forward evidence of union discrimination, of unfair practices, and of unemployment.

The work, so auspiciously begun, soon collapsed into a comedy of errors. An official of the NRA Labor Advisory Board came to Chicago to collect evidence, but he ended up at the headquarters of a rival splinter union. Gorman meanwhile turned over a bundle of affidavits on union discrimination to a minor NRA official who shortly thereafter left the service without bothering to forward the documents to Washington. In early December, an impatient query came from the Labor Advisory Board. It took several more weeks to track down the evidence of code violations. Then the board asked the Amalgamated for another survey. Thoroughly disgusted by the confusion in Washington, Gorman refused to take any further trouble. The NRA had to withdraw its attempt to impose a code for lack of evidence.[19] That sorry note ended the Amalgamated hope in the Recovery Act several months before it received its quietus at the hands of the Supreme Court.

The Amalgamated was meanwhile retrenching. At its meeting in January 1935, the executive board withdrew the waiver of local dues for packinghouse workers: the international would no longer finance their local unions. The number of organizers was drastically reduced. The international, Lane decided, would have to live within its income. On May 14, 1935 the regular salary scale of the vice-presidents was restored, increasing their earnings from $62.50 to $75 a week. "We are now getting back to a more normal basis of operation." [20]

The packinghouse gains quickly melted away. The Amalgamated paid per capita to the AFL for November–December 1934 for 22,625 members; for December 1935, 14,769 members. The most serious losses came from unfortunate strikes that undermined strong points in the industry. A "setdown" began a violent two-year struggle at the Morrell plant at Sioux Falls, South Dakota. The Amalgamated used its resources to the utmost in the fight, including a national boycott of Morrell goods. The bitter contest ended finally on March 10, 1937 with the company backing down on the principal issue, namely, the reemployment (although on a gradual basis) of all the sit-down strikers. But, despite the tenacity of some of the strikers, the Amalgamated had by then lost its grip on the Morrell employees. Local 89 of Oklahoma City entered a strike on June 4, 1935 against Wilson & Co. that lasted for seven months. A respectable settlement was finally reached, but the weakened union did not long survive. Other strong unions were lost through secession or undermined by disruptionist activity, and there was a general withdrawal of membership from the weaker organizations.

In the end, the Amalgamated had little to show for its exertions. Gorman and Lane noted in May 1935 that while "there were many thousands joined, yet there were [sic] but a small percentage . . . that remained as dues paying members. The greater portion of this influx came from the packing industry." The statistics were inescapable: 60,889 people had entered the Amalgamated between 1930 and June 1936; yet the increase since the start of the NRA was 5480, the majority of whom came from the retail trade. [21] At best, a mere few thousand of the 120,000 in meat packing had been gained during the NRA period.

The Amalgamated experience had an ominous new aspect. The primary cause for failure was not different from the past: the resistant powers of the packers. But for the first time the exclusive role of the Amalgamated in meat packing was called into question. The challenge arose not from within the labor movement, but from within the industrial ranks. Secretary Lane himself observed privately "that the International Union made a real earnest effort to help the packing house workers organizing and . . . they did not take advantage of what we were doing." [22] The dangerous feature was that, unlike earlier periods of union upsurge, there were now appearing packinghouse unions outside the Amalgamated.

The most vigorous independent sprang up at Austin, Minnesota, in July 1933. Angered by payroll deductions for a new pension plan, the employees of Hormel & Co. spontaneously formed an organization called the Independent Union of All Workers. The turning point came in November. Deciding to strike, the men "cleared out" the foremen and executives and took complete possession of the plant. National Guard units were mobilized, but Governor Floyd B. Olson was able to avert the crisis by persuading both sides to accept arbitration by the state industrial commission. Thereafter, Morrell and the IUAW entered into an unusually fruitful relationship, and the independent union assumed a permanent footing. Strongly influenced by radical ideas ("We recognize that we are under a system which perpetuates wage slavery"), the IUAW at first took in workers from other trades, and it organized actively in other packing centers in Minnesota and Iowa. Its success outside Austin was not conspicuous, but by 1936 the IUAW did have small groups in Albert Lea, Faribault and South St. Paul, Minnesota, and in Mason City and Waterloo, Iowa.[23]

The other notable independent center at the close of the NRA period was Cedar Rapids, Iowa. The Amalgamated local in the Wilson plant, which was strongly organized, had seceded in January 1935. This was the result partly of dissatisfaction that began when the international refused to pay death benefits for men delinquent in dues. The more important cause resided in the ambitions of the local secretary, Lewis J. Clark. He had urged the expansion of the Iowa State Branch, in which he played a leading role, into a "Midwest State Branch" with authority to direct an organizing drive in the packing centers. This projected body, which would have been virtually autonomous, was refused a charter by the international. Clark thereupon reconstituted his organization as the Midwest Union of All Packing House Workers. Despite its resounding title, the dual union was limited to Cedar Rapids, but, like the Austin independent, it actively sought to extend its influence into adjacent areas.[24]

Chicago also had experienced considerable independent activity. The Communists had organized the Packing House Workers' Industrial Union in the yards, as well as engaging in organizing activity elsewhere through the Trade Union Unity League.[25] The important rival union in Chicago, however, was the Stockyards Labor Council. The organization carried the same name as the dissident body of the world war period and had the same president, Martin P. Murphy. But, unlike its predecessor, the SYLC was entirely independent, having no ties with the AFL or the Chicago Federation of Labor. Initially quite strong, the SYLC was weakened by the dismissal of forty active members from the G. H. Hammond plant on December 28, 1933. When the Stockyards Labor Council collapsed in late 1934, the remnants drifted into the Amalgamated.

The SYLC men proved more dangerous inside the Amalgamated than outside. They quickly became the nucleus of a Chicago group opposed to Amalgamated policy. Against the wishes of the international leadership, delegates from four Chicago locals in the summer of 1935 requested the aid of the Chicago Federation of Labor for an organizing drive in the stockyards. Despite some success in the small houses and the Armour pork cutting department, the campaign soon petered out. But the dissident potential was clearly present in Chicago.[26]

The independent movements held a very minor place in meat packing in 1936. But they were significant for two reasons. First, they were a sign that the years of failure had consumed much of the credit of the Amalgamated among the packinghouse workers. And, of more concrete importance, the independents could constitute the nucleus of a major rival union in meat packing at the conjunction of the CIO and the Wagner Act in American labor history.

The threat did not materialize for many months after the formation in November 1935 of the Committee for Industrial Organization.

The Amalgamated had no quarrel with the views of John L. Lewis. The unskilled had found a place in the union almost from the start. And its charter, as Secretary Lane observed, gave it jurisdiction over "wage earners in any way connected—that is pretty broad— . . . with slaughtering and packing establishments." It was true that, in contrast to the Brewery Workers and Mine Workers, the Amalgamated had not opposed the AFL policy of awarding mechanical workers and teamsters in the stockyards to other internationals. Even that accommodation, however, was altered as a result of the strike of 1921. At the Amalgamated convention several months afterward, delegates had agreed that the craft division of the labor force had made "the necessary solidarity of the workers an impossibility." The strike itself had suffered from the treachery of the Teamsters, Stationary Engineers, and Firemen. The convention favored industrial unionism, but took no actual measures. To demand a transfer of all packinghouse jurisdictions, Lane prudently argued, was unnecessary because of the broad charter of the Amalgamated and foolhardy because of the union's weakened state. The consensus was first to organize the industry "and then by reason of the strength that would accompany such an organization, take and retain control over all men of whatever craft employed in the industry."[27]

Therefore, at the fateful AFL convention of 1935 the Amalgamated, while it had not itself directly challenged the federation's craft policy, supported the minority report which became the CIO platform. (Actually, the meat cutters' delegation was split, casting 149 votes for and 40 votes against the minority report.) In the ensuing controversy, the Amalgamated ranged itself on the side of the industrial unionists. In March 1936 Secretary Lane assured CIO director John Brophy that, "with slight exception, the members of our International Executive Board are in full accord and sympathy with the program of the C.I.O." The international convention of 1936 formalized the Amalgamated position that industrial unionism, as advocated by the CIO, "is the only effective means of establishing permanent organization among the wage earners of the meat industry." Gorman had already expressed that view to William Green, evoking from the AFL chief his standard defense of the feasibility of the AFL program in mass-production industries.[28]

Events at the time lent substance to the Amalgamated espousal of the CIO view. The union had recently organized all the employees of six Philadelphia plants. A strike broke out in late February 1936. The Brotherhood of Teamsters, Lane reported to John Brophy, "went into those packers and signed an agreement when they did not have a man connected with their organization involved in the trouble . . . and put other teamsters on the trucks." Following the dispute with interest, John Brophy sympathized with Lane: "I agree with you that when craft unions project themselves into a strike situation . . . inevitable confusion and demoralization is [sic] bound to occur."[29]

The Amalgamated continued to support the CIO as the split deepened within the labor movement. It was among the few organizations to urge a modification of Green's demand in February 1936 for the dissolution of the committee. At the AFL convention of November 1936, the Amalgamated delegation abstained from the vote sustaining the suspension of the CIO unions two months before. The union's monthly publication registered more forthright disapproval of that act. When John P. Frey of the AFL Metal Trades Department criticized the sit-down tactics used against General Motors, Gorman took the opportunity to answer that the labor movement could not "stick to horse and buggy days in its effort to unionize . . . industrial organizations producing on a mass production basis." The Amalgamated stand had repercussions. The AFL Executive Board in March 1937 rejected its application for jurisdiction over fish workers. "This decision," Gorman believed, "would not have been made if we had not been so outspoken for the C.I.O." [30]

The Amalgamated consequently retained the good will of the CIO for a time. Friendly letters passed between the two organizations, and John L. Lewis was invited to speak at the Amalgamated convention. John Brophy consistently turned aside the overtures of the independent unions in the packing field. "Our advice has been that you affiliate with the AMCBW and work out your problems with them as this organization favors industrial unionism." The Butcher Workmen, Brophy told the Austin and Cedar Rapids independents, "were in sympathy with the position of the committee." [31] By early 1937, however, the CIO began to demand more than verbal support.

For one thing, meat packing presented an obvious field for organization. Other mass-production industries were undergoing rapid organization, but the Amalgamated, having expended itself during the NRA period, remained relatively inactive. "What are the prospects for the extending of union activities in the packing industry?" Brophy asked pointedly in December 1936. "I receive communications from different individuals urging action and requesting cooperation." Secretary Lane promised a renewed effort. "We expect soon to have a complete force organized to make a drive in all of the packing centers at the same time," he assured John L. Lewis on March 15, 1937. Lane wanted "very much to have an opportunity sometimes in the near future of sitting down with you and discussing a program." [32]

But something else was now in the mind of the CIO chairman. Along with several other CIO supporters such as the Bakery Workers, the Amalgamated had not joined the committee. The Amalgamated resolution of 1936 that praised the CIO program also rejected any split from the AFL. Gorman had then predicted that "secession or separation from the A.F.L. . . . is remote and not contemplated by any of the splendid leaders fostering Industrial Unionism." That hope was quashed by the suspension of the CIO unions, and by the General Motors and U.S. Steel victories. John L. Lewis became increasingly committed to his own industrial-union federation. In March 1937, the CIO executive officers were authorized to issue certificates of affiliation to internationals, locals, city centrals, and state bodies. Thereafter, the pressure mounted sharply for the Amalgamated to leave the

AFL. Hitherto scrupulous in its regard for Amalgamated jurisdiction, the CIO became active in meat packing. Gorman complained to John Brophy on April 19, 1937 of "many communications that C.I.O. organizers and organizations are attempting to enroll packing plant workers with the C.I.O." [33]

The alternatives crystallized in the next weeks. Amalgamated leaders met twice with John L. Lewis. President Gorman told him at the first encounter "that it was perhaps an unfriendly jesture [sic]" for the CIO "to enter the field against an organization that had so consistently supported the industrial form of organization." His aim, Lewis retorted, was to organize the mass-production industries; he had abundant evidence that "the sentiment among packing house workers was that they didn't want to join the American Federation of Labor." Lewis made his position unmistakable at the second meeting in early May: "You must stand on one side or the other—these men are going to be organized—you have got to stand with the C.I.O. or with the American Federation of Labor." That was the nub of the matter: either quit the federation or the CIO would issue charters in the packing centers. [34]

The executive board met on May 10, 1937 to reach a decision. Everyone present recognized, as Gorman said, that "this proposition is perhaps the most paramount one the Executive Board has had to deal with in a good many years." The CIO threat was not underestimated. Polled by Gorman, the vice-presidents one by one acknowledged the popularity of the CIO among the packinghouse workers. Earl W. Jimerson of East St. Louis, in fact, judged that in his district "it is doubtful whether or not we can organize the big houses other than C.I.O., as they all seem to be C.I.O. minded." The board nevertheless voted to reaffirm the position of the Memphis convention in favor of "industrial organization in the meat industry but with no thought in mind of any separation or secession from the American Federation of Labor." On May 12, Secretary Lane notified John Brophy. "We have decided to continue our status as is . . . If the C.I.O. must enter our field, then I suppose that's what it will be." [35]

Foremost among the determinants of that fateful decision was a deep-seated abhorrence of secession. The Amalgamated itself had suffered from repeated internal divisions. Gorman and Lane, moreover, respected the legitimacy invested in the AFL. "We are not secessionists," Lane told Brophy, "for we have always fought secession and are firm believers in settling disputes within our own household." It was a matter of principle. The Butcher Workmen would leave the AFL only at the latter's initiative. That, after all, had been how the original CIO unions had split from the labor movement. "If we get kicked out of the Federation, then that will be another thing—It will not be that we leave thru the avenue of secession." In his efforts to placate John L. Lewis, Secretary Lane had intimated that this was a real possibility.[36] Actually, the Amalgamated had not given the AFL any occasion for such action.

The principle of unity, compelling as it was, was not the only factor in the decision of the Amalgamated Executive Board. There were also good practical reasons for remaining within the AFL.

Embracing the CIO, for one thing, involved hidden risks. An influx of packinghouse workers under CIO auspices might well engulf the entrenched Amalgamated administration. Identified with earlier failures, Dennis Lane was not a popular figure among packinghouse workers. His policy of caution, as it had in the past, would surely cause dissension. Among the newcomers were certain to be ambitious leaders, such as Lewis Clark of Cedar Rapids, and Frank Ellis and Joseph Vorhees of Austin, Minnesota. The experience with the Stockyards Labor Council after World War I had been bad enough. The potential danger would be very much greater with packinghouse organization rapidly developed through the CIO. Gorman reported an offer by Lewis at the first meeting for the CIO "to organize the packing house workers for us and then at the opportune time he would turn over this entire membership . . . Of course I frowned upon that proposition as we want to do our own organizing." [37] An alliance with the CIO clearly gave cause for hesitation.

There were, on the other hand, positive reasons for remaining in the AFL. The Butcher Workmen, unlike the Amalgamated Association of Iron, Steel, and Tin Workers, was not limited to a mass-production industry. The retail trade gave the Butcher Workmen an independent base. And the Amalgamated was making rapid headway here at the very time of the CIO crisis: the paid-up membership increased from 16,500 to 34,700 in the fourteen months after April 1936. The AFL affiliation provided important advantages for retail organization. Dealings with the top management of the chains benefited by the AFL connection as they would not have under the CIO. In addition, the support of the Teamsters, local although it then was, was very important in the retail field. The meat cutters' unions would not want the Teamsters as enemies. Finally, the vital consumer boycott depended on the AFL connection. In comparison, the CIO city centrals and state bodies were in a rudimentary stage and the membership concentrated in fewer areas. The commitment to the retail trade thus influenced in an unremarked fashion the Amalgamated decision.[38]

The break with John L. Lewis soon became decisive. The CIO not only began to issue local packinghouse charters, but also to send out circulars urging Amalgamated unions to shift affiliation. Gorman sent a pained letter to Lewis on June 16, 1937. "I didn't feel that this was true to the policy of the C.I.O. to organize the unorganized. I was wondering if you knew that Brother Brophy has sent such letters." In response, Lewis pointed to the intensifying conflict within the labor movement. The AFL had "declared war on the Committee for Industrial Organization."[39] The CIO was entering the packing industry in earnest.

A narrow base already existed for a rival organization. The strong independents at Cedar Rapids and Austin had long sought affiliation with the CIO.[40] Rebuffed, the two unions nevertheless persisted. They formed a loose alliance, holding conferences and exchanging speakers. They also implied a connection with the CIO in their organizing work in nearby centers. On December 5, 1936 a conference was held at Des Moines of representatives of the two independents and also of workers from Des Moines and Sioux City. The outcome was an agreement to establish a committee to begin an organizing drive and to consider forming an international union.[41] (The CIO, among other reasons for putting off the independents, had explained that it was composed only of international unions.) Before long, the independents began to receive the first signs of encouragement from the CIO.

An insurgent movement was meanwhile developing in Chicago. Excited by the CIO successes in steel and automobiles, the militants in the Amalgamated locals there began to press for action in early 1937. Following a meeting of a small group in the backroom of a saloon near the yards, a committee of three was sent to see Secretary Lane. He discouraged any immediate action until the impending Supreme Court decision on the Wagner Act. The militants then turned to the CIO. A committee visited the Chicago office of the Steel Workers Organizing Committee. Greeted without enthusiasm, the packinghouse representatives were permitted to go ahead on their own. Constituting themselves volunteer organizers, the militants began to pass out "C.I.O. membership cards" which had been printed at their own expense. The initial success came in the Armour plant. On March 27, 1937, the employees of the pork department formed a local union

with Arthur Kampfert as president. The rank-and-file campaign continued into the summer, despite the discharge of Kampfert and another key
man. A CIO nucleus thus existed in Chicago when the break came between John L. Lewis and the Amalgamated.[42]

Packinghouse organization was beginning at many places during the
late spring and summer of 1937. In East St. Louis, the Mine, Mill, and
Smelter Workers were active, in fact giving their own charters to locals at
the Armour and Hunter Packing plants. The Auto, Rubber, Mine, and
Steel Workers did organizing work elsewhere, as did the packinghouse independents in Minnesota and Iowa.[43] A few Amalgamated locals came
over and, occasionally, weak locals were displaced by CIO unions. In addition to its own staff, the CIO appointed a few men to work directly in meat
packing. Arthur Kampfert became regional director in Chicago on July 28,
1937, with authority to employ four or five organizers in the Union Stock
Yards.

The final step occurred at a Chicago meeting of local delegates on October 24, 1937. Van Bittner, the Chicago SWOC district director, announced
the formation of the Packinghouse Workers Organizing Committee.
Hitherto local industrial unions attached directly to the CIO, the packinghouse bodies would now be transferred to the PWOC. The CIO appointed
Bittner as chairman; Don Harris of the Hosiery Workers as national director; and later, Henry Johnson, a Negro who had been on Kampfert's
Chicago staff, as assistant national director. The majority of the original
nineteen regional and subregional directors were CIO officials for whom
this was merely an additional responsibility, but there was also a number
of packinghouse men: Kampfert in Chicago, officials of the two former independents at Austin, Cedar Rapids, Omaha, and Waterloo, Iowa, and several local leaders at other Iowa points.

A rival union thus resulted from the Amalgamated decision to stay in
the AFL. A national organization was in operation under the aegis of the
triumphant CIO. The PWOC, of course, had made only a bare beginning.
But the CIO base was firm, if narrow, in meat packing at the close of 1937.
The Amalgamated would face a formidable rival for the opportunities
opened by the Wagner Act.

CHAPTER 9

THE THREE-CORNERED CONTEST

Packers and unions competed for the allegiance of the packinghouse workers. Theirs, however, was the choice to make freely through majority rule. The implementation of that fundamental principle of the Wagner Act created a new framework within which management and the rival unions had to operate. The consequences were decisive on all sides.

The Wagner Act put an end to effective employer resistance to union growth. Years later, James D. Cooney stated the fact bluntly for his company: "The unions would not have organized Wilson [& Co.] during the Thirties if it had not been for the Act."[1] The judgment held for the entire industry.

The impact was not felt for almost two years because the major packers ignored the law of the land until the Supreme Court's decision in April 1937. Even then, they harbored the hope, as the *National Provisioner* put it, "that no abrupt change will be necessary, and that some satisfactory solution can be worked out which will retain the benefits of the existing system."[2]

The key was the company union. The company union patently violated the prohibition on employers, under Section 8(2), "To dominate or interfere with the formation or administration of any labor organization or contribute financial or other support to it." But the expectation remained that "inside" organizations could be reconciled with the law. Following a standard course of action, company officials called together employee representatives and informed them that the Wagner Act required the disbanding of the assemblies and councils. The superintendents of the Swift plants, for example, read a statement from the Chicago office:

It is Swift & Company's intention to comply with the law as the court has now construed it and it is not possible to continue the present Representation plan.

Whether you wish to establish an employees' representation plan for collective bargaining, that will comply with the terms of the law, is a matter for you to decide . . . The company earnestly desires that the understanding growing out of our relationships during these past many years will be the basis upon which the continued good relations between employees and the company will be maintained.[3]

The president of the Cudahy Packing Company made a franker appeal:

In these times of change in the general conception of the relations between worker and employer we, of the Cudahy organization, will do well to keep *to our chartered course of industrial relations by which we have progressed so steadily for so many years.*

Further, it seems to me that until the rules and regulations embodied in the Wagner Labor Relations Act are known and understood thoroughly, it is the part of good judgment to *maintain our accustomed way and not be unduly influenced by developments of the moment.*[4]

Acting generally in the same meetings ending the old plans, the representatives constituted themselves the nuclei of independent unions in which the employer did not participate (as they did in the employer representation plans). The objective was, as a Swift employee representative at the Nashville plant said, to avoid having "the C.I.O. or the A.F. of L. crammed down our throats."[5] Organizations were thus quickly established under a variety of titles: at Swift plants, as Security Leagues; at Cudahy, as Packing House Workers' Union of Sioux City, Omaha, and so forth; at Armour, as varying titles such as Employees' Mutual Association (Chicago), Employees' Mutual Bargaining Association (Denver), Employees' Industrial Organization (St. Joseph).

The company unions of course experienced no difficulty with management. They were accorded recognition as bargaining agents as soon as they claimed a majority. Collective bargaining began, but the reality was little different from before April 1937; there was rarely genuine negotiation or signed agreements, and activities were still limited mainly to grievances.

Cudahy was somewhat of an exception. Delegates from ten plant unions on May 24, 1937 attended a conference which resulted in a signed contract with the company. The principal issue involved overtime pay. The delegates pleaded for a real concession. The spokesman from Kansas City warned:

I have signed up over 51 per cent of the men. How long we can keep them I don't know—they are switching right now . . . I would like to see this put through, just in order to keep these men who are backsliding . . . There won't be 3 per cent of the men in the plant who won't go over to the C.I.O. You are saving an entire plant—you will have us bargaining with you.

The argument impressed the reluctant E. A. Cudahy, Jr.

I think as you do from the standpoint of the advisability of your going home and satisfying them 100 per cent. We want to go along with you. We don't want some A.F. of L. or C.I.O.

Compromising, the company agreed to pay overtime rates after forty hours for ninety days, or, if it came earlier, until the passage of Federal wage and hour legislation. The delegates and company officials were in unanimous agreement, Cudahy observed, "principally on account of our past," that dealing "among ourselves" was preferable to having "any outside affiliation." [6]

The hope of the *National Provisioner* seemed fulfilled that the past representation plans could "lead in many cases to an easy and almost automatic transition into wholly independent groups, which will be equally cooperative in the future." [7]

But the new organizations could not survive the scrutiny of the National Labor Relations Board. The packers had misjudged both the intent and the enforcement of Section 8(2). It was not enough merely to withdraw management representation from the employee organizations. The NLRB demanded the complete absence of employer domination and, from the nature of things, could not find this to be the case in the packing houses. Invariably, the board condemned the circumstances attending the start of the new independent unions. The initiators and subsequent officials were almost always the employee representatives of the earlier plans. Solicitation of members went on during working hours, under the eye of the foremen, and often with their open assistance. Company support or domination continued after the launching of the independent unions. They frequently used plant equipment, were housed on company property, and received thinly veiled financial aid. Cudahy, for instance, picked up the sizable hotel bill for the Omaha conference in May 1937. Two unions at South St. Paul hired a lawyer who was already retained by the Cudahy Packing Company. The constitution of the union at Wilson-Chicago plant could be amended only with the "approval of the management."

Disestablishment of company unions therefore resulted at more than fifteen major plants of the Big Four, as well as at independent houses. In a number of instances, the packers resorted to the Federal courts. While gaining postponements, this course never prevented the ultimate enforcement of the NLRB's disestablishment orders. Recognizing the inevitable, Swift & Co. withdrew recognition from the Security Leagues by stipulation with the board in 1940 and 1941.[8] Plants were particularly vulnerable following disestablishment orders, because company unions could not participate in elections immediately after losing recognition. But the independents were defeated, with notable exceptions, even when they did take part in elections. The fact was that employers had lost the means of control.

The packinghouse workers, as never before, were becoming free agents under the Wagner Act. Intimidation and discrimination were vigorously prosecuted. In a number of instances, union men were reinstated with back pay by order of the NLRB. In addition, stipulation cases required the posting of notices such as the following at a Sioux City plant:

> The Cudahy Packing Company wants it definitely understood that in the future the company, its officers, and supervisory staff will in no way interfere with its employees' right to organize. No one will be discharged, demoted, transferred, put on less desirable jobs, or laid off because he joins Local No. 70 or any other labor organization . . . Union membership and union activity will in no way affect the jobs or rights of Cudahy employees.
>
> . . . .
>
> If the company, its officers, or supervisors have in the past made any statements or taken any action to indicate that its employees were not free to join local No. 70 or any other labor organization, these statements and actions are now repudiated.[9]

The weapons of coercion were mostly nullified by the Wagner Act.[10]

Nor could positive policies counteract the union appeal. Paid vacations, pensions, group insurance, seniority plans, safety programs, and the thirty-two-hour weekly guarantee were common in the industry. Wage increases of 7 per cent in October 1936 and 13 per cent in March 1937 brought the minimum for male labor to $62\frac{1}{2}$ cents an hour—$26\frac{1}{2}$ cents over the 1932 level.[11] When the PWOC opened a drive at the Rath Packing Company in 1941, the company passed out stickers: "Mr. Rath, I am Satisfied." But most packinghouse workers were not satisfied. They flocked into the outside unions. Ultimately, the company unions disappeared from the plants of Armour, Wilson, Cudahy, and most of the important independent firms.

Swift & Co. was the major exception. Its employee relations had long been the best in the industry. "Swift's labor policy was shrewd," acknowledged a PWOC official, "and its working conditions were slightly better than the other big packers." Its company union, Van Bittner told the CIO

Executive Board, was "one of the most effective in the country." [12] The company wisely voluntarily withdrew its recognition of the Security Leagues after 1939. And the full force of the union drive came last to Swift. As a result, there emerged independent unions that were acceptable to the NLRB, and able to compete with the outside unions. A series of tough contests took place from 1941 through 1943. The independents lost twelve elections, but they were certified in a number of Swift plants, including St. Joseph, St. Paul, Kansas City, and Fort Worth. [13] As early as 1939, the independents had formed themselves into the International Brotherhood of Swift Employees (later changed to the National Brotherhood of Packinghouse Workers).

The bulk of the industry, however, was destined to move into the main line of AFL or CIO organization.

The Wagner Act neutralized the antiunion weapons of the packers. But the legislation imposed new requirements on the unions. They received representational rights only if they were the free choice of the majority in a plant. The right to choose was greatly magnified in importance by the existence of two vigorous unions in meat packing. The onset of depression in late 1937 had hindered the activities of the PWOC for a year. Thereafter, its campaign in the packing centers mounted in force. The Amalgamated had at first avoided direct competition with the PWOC. "We have so far remained out of any fights that the C.I.O. was engaged in," Patrick Gorman commented in October 1938. [14] From that point, the AFL union entered the ring without reservation. These two facts—the Wagner Act and union rivalry—made decisive the ability to win the vote of the packinghouse worker.

The contest was on fairly even terms in some respects. The availability of resources gave neither union a great advantage. The Amalgamated could expect little outside help, although it did benefit from the work of the expanded AFL staff, particularly in Chicago in 1938–39. But it had to rely on its own income. The rise in dues-paying membership—from 43,422 in December 1937 to 70,591 two years later—resulted in greater income; Secretary Lane reported assets of $753,255 on January 31, 1940. The Amalgamated thus had adequate means to support organizing work in meat packing. The CIO connection, on the other hand, provided relatively more outside help for the PWOC, but never in massive amounts. The five organizers first assigned were all withdrawn in November 1939 for lack of funds, as were other men a year later. The total CIO investment in the

Packinghouse Workers, including the services of organizers, amounted to $92,615 by September 30, 1941. CIO affiliates, particularly the Mine Workers, also donated men from time to time, but this help was substantial only at critical points. The PWOC was itself always in financial straits. The CIO campaign in meat packing never had the generous infusion of money that moved the steel drive.[15]

Nor were there significant structural differences. It was true that, as earlier, the Amalgamated found it expedient to bow to insistent claims of craft unions. When a joint drive was started in the Union Stock Yards under the auspices of the Chicago Federation of Labor, Secretary Joseph Keenan reported that all the recruits would be taken into the Amalgamated "and later these workers would be transferred to their respective international unions."[16] Nevertheless, there was little actual difference between the Amalgamated and PWOC. For one thing, packing houses had relatively few maintenance and mechanical workers and, more often than not, their appropriate unions did not lay claim to them (at least in NLRB elections). The NLRB also narrowed the freedom of the PWOC on this issue, because the *Globe* doctrine permitted craft workers to choose between a separate union and inclusion in an industrial unit. Finally, the PWOC was not itself entirely faithful to the doctrine of industrial unionism. The prospect of a close election sometimes prompted a desire to exclude unfriendly craft groups. The PWOC also usually omitted the teamsters. In the Swift case in New York City, the PWOC local claimed "an agreement with the International Brotherhood of Teamsters . . . that the United [PWOC] will not attempt to organize truck drivers." No less than the Amalgamated, the PWOC valued the good will of the strategic Teamsters.[17]

The structural issue had another side. Ever since its beginning, the Amalgamated had drawn the packinghouse locals on "craft" lines—that is, giving each local jurisdiction over a single department in all the plants in a meat center. Just before the open break with the CIO, the international adopted an "industrial" arrangement. Secretary Lane informed the executive board at the decisive May 1937 meeting that he had already inaugurated "a program of organization which he believed would be effective, that is, an Armour local, a Swift local, etc., and all workers in each house would be taken into the local union representing that plant."[18] This plan was not forced on established locals, but it was applied to plants organized after May 1937. No substantial difference existed, therefore, in the scope or local union structure of the CIO and AFL organizations in meat packing.

The gap between the rivals began with their reputations in the industry.

The Amalgamated emphasized, as a circular said, that it was "the OLD, RELIABLE and THE ONE ALERT, METHODICAL, MILITANT and THOROUGH BUSINESS–LIKE organization for the employees in the Butchering trade." The PWOC was led by outsiders with no knowledge of the industry. Chairman Van Bittner, the *Butcher Workman* taunted, "cannot pass through the killing department of a a packing plant because he becomes nauseated." [19] But the identification with the past was more a liability to the Amalgamated. It bore the heavy onus of remembered defeat and bitterness. "A Stock Handler" could address an open letter to Gorman in the PWOC journal:

Yes, we can look back over the last 25 years and see with remorse what your organization has done for us poor slaves of Packingtown. We can see what was done for us in 1921, 1922.

Shortly after the decision to organize local unions on plant lines, the St. Joseph organizer reported a meeting of Swift employees "to form a unit of their own": "but upon finding that they would have to get their charter from the Amalgated [sic] then there was a decided change in affairs." An independent union was formed instead.[20] In contrast, immediate enthusiasm greeted the PWOC. It benefited from the immense popularity of John L. Lewis and the CIO, from the reports of extraordinary success in steel, autos and elsewhere, and from the mistaken belief that it alone offered an industrial form of organization. Other things being equal, the packinghouse workers preferred the CIO organization.

Certain limitations were built into the Amalgamated. The union was unprepared by experience or outlook for the special demands of organizing a mass-production industry in the late 1930's under the Wagner Act.

The recruitment of effective organizers proved difficult. Men trained in the retail field and as business agents were of little value in the big packing centers. There were numerous dismissals for "lack of results." Secretary Lane explained the situation to a discharged organizer at St. Joseph:

We do not continue a person or persons in the capacity who are not getting results . . . We are not mind readers, nor prophets, therefore we cannot pick with a certainty those who are going to make good. We try our best by selecting the fellow who has been most active, believing that he would have a following but after a fair trial and results are not forthcoming, a change is made by another selection.

One indicator of the Amalgamated difficulty was the employment of several men of shady reputation in the Chicago yards.

The PWOC, of course, had its own problems. Organization in Wichita, for example, was long retarded, first by an organizer who absconded with the funds and then by a Communist whom the workers, "when they found out," dropped "like a hot potato." [21] But, in general, the PWOC recruited energetic men who could handle large crowds and master the hard techniques of organizing in meat packing. The men assigned from the CIO already had experience in mass-production fields. The formerly independent unions proved a rich source of aggressive organizers. Others of this militant variety turned up in the course of organization.

Nor was the Amalgamated leadership sensitive to non-economic aspects of organizing in the packing centers. Ever since World War I, the industry had employed large numbers of Negroes. The 1930 census indicated that they constituted 13 per cent of the total semi-skilled and 22 per cent of the common labor force; and the concentration was higher in Chicago and the river points. Amalgamated officers understood the importance of the Negroes. Several were on the international organizing staff. But no special appeal was made. Although segregation existed at a few points (including the retail field in Chicago), the Amalgamated was not among the discriminatory AFL organizations; it had followed a policy of open membership from the beginning. But the Amalgamated would not crusade on the race issue.[22]

This the PWOC did. An early achievement of the grievance committee at the Armour-Chicago plant, for instance, was to end the practice of "tagging" the time cards of colored employees; "the Stars will no longer offend the Negro workers of Armour & Co." When three colored workers were sent out of sight while a tour was passing through the lard refining room of the Wilson-Chicago plant, the PWOC representative demanded a public apology from the company. He also seized the opportunity to protest "Wilson's whole policy of racial discrimination." The first informal agreement at the Swift-Chicago plant included a company pledge to hire Negroes in proportion to their numbers in the Chicago population.

Militancy raised certain problems. The Iowa and Minnesota centers, which had relatively few colored workers, evidenced some hostility toward the Negroes on the organizing and administrative staff. And concessions had to be made to southern sentiment. The director of the southern district declined the offer of a Negro member to attend a Chicago conference for the district: "To have it advertised through the South that we from the South were represented by a colored man would do more harm than good. We have enough of a fight as it is without sticking our necks out in this manner." Nevertheless, its aggressive racial stand won for the PWOC the widespread support of Negro packinghouse workers.[23] The union was similarly effective in its approach to ethnic minorities and the women in the industry.

PWOC officials also saw the value of communal activities. They cultivated the good will of church and civic leaders of Negro and immigrant groups. They pushed vigorously for relief and housing measures, and played an active part, for example, in the "back of the Yards" movement in Chicago. They took an interest in local politics. PWOC support for the Democratic administration in Chicago paid handsome dividends during controversies in the stockyards. Construing its function narrowly, the Amalgamated was remiss in this important area. The PWOC owed its Chicago successes in large part to the support of the larger community.[24]

The main task of course resided within the confines of the stockyards. Unremitting effort and a mastery of the organizer's techniques were the main ingredients of success. The PWOC district director at Oklahoma City outlined his program for the Armour plant: "Home to home contract work, meetings once each week, little social gatherings at someones [sic] home, and a little later on I will take some of the Boys from Denver down to one or too [sic] of their meetings." [25] Organizers massed their supporters for picketing and demonstration purposes, and they did not back away from the rough-and-tumble of union rivalry. The PWOC excelled in ruthless effectiveness on the organizing field.

Beyond the standard methods, two factors drew the packinghouse workers into the CIO camp. First, the PWOC emphasized rank-and-file participation. An official instructed Fort Worth unionists on how to organize the Armour plant.

> It takes Organizers inside the plant to Organize the plant.
> The Committee that organized the Oklahoma City plant was a voluntary committee, established inside the plant.
> You cannot wait for the National Organizer to do all the work; you have capable workers here in Ft. Worth. . . . You people here can have a Union, but you will have to work to build it.

The standard first step of a PWOC organizer was to win over some of the natural leaders and constitute them a "volunteer organizing committee." A departmental steward system was next erected, first to complete the organizing process and then, as the union paper said, to be the means by which "more men are given responsibility, and our organization becomes more powerful and more closely knit." The major plants had large numbers of elected officials: chief and co-stewards in departments, grievance committees, divisional representatives, local officers and, in some unions, educational directors. The PWOC eschewed the "bureaucratic" control of which it accused the Amalgamated. The rule of the packinghouse unions, asserted an organizer, should not be in the hands of a few, but in "the whole body, in one, acting as one. All of these collectively comprise your leadership." That concern for broad participation, inspired partly by the leftist orientation of some officials, clearly contributed to the strength of the PWOC.[26]

Its aggressive stance—the second element—was equally important. The packinghouse workers found in the PWOC the vehicle for releasing their accumulated resentment. Organizers understood this. "It seems to me that we are not having enough troubles in the plant," observed an official of the Wilson local in Faribault, Minnesota. "When something goes wrong they pay their dues better." The aggressive handling of grievances was necessary partly to offset the long delays before the achievement of formal negotiations and signed contracts. But the immediate object was, as Arthur Kampfert said, the "psychological effect among the workers." They flocked into the organization when they saw "how the C.I.O. was fighting to protect workers' rights." [27]

Direct action also expressed PWOC militancy in the early period. Slowdowns and sudden stoppages became a standard means of pressing unresolved grievances. After a brief strike, the Armour local at Denver was

firmly convinced of the truth behind the old saying, "Spare the rod and spoil the child." After this 35 minute punishment the management has promised to be very good.

New confidence grew from this common experience. Blowing their whistles precisely at 9:00 A.M., union stewards completely halted operations for 50 minutes at the Cudahy-Sioux City plant on December 31, 1938 to protest the slow grievance process. These actions invigorated the growth of PWOC organization. The stoppages at the Armour-Chicago plant, asserted a unionist, "demonstrated to all, union members and non-union members, that the C.I.O. had plenty of stuff on the ball and that there was no such thing as waiting for something to happen." [28]

Such tactics, however, entailed serious risks. Besides a number of stoppages lasting a half-day or so, packinghouse locals became involved before the end of 1939 in serious strikes in six cities. Only the Buffalo and Chicago stock handlers' strikes ended in union victories. In fact, such strikes were foolhardy. This was a period of unemployment; the packers were multi-plant firms; and, as a result of the Wagner Act, they could vent their animosity only during a strike. The *Fansteel* decision of 1939 increased the hazards of sudden stoppages, for the Supreme Court ruled dismissals of sit-down strikers not an unfair labor practice. The NLRB subsequently refused to order the reinstatement of 154 strikers at the Swift plant at Sioux City.[29] When rank-and-file enthusiasm spilled over into strikes, local strength could be dissipated, not to speak of the money diverted from more fruitful work elsewhere.

The Amalgamated did not emulate the CIO penchant for "action." Informed of a strike threat in the Cudahy plant at Jersey City, Secretary Dennis Lane answered with the voice of experience:

The history of the labor movement in the meat industry does not point out any degree of success gained through strikes but, to the contrary, all material things that the butcher workmen have gained through organization was accomplished across the table and not by strike. In fact the packing house workers were always beat down and their organizations broken up when they resorted to strike. [Let us] avoid the pitfalls that destroyed us in the past and use better tact in the handling of our problems from now on.

In the summer of 1939, a stern order was issued to the packinghouse branch. "We cannot tolerate local unions, even though they may be newly organized, to jeopardize the good name of the Amalgamated by stopping work promiscuously and over matters that are minor." Gorman and Lane urged "all locals unions to put in place any of the local leaders who will urge strikes . . . without first seeking the aid of the International office to adjust the controversy peaceably." The Amalgamated did not shrink from necessary fights, but it rigidly opposed walkouts caused by "undisciplined local membership" and "mob spirit."[30]

The PWOC leadership came to the same conclusion. On January 16, 1939, as the costly Swift–Sioux City strike was drawing to an unfavorable close, the local unions were ordered thenceforth not to strike without the authorization of the national office; never to permit stoppages in plants under contract; and to settle grievances through the machinery provided in agreements. "Our word is our bond, and our union must respect these wage agreements in letter and in spirit." The enforcement of this directive gradually became evident. When Cudahy complained in July 1941 of a slowdown at its Kansas City plant, National Chairman J. C. Lewis assured the firm "that we do not approve, nor will we condone slowdown and the officers of the Cudahy Local Union have been advised that . . . it is contrary to the fundamental principles of the trade union movement and . . . our National Union condemns such practice." [31] The PWOC thus came to the approximate position of the Amalgamated on slowdowns and strikes.

Yet the CIO body continued to exploit rank-and-file militancy. The reputation for "action" remained. Despite the announced policy, the national leaders still talked tough and, as a result of the effort to get national contracts, with many references to impending strikes. The PWOC, moreover, in the early period did not, or could not, rigorously enforce the order of January 1939. And the local leadership was more aggressive, for the fighting spirits had gravitated to the CIO. Despite the constraint of practical considerations, the PWOC remained the beneficiary of the angry sentiment within the industry.

The Amalgamated did its utmost to counteract the impact of its rival. Organizers pointed to concrete gains won by the Butcher Workmen. They warned of the losses incident to repeated slowdowns and strikes under PWOC leadership. They cautioned against "the pinks and reds who swamp the packinghouse workers with their own brand of radicalism and communism." [32] Just before the second Armour election in Chicago, in fact, Martin Dies of the House Un-American Activities Committee arrived to investigate the Communist connections of several PWOC officials. But the packinghouse workers seemed indifferent to "regular and bona fide" trade unionism with "fine and responsible" leadership.

Nor did the Amalgamated succeed in bringing into effective play the conservative elements in the stockyards. One possibility was to step into PWOC strikes. The stock handlers' strike in Chicago in November 1938 offered an early opportunity. PWOC Local 561, led by President Ben Brown, had come over from the Amalgamated. Although the CIO local had won NLRB certification, the Amalgamated continued the charter of the original Local 517 and was claiming a majority of the stock handlers in the fall of 1938. On November 21, the PWOC local was locked out as the result of a work stoppage pulled by the men. At the end of the week, the Amalgamated local announced that its members had voted to return to work. And the Union Stock Yards and Transit Company said it would resume operations with the AFL men on Monday morning, November 28. Faced by a PWOC threat of violence and a sympathy strike, the back-to-work movement failed to materialize. The way was then open for a settlement of the strike on December 4, 1938. [33]

The incident made evident the dangers in taking advantage of PWOC trouble—first, because of the unavoidable violence and sympathetic walkouts and, more important, because of the stigma attached to the label of "strike-breaker." The PWOC, in fact, made good use of this charge, whatever its truth, against the Amalgamated in Chicago. There were occasions in which AFL adherents ignored PWOC stoppages, for instance, during the sudden ten-hour strike at the Cudahy plant in Kansas City in September 1939. But the national policy was to keep clear of PWOC strife with the packers.

Company unions seemed another way of using conservative elements against the PWOC. The AFL sponsored a conference on August 9, 1938 to consider launching a joint drive in the Chicago stockyards. In his report afterward, President Green mentioned

some substantial company unions that . . . were making overtures to the Butcher Workmen to come over to that organization and become a part of it . . . If that was brought about it would form the nucleus of splendid organizations . . . It was agreed that responsible representatives would continue to carry on negotiations with these independent organizations for the purpose of working out a plan by which they would come over in a body.

These hopes materialized. The Amalgamated affiliated the Live Stock Handlers' Benevolent Association under the charter of Local 517, and in late November the Armour Employees' Mutual Association as Local 661.[34]

The drawbacks of this maneuver soon became obvious. For one thing, both independents had suffered major defeats at the hands of the PWOC in NLRB cases before entering the Amalgamated. Nor was it able to keep entirely intact the independent organizations in the transitional period. Finally, the action laid the Amalgamated open to the crippling accusation of attempting "to play company union for the bosses." The union president at the Armour plant, indeed, complained bitterly that the PWOC was labeling "the whole Local No. 661 as a 'Transformed' Employees Mutual Association and a company union." Not even the defection of the PWOC local president, Al Malachi, could invigorate Local 661. It lost the second NLRB election in December 1939 to the PWOC local by a four to one margin, and, as a further sign of the dubious benefit derived from the company union, received fewer votes than went into the "neither" column.[35]

Having learned its lesson, the Amalgamated avoided the reconstituted employee representation groups, even where the absence of the PWOC might have permitted collaboration in relative safety. Both unions, of course, sought the widest base of support. Gorman put the matter succinctly:

When we are in the fight we won't chase away from the ballot box anybody who wants to vote for us, even though he might be a stool pigeon for the company.

However, we know the C.I.O. is not so holy that they wouldn't welcome every stool pigeon of the company in their organization if through that they thought they could win.

Gorman's appraisal of the opposition was accurate enough. When the Amalgamated defeated the Swift Employees' Association in an NLRB election at the East. St. Louis plant, the PWOC began to enroll the company unionists. The PWOC organizer questioned the ethics of taking them without initiation fees and thus "scabbing on the A.F.L.," but he was ordered to go ahead if the Amalgamated could be beaten in that way. There were, in fact, instances—such as at the Rath and Swift-St. Paul plants—where the PWOC worked on company union leaders to bring their followings over *en masse*. The PWOC actually had greater freedom in this sphere, for its militant image rendered less credible the charge that it was a company pawn. Whatever its rival did, in any event, the fact remained that the Amalgamated could not use the company unions as such during the peak competitive period.[36]

The PWOC could draw into its organization the mass of packinghouse workers. But there remained the parallel task of forming the green ranks into a viable union. That had always been a tougher problem. The failures of 1904 and 1921 were, in part, ascribable to the Amalgamated inability to maintain effective control over the packinghouse membership. The problem was, if anything, more acute in the 1930's, for rank-and-file militancy was then of greater intensity. The PWOC could not avoid a turbulent period of divisiveness.

The course of dissension took its direction from the governing arrangements imposed on the packinghouse organization. Following standard procedure, the CIO had appointed an "organizing committee" to direct the drive in meat packing and to administer the new union in its formative period. The CIO official who became the first chairman of the Packinghouse Workers Organizing Committee, Van A. Bittner, had powers equivalent to those of the president of an international union. But he was answerable to the CIO Executive Board, not to a convention representative of the rank and file.

This system raised popular resentment. By 1940 the demand was widespread, as a loyal district director observed, for "a packinghouse workers' international union free from outside influence . . . an organization of, by, and for the packinghouse workers." [37] Inevitable under any circumstances, this sentiment was magnified by the situation within the packinghouse unions. They were just emerging from the heroic organizing period. Their fierce factionalism was the internal reflection of the fight against the packers. One Austin official expressed the democratic lesson learned from dismissals, picketing, and sit-downs: "The workers themselves are realizing today, as they never have in the past, that they, themselves, are responsible for their own organization." [38] How could that view be squared with the imposed regime of the PWOC? Such rule, moreover, ran counter to the ambitions of local leaders. From the moment of its birth, in fact, there had been resistance to the PWOC because it lacked autonomy. That sentiment was bound to increase with the passage of time.

For his part, Chairman Bittner did little to smooth his own way. Since the other members of the first committee were inactive, he ruled alone until June 1939. His vice-chairmen thereafter—Nicholas Fontecchio and then John Doherty—were, like Bittner, Mine Workers and SWOC officials. No packinghouse worker was on the committee before May 1941. Bittner, moreover, was something of an autocrat, keeping his own council, refusing to divulge financial data, and dealing with directors and organizers at will. "The present set-up is a machine," one Austin man charged. "They hire and fire, and hire who they please. It is by the national office and not the packinghouse workers." [39] The number and scope of packinghouse conferences gradually increased. But that could not disguise the fact that the locus of power remained with Bittner. As a final affront, he was regularly absent to attend to other duties.

Collective bargaining strategy created another area of conflict. Bittner was an old hand who played a cautious and patient game. But, as in previous periods, the rank and file pressed for action and quick gains. Bittner's subordinates and the local leadership were green and volatile—a dangerous combination. The PWOC national director and assistant director, a Federal mediator noted during an Armour controversy in early 1939, "seem entirely at sea . . . without Bittner's counsel or direction." But their inexperience did not restrain them. "It is difficult to hold this group down," the mediator continued, or to "know which way this was going to swing." [40] The inevitable clash over strategy came several months later. In July 1939, a conference voted to strike Armour if the company did not sign a national agreement. At the height of the crisis, Bittner accepted—to the outrage of his subordinates—an Armour offer to negotiate on a plant basis. He "sold us out" by taking less than the "Blanket Contract as promised all of us Delegates," an Armour local leader later charged. "The only way we will ever get our Union firmly established is to strike the entire Armour chain." Bittner resisted such rashness in 1939 and later. And the hot-heads, in turn, charged that "this dictator maniac, Bittner . . . would sell us out under any terms if he could retain control of the unions." [41]

To counter this rising opposition, Bittner deliberately constructed an administration party. His later appointments fell into the categories of, first, seasoned CIO officials and, second, packinghouse leaders with a large regional following. Foremost among the latter group was Lewis J. Clark of Cedar Rapids. Business agent of the union at the Wilson plant since its creation in 1933, Clark was appointed to the PWOC staff in April 1940 and five months later raised to the directorship of District 3. This Iowa-Nebraska territory consequently became a Bittner stronghold. An ambitious and wily leader, Clark understood the realities of union politics. Thus his advice to win over the dissident District 2 in November 1940:

> For instance, there are a number of men in the Austin Local seeking power. If one was elevated to Assistant Director I feel that he would realize . . . his individual support . . . would naturally be in District Number 2, and he, of course being understood, must follow the national policy, would bend every effort to see that his own district was straightened out and cooperation on a unity basis would follow the same there as in other districts. Then too . . . [he] would be in a position to assist in selecting a District Director for that district, and then the local unions could not say that they were not allowed to participate in selecting their leaders.[42]

Although another choice was made, the logic of administration strategy was evident. Beginning with the selection of Arthur Kampfert of Chicago to replace Don Harris as national director in December 1939, appointments had been designed to create a dominant administration party.

The substantial success of this policy could not prevent growing factionalism. Shut out of the central PWOC structure, the dissidents retained local bases of support. The joint councils became "a breathing [sic] ground for disruption." [43] Moreover, summary dismissals tended to focus opposition sentiment and to create new issues. Bittner hesitated to fire certain Negro leaders for fear of raising the cry of race prejudice—as indeed did occur at their dismissals." [44] Finally, an administration party lent substance to the suspicion that Bittner, his denials notwithstanding, aimed at permanent control of the organization.

The diffuse factionalism crystallized into a coherent opposition in the aftermath of the presidential election of 1940. About thirty dissident locals led by Henry Johnson, who was still assistant national director, followed John L. Lewis in support of Wendell Willkie. This defiance of the regular

CIO and PWOC position also provided cover for an attempted coup against Bittner. On November 3, Johnson telegraphed the locals of the Nebraska-Iowa district:

Please be advised that our great leader John L. Lewis has given back to us our fighting brother Don Harris. You may expect him in your city this week end. I as personal representative of John L. Lewis in charge of the P.W.O.C. would appreciate your giving him your full cooperation and support.

Director Lewis Clark fought off "the Blitzkreig [sic] tactics." A resolution was adopted urging Harris (who may not have been privy to Johnson's plan) "to get here immediately as . . . we haven't seen anyone swim the Missouri River for quite some time." [45] The abortive revolt ended Johnson's career as a PWOC official, as well as that of his supporters on the staff. The dismissals in turn stiffened the opposition. Delegations of packinghouse workers pressed demands for an autonomous international union and the reinstatement of Johnson at the CIO convention in Atlantic City later in November and at the CIO Washington headquarters in early 1941.

By then, the PWOC was split wide open. The dismissed officers actively fomented opposition with personal visits and circulars that excoriated the national leadership and demanded thoroughgoing reform. The Johnson party controlled the joint councils of St. Paul and Chicago; and the rebellion spread to Kansas City, to East St. Louis, Fort Dodge, and some lesser points. The insurgent unions represented a considerable part of the total membership. In some cases, PWOC representatives were barred from meetings, and two of the locals boycotted the important Armour policy conference in March 1941. Worst of all, per capita was being withheld from the national office. A severe internal crisis gripped the PWOC in the spring of 1941.

The CIO had to intervene. Philip Murray appointed a committee—Allan Haywood, James Carey, and Powers Hapgood—to investigate the controversy. The outcome of lengthy hearings was the resignation of Chairman Bittner and Vice-Chairman John Doherty in May 1941. The new full-time chairman was J. C. ("Shady") Lewis. A Mine Workers' official like Bittner, Lewis combined the advantages of packing-union experience and uninvolvement; his connection with the PWOC as director of District 3—Clark's predecessor—had ended before the rebellion. A trusted John L. Lewis man, his appointment may also have been designed to reconcile the UMWA chief to the fact that the dissidents, who had been acting in his name and in support of his political position, were not reinstated. No immediate reforms altered the PWOC government, nor did any concession open the way to an international union. But Lewis Clark was appointed vice-chairman, placing a packinghouse worker—albeit of the loyalist group—in a policy-making position for the first time.

The revolt gradually subsided. Chairman Lewis and his subordinates toured the dissident centers, urging the resumption of per capita payments and promising reforms. Not having been restored to their posts, the insurgent leaders lacked a base to continue effective opposition; and others still on the staff, such as Vernon Ford, district director at Kansas City, were soon dismissed. In Chicago, critical local elections were won by the unity group. The PWOC, moreover, was entering a decisive stage in its relations with Armour. "This split within the PWOC," Lewis warned the Kansas City unions, "will seriously hamper progress, especially negotiations, and I am of the opinion the Packers paid the freight in promoting this split." [46] The national leadership gained support from the exigencies of collective bargaining, and then from the successful conclusion of agreements with Armour and Cudahy in September and October 1941. By then, ranks had fairly well closed.

The turbulence of immature organization, hampering although it was, lacked the fatal effect of earlier internal troubles of packinghouse organization. Two changes had occurred. The lack of autonomy, itself a cause of dissension, now permitted a resolution of the internal strife. Recourse was possible to a higher authority with binding powers—that is, the CIO—at the point of crisis. Second, the Wagner Act gave organization a new permanence. The CIO leadership did not have to choose, as did Michael Donnelly in 1904 and Dennis Lane after World War I, between surrendering to rank-and-file pressures and seeing the local unions melt away or break up in factional conflict. NLRB certifications gave the PWOC a hold on packing plants that could surmount divisiveness and temporary disaffection. Notwithstanding the intensity of the fight, not a single certified union was destroyed during the struggle. The packinghouse union fostered by the CIO had the capacity, not only to organize, but to survive.

"Our road in developing substantial unions among the Fig Four has been rather hard," Patrick Gorman confided in June 1940.[47] The utmost efforts of the Amalgamated were falling short in the major centers. The packinghouse workers there seemed ineradicably "CIO-minded." But the historic interest in meat packing could not be surrendered. The question then remained whether, given its lack of mass appeal, the Amalgamated could gain its ends in other ways.

Two advantages were available. Unlike its CIO rival, the Amalgamated had a decentralized structure and was therefore better equipped to reach the dispersed segment of the packing industry. The Amalgamated, secondly, could use employers for organizing purposes. There was, in principle, no difference between the rivals on this score; the PWOC was not averse to support from employers. But the Amalgamated was far more effective. The AFL union had a stronger arsenal of weapons to pressure employers, and, over the years and out of the retail experience, Patrick Gorman and his colleagues had become adept at dealing with management. These advantages, which were related, could not prevent the CIO successes in the big centers. But they did open to the Amalgamated a lesser—and increasingly important—portion of the field.

Fear of the CIO initially fostered the prospect that the Amalgamated would be the preferred union of the packers. "When dealings are had with A.F. of L. organizations like the Amalgamated Meat Cutters there seems to be a much better understanding of meat plant conditions, and a much more reasonable attitude on the part of the organizers in many cases," the *National Provisioner* observed during the first months of the CIO drive. PWOC strikes and threats reinforced the concern of the packers.[48] In the early period, the PWOC clearly seemed an uncertain and dangerous quantity to the industry.

For their part, Gorman and Lane stressed the business-like virtues of the Amalgamated. "If we are to sell our organization," Gorman told the executive board, "we have to prove that our organization is sound and our contracts are going to be religiously lived up to." The prohibition of unauthorized strikes was necessary to protect "our enviable reputation and our organization's integrity, sincerity and trustworthiness." Secretary Lane, for example, forbade a threatening walkout over unresolved grievances at the Cudahy plant in Jersey City.

We must be very cautious and not do anything that would cause the management to lose confidence in our ability to carry our part of our relationship through in a businesslike manner. If we were to lose the confidence of the management you may then find them doing things that would lend encouragement to the C.I.O.

Later, President Gorman warned the Cudahy local that "these threatened stoppages of work . . . weaken the possibility of our organization securing friendly contacts with the company at points away from New York City."[49] To that end, the packinghouse locals had to eschew contract violations and unauthorized strikes.

In addition, demands had to be reasonable. Seeking a first contract with Kingan in June 1937, Gorman sent the company a copy of a recent agreement with Oscar Mayer and Company: "There are so many advantageous features to this contract that I felt you might be interested in looking it over. Under its provisions there will be, we are sure, an absolutely harmonious relationship . . . during the life of the contract." Several months later, Gorman was using the Kingan agreement, now in effect, in the same way with Cudahy officials, "in hopes that you might be able to work out some understanding along similar lines." When the membership at the Jersey City plant made extreme demands in November 1938, the Cudahy general superintendent asked Gorman whether the proposed contract was "in accordance with your ideas as expressed last week." After securing a copy, Secretary Lane sharply rebuked the local. "You are drawing a contract that two sides must live under and unless it is reasonable you cannot expect mutuality that would create cooperation and friendship." [50] Amalgamated chieftains thus sought employer approval and, in due course, organizational gains.

These efforts, by themselves, had only limited success. The major companies, initially committed to fostering independent organizations, found any form of outside unionism suspect. Before this resistance could be overcome, the PWOC quieted the misgivings of the packers; public militancy did not interfere with realistic negotiations. Nor could the Amalgamated always implement the policy of moderation and restraint, particularly among the established packinghouse organizations, which enjoyed considerable local autonomy. Actions were taken, such as the Armour boycott in Seattle in 1940, which Gorman regretted but could not prevent. A number of serious strikes were fought over contracts in the late 1930's. An organizer during a Pacific Coast strike in 1939 thus remarked that "the Companies naturally concluded, while the C.I.O. always talk of striking they haven't had very many in the past months, our organization simply goes on strike and thats [sic] all there is to it." [51] Arguing against unity negotiations with the AFL in 1940 before the CIO Executive Board, Van Bittner claimed that both Armour and Swift preferred the PWOC to the Amalgamated.[52] In any event, by 1940 the Amalgamated had lost any real advantage over its CIO rival in the view of the packers.

But the Amalgamated was not dependent only on blandishments. The organization of the retail field, well advanced by 1940, made a formidable threat of the secondary boycott—that is, the refusal by union meat cutters to handle unfair meat. This tactic, as well as the regular consumer boycott, had been employed with considerable success in the two-year fight that be-gan in 1935 against Morrell & Co. Its willingness to accept a compromise agreement long after the actual end of the strike indicated the force of the boycott.[53] The weapon was used with increasing frequency and effectiveness in subsequent years.

The boycott was, of course, subject to a variety of limitations. The inter-national, to begin with, was reluctant to see retail disputes grow out of the boycott. One safeguard was a contract clause requiring retail employers to respect boycotts. But, in exchange, the unions had to accept restrictions, fol-lowing the general rule of no boycott unless alternative suppliers were avail-able to retailers.[54] In addition, boycotts were not easily mounted. They were mass efforts requiring, for maximum results, disciplined membership and strong leadership. The unionization of the chains somewhat lessened this problem. Considering a boycott against the Kentucky Independent Packing Company, President Gorman asked a chain executive "if you can consist-ently do so, inasmuch as the Kroger Company is a customer, you make some mention of our intention when you next confer with . . . [the] President of this concern." This could have a marvelous effect on a recalcitrant packer. Chain managements, however, were not eager to cooperate, nor the Amal-gamated usually willing to alienate them for the sake of a packinghouse union. Achieving their support was a delicate matter involving the circum-spection and diplomacy at which Gorman excelled.[55] And the boycott still en-tailed legal risks, especially in intrastate cases.[56]

The organizing boycott had a decisive effect within limited, well-organized markets. Packing houses in New York City, for instance, were ideal targets because their output went mainly to the thoroughly unionized Jewish trade of the metropolitan area. On December 3, 1938 an Amalgamated organizer visited the Wilson plant in New York City to request a contract. Otherwise, Wilson Vice-President James D. Cooney complained to Gorman, "the A.M.C. would declare a boycott against the meat produced at that plant." The Amal-gamated man "appeared to be very impatient and arbitrary about the mat-ter and unwilling to discuss any question other than the immediate signing of the proposed contract." He had some cause for confidence; the Armour and Cudahy plants in the metropolitan area had both been organized in this fashion. A local official later explained to Secretary Lane the events that led to the boycott on Wilson:

New York Butchers (Armour) never did pay dividends as you know after spending a whole year before the National Labor Board courts and meetings to bring the men around, but they would not respond. . . . Not until International

Vice President Belsky put the company on the spot with a threatened kosher boycott was a contract signed and then the members came into the New York Butchers Local 334 . . . Cudahy fought us for a period of nine months . . . but Vice President Belsky again put pressure with a kosher boycott and again a contract. Wilson was next, pressure . . . was laughed at, so we had to really apply the boycott, as our bluff was called so the fight was on.

In this instance, the Amalgamated did not prevail, but clearly the boycott was a powerful weapon.[57]

This did not apply to the major plants in the western packing centers, nor to the entire output of big multi-plant firms. The lesson, learned at the turn of the century, remained binding 40 years later. The Amalgamated leadership turned aside requests for national boycotts against any of the Big Four.[58] Still, the boycott strengthened the Amalgamated in an important part of the packing industry.

So the Amalgamated labored to enlist the employers in the unionization process. They cooperated because of a preference for the Amalgamated, or fear of its boycotting power, or, most commonly, a subtle combination of the two. The small independents were very vulnerable to Amalgamated tactics, and they responded accordingly. A Milwaukee sausage manufacturer, for example, observed that "it was thru the splendid cooperation" of the local Amalgamated officials and himself in May 1937 "that we got all the rest of the plants 100% (all but the big meat packers). We have 21 such plants here in Milwaukee." [59] The local meat and provision trade, supporting many small firms in every urban center, was the prime area of Amalgamated success in the packing industry.

Among the major independents, Kingan & Co. provided the most notable example of employer cooperation. Through the intercession of an Indianapolis railroad brotherhood official who had served with a high Kingan executive on the NRA Regional Labor Board, Gorman and Lane were enabled to visit the Kingan offices in April 1937 to explain the advantages of dealing with the Amalgamated. The response was favorable. Vice-President Jimerson told the executive board on May 10 that "the C.I.O. is active . . . However, Mr. Kingan would offer no objection to his employees going into our organization." The Amalgamated chartered Local 165 and began organizing work. Gorman wanted "to, as quickly as possible, reach an understanding with your Company, because of the very unsettled industrial conditions prevailing quite generally at present." The outcome was a contract for the Indianapolis plant on June 19, 1937, followed in the next few months by agreements for the company's smaller plants at Storm Lake, Iowa, and Richmond, Virginia.[60]

Of the Big Four, the Cudahy Packing Company seemed receptive to the Amalgamated. Promising conversations with company officials prompted President Gorman to tell the executive board in March 1938 of his expectation that Cudahy would "absolutely treat with us and move from one plant into another if we can get the members of the Board and the local unions in their respective districts to cooperate in this program." That summer, Gorman mentioned a clause in "the agreement we are negotiating with Cudahy and Company that would pledge . . . all possible cooperation to the Union to bring about organization of the employees." The Cudahy arrangement, however, was never fully realized, in part because of the departure in mid-1938 of General Superintendent R. E. Yokum.[61] In the end, the company actually responded only in plants—in New York, Los Angeles, and Portland, Oregon—where the union was able to exert effective boycott pressure. That held true also for Armour; but Swift and Wilson were more resistant. On balance, nevertheless, employer cooperation had a substantial role in the Amalgamated organizing effort.

The Wagner Act raised a barrier to this strategy. Swift, for example, complained in 1939 that in Los Angeles the Amalgamated was "asking that we violate the law": "the sole reason of the boycott is that we have declined to force our employees to join your Union." In exchange for lifting the boycott, Gorman could get no more than an assurance that the company "certainly would not place any obstacle in the way of the employees' . . . right, as provided under the National Labor Relations Act, to become affiliated with the Union." (A year later, the PWOC received NLRB certification for the Swift plant.) For its part, the union acknowledged the limitations imposed by the law. During the contest with the PWOC at the Cudahy plant at Kansas City, a suggestion was made to persuade the company "to coyly let [the] threat be circulated among the employees" that a CIO victory would cause the transfer of several departments to other plants. Gorman demurred on the ground that such statements "would make a labor case . . . as coersion [sic] on part of the company toward a favorable vote for our organization."[62]

Still, scope remained within the bounds of the Act. A New York organizer brusquely dismissed the possibility of action by the NLRB: "We are not children and we know that if you sign this contract and go out and post a notice in the plant you have signed we know what all the employees will do about it." Gorman disapproved this crude approach, but he felt the employer could not be entirely passive. He answered the complaint of Wilson Vice-President Cooney (who had quoted the organizer):

I am not unmindful that your company, under the Wagner Act, is not in a position to force employees to join any union, but, Judge Cooney, employers are faced with employee relations problems most every day. It is generally conceded that union labor is here to stay. That being true, isn't it better for us to meet these problems in a way that will prove most advantageous to both the union and the company involved. Surely we can keep within the law in accomplishing our purpose.[63]

Since employers could not be absolutely neutral, the Amalgamated wanted to be the beneficiary, not the victim, of managerial influence.

This took many forms. A packer could be asked to stop the "undue opposition" of a plant manager; or direct him to "cooperate with us at least to the extent of our privileges under the National Labor Relations Act"; or "pass the word down to the superintendent . . . that the boys immediately make application to Local 537"; or order no delays when the union desired a quick election and the consummation of an agreement.[64] Rarely did these maneuvers lead to unfair labor practices action by the NLRB. Employees were not "turned over" to the Amalgamated. Employer cooperation was in conjunction with organizing work already in progress by the union. Generally, too, it was a situation in which the CIO posed little or no threat. Essentially, the Amalgamated was seeking, by employer approval as well as by threatening employment through the boycott, to create a condition whose logic was for employees to enter the union.[65]

The effectiveness of the Amalgamated tactics could be seen at the Kingan plant in Indianapolis. The CIO was already active there—in this respect an atypical situation—when contact was made with the Kingan management. The talks having gone well, the Amalgamated dispatched Vice-President J. P. McCoy to direct an organizing drive. In June 1937 a contract was signed. The CIO at Indianapolis, of course, vehemently protested this sudden turn of events and lodged charges with the NLRB. The outcome was a consent election scheduled for April 15, 1938. By then, Amalgamated Local 165 had entrenched its position. The Amalgamated had the concrete advantage of a signed contract which, although not changing wages, provided gains in paid vacations, seniority, and overtime. (The PWOC local in the nearby Armour plant, on the other hand, still lacked a negotiated agreement.) The company attitude also had intangible benefits for the Amalgamated. Among other things, a rumor persisted that Kingan would close the plant if the CIO won the election. The result was a resounding victory for the Amalgamated, 1198 to 420 for the PWOC. A union shop in the next contract gave the union the apparent final safeguard for permanent control.[66]

Such gains, however, could not always be held. Designation as a bargaining agent under the Wagner Act was a contingent status; it lasted only while the union retained the support of a majority of the employees in a unit.

Contracts, moreover, proved pregnable as defenses against PWOC raids. The Amalgamated learned the painful lesson at an early point. Its national boycott against Morrell had finally resulted in agreement on March 9, 1937. The gain vanished when an independent won an NLRB election in the Sioux Falls plant four months later. Answering an Amalgamated charge that this cancelled a valid contract, the regional director cited the *New England Transportation* case: an existing contract did not bar an election; the contract remained in effect even if the bargaining agent changed. The boycott had been waged to no apparent purpose.[67] Union-shop contracts offered greater protection. Although its position still was undefined by 1940, the NLRB skirted the issue in Amalgamated cases by finding that the contracts had ended or been renewed after a representation question had been raised.[68] A possible countermeasure was to put a reopening clause, rather than a termination date, into contracts. The Amalgamated did this with Kingan in the 1939 agreement; and the NLRB ruled that the union-shop clause was a bar to an election the next year. But, despite a growing CIO threat, the union had to ask for a reopening in 1941 because of rank-and-file pressure for contract improvements. The fact was that the PWOC could not be prevented from testing the Amalgamated at its weak points.

Amalgamated vulnerability was part of the price for employer support. The Amalgamated could not respond to rank-and-file pressures for "action" and extreme demands; sometimes, in fact, the union had to permit gains to be surrendered. For example, the Kingan contract of May 28, 1938 permitted wage reopenings which could go to arbitration. Within a month of the signing (as Gorman had then suspected), the management requested a 20 per cent reduction. In August 1938 an arbitration board awarded the company an adjustment of $6\frac{1}{2}$ cents an hour for men and $4\frac{1}{2}$ cents for women. One man, writing to Secretary Lane, expressed the reaction of the membership:

The Union was all right, we took a vote on the cut, 20 out of 1700 voted for the cut. Well what did the Union do . . . They had an attorney and meet [sic] in a big building here in the city and all there was to that was Kingan and the Union got together and the Union sold out to Kingan. It is plain as the nose on your face. We took a 6½ cent cut, that is what we got. . . . Before I pay any more dues, they will have to show me they are not a bunch of crooked half wits.

The international stood its ground against pressure for a change in the contract "to gain the privelege [sic] of striking to use as a weapon in arbitration for regaining our cut in wages." President Gorman emphasized the sanctity of contracts: "If we would do this we would not deserve the confidence of a single employer throughout the United States, and . . . it would be a clear demonstration that your membership will not live up to an agreement entered into in good faith." [69] The Amalgamated barely managed to save the Kingan local in the subsequent election in 1939. In the early period, the conciliatory policy of the Amalgamated tended to make its plants vulnerable to the rival union.[70]

The PWOC brought to bear on these weak points its full appeal. In its four-year campaign at Kingan, the PWOC accused the Amalgamated local of being a "company induced union," of "sell-out" on the wage cut, of timidity in grievance cases, and of dictatorial leadership. In contrast to segregated dances for Kingan workers in Indianapolis and segregated locals in the Richmond-Kingan plant, the PWOC publicized its policy of racial equality. The quiet defection of the Negro vice-president of the Amalgamated local probably determined the preference of the colored workers. On July 17, 1941, the PWOC finally took the Kingan plant by a vote of 1096 to 847. Having repeatedly castigated the "pro-company policy" of the Amalgamated, the PWOC now proceeded to negotiate precisely the same agreement with two minor changes. But that fact did not weaken its hold; nor did the dissension and venal leadership that soon gripped the local union. Election losses by the Amalgamated in 1942 and 1944 showed that the Kingan plant was irrevocably in CIO hands.[71] In a number of other instances, most notably in Los Angeles and New York City, the CIO appeal similarly overbore early organizational gains of the Amalgamated.

The great period of organizing had come late to meat packing. The union drives had started only in 1937 and had then been impeded by the 1938 recession. Thereafter, the organizing pace quickened under the stimulus of a defense economy and collective-bargaining successes and reached a climax in 1942–43. The bulk of the industry was organized before the end of the war. The small independents of the south, as yet not a significant factor, remained largely unorganized. Of the important plants, only the Wilson house in Oklahoma City was not recorded on the list of NLRB certifications.

The major plants of the industry—both of the Big Four and the important independents—fell mostly to the PWOC. This area responded best to the CIO appeal and tactics. It was also where the PWOC concentrated its efforts. "We felt if we could organize the Big Four," commented one official, "the others would fall into line." [72] The initial successes were with Armour & Co., followed by Cudahy and Wilson. Swift was the last of the giants to be penetrated by the CIO. (A subsidiary, the Plankinton Packing Company of Milwaukee, had been organized much earlier.) The first Swift certification came at Los Angeles in December 1940, followed by many others, usually by the defeat of independent unions. Fifteen Swift plants were in the PWOC fold by May 1943. But the National Brotherhood retained nine Swift units. The major plants owned by the leading independents, such as Morrell, Hormel, Rath, and Hygrade were also taken by the CIO.

The Amalgamated had minimal success in this area. The only major unit in the Big Four chains with AFL certification was the Swift plant at East St. Louis. A fierce three-sided fight had raged here in 1941. The PWOC was eliminated by the NLRB after two ballots, leaving the field to the Amalgamated and the Swift Employees' Association. Effectively led by John Powderly, Local 78 defeated the SEA by a narrow vote in December 1941 and won the big plant for the Amalgamated.[73] After an earlier defeat, the union was also certified in 1939 to represent the 1800 workers at the Sioux Falls plant of Morrell & Co. The Amalgamated profited here from the persistent work of Sam Twedell, the local leader, and from its reputation gained from the recent strike and boycott against the company. The Amalgamated also had success with the substantial plants of the medium-sized independents, such as Oscar Mayer, Mickelberry's Foods, and E. Kahn's Sons.

The prime AFL achievement, however, was in the smaller production units of the industry. The master agreement with Armour after the war covered twelve such plants, all of them away from the major centers: Baltimore, Columbus, Grand Forks, Huron, Peoria, Pittsburgh, Portland (Oregon), Prairie du Chien, Reading, San Francisco, and Spokane. The Amalgamated had sixteen plants of the other members of the Big Five (including Morrell), and represented an estimated 15 per cent of their total labor force. Greater success was experienced among the independents, particularly the many small firms serving local markets. According to Gorman, more than 500 independent plants were under Amalgamated contract in June 1945.[74] Despite its concentration on the big plants, the PWOC also worked the dispersed section of the industry. Canada, moreover, was a relatively clear field for the PWOC, since earlier failures had resulted in an Amalgamated decision to stay out of the

Dominion. The PWOC consequently built up a substantial roster of unions outside the packing centers, reportedly with agreements covering 65 independent plants in October 1941.[75]

Only a rough approximation can be made of the division of packinghouse workers after the main surge of organization. In May 1943, the U.S. Bureau of Labor Statistics reported the labor force at 154,000 in slaughtering and meat packing. The PWOC paid per capita to the CIO in May for 53,505 members. The number represented by the PWOC was of course still larger, since it then had few union-shop contracts. Probably 90,000 packinghouse workers were employed in plants which were certified to the PWOC. The Swift Brotherhood was bargaining agent of nearly 10,000. The Amalgamated paid per capita to the AFL in May 1943 for 95,000 members, of whom no more than 25,000 fell within meat packing proper; several thousand additional non-members were employed in plants for which the Amalgamated was bargaining agent.[76] In sum, then, the CIO represented about 60 per cent of the packinghouse workers, the Amalgamated 20 per cent, and the Brotherhood 7 per cent at the end of the great organizing period in the middle of World War II.

This was not a fixed relationship, for meat packing was changing. The industry was proceeding away from the characteristic of concentration—in plant size, in the domination of the great packing centers, and in the relative importance of the giant firms. The transformation would revise the prospects of the competing unions. No one could foresee in 1943 the dimming future of the triumphant CIO in meat packing.

CHAPTER 10

THE FIGHT FOR COLLECTIVE
BARGAINING, 1938–1945

Organizing the packinghouse workers carried the union task only to the halfway mark. There remained the achievement of authentic collective bargaining. In the past, that had been the point of fatal failure in the unionization process. Now success was within reach. But the campaign would be long and arduous. Only at the end of World War II was a viable system of collective bargaining in sight; and only then could the trade unions become confident of their future in meat packing.

The accustomed hardness toward trade unionism did not subside when the big packers realized that they could not prevent the organization of their plants. Rather, their opposition was channelled against collective bargaining. Historically, in fact, that had been the decisive area, for there had been periods—1901–04, 1917–19, 1933–34—when the organizing process had been irresistible. After 1904, the packers had developed the policy of refusing "recognition," that is, to have any union dealings. Trade unionism had been broken by that measure. Its use in pure form was now forbidden. Section 8(5) of the Wagner Act required employers to negotiate with unions which were certified bargaining agents for their plants. Even the most hostile managements, therefore, were prepared to *talk* to union representatives. The law also forced them to discard the practice, common during the NRA period, of treating unions as spokesmen only for their own members rather than as the exclusive agent for a bargaining unit.[1] Still, it seemed possible to withhold from the unions any meaningful role as bargaining agents and hence ultimately to eliminate them.

Resistance took a variety of forms. For instance, after several meetings the Neuhoff Packing Company of Nashville, a Swift subsidiary, in July 1937 rejected all demands of the Amalgamated local, refused to make counterproposals, or even to incorporate into an agreement the existing terms of employment (on the grounds that these were subject to change). In the early union period important alterations generally resulted from unilateral action rather than negotiation. The packers also withheld the forms of collective bargaining. They refused to reduce understandings with unions to contractual form. Armour & Co. simply issued "Statements of Policy." Wilson made only verbal agreements, so that, as unionists complained, they "have to be fighting back and forth and discussing things over that was [sic] already settled times before." Even when agreements were in written form, none of the Big Four would affix a signature to the contract.[2] These evasions of the intent of the Wagner Act were gradually remedied. The NLRB and the courts ruled against a variety of employer strategems, including the failure to make counteroffers, to sign agreements, and to consult the union before changing the terms of employment.

The packers still had ample leeway, for the Wagner Act was a less effective guarantee of collective-bargaining rights than of the right to organize.[3] To begin with, the former involved the commission of, rather than omission of, actions. Delays and postponements were more damaging, and there were difficult problems of definition and proof. The law, as it was interpreted, required employers to bargain "in good faith." But how was lack of good faith to be proved? From the union standpoint, moreover, the Wagner Act had a limited range. "From the duty of the employer to bargain collectively," the Supreme Court noted in the *Sands* case, "there does not flow any duty . . . to accede to the demands of the employees."[4] This applied not only to substantive issues—wages, hours, and conditions—but to such vital questions as union security, arbitration of grievances, and master contracts. The unions quickly concluded that, while the NLRB had some value, it could not be the basis for achieving satisfactory collective bargaining.

Resistance did not extend significantly to the lesser firms. They generally entered signed agreements with the Amalgamated or the PWOC, including provisions for the union shop and in occasional instances the check-off. The boycott power of the Amalgamated permitted progress even against certain Cudahy, Armour, and Swift plants. The terms of agreement were reduced to writing and sometimes were accompanied by an exchange of signed letters. Finally, the Amalgamated did get signed agreements for Big Four plants on the West Coast. Some advance was also made toward union security. The

Amalgamated failed to carry this point against Swift in San Francisco in an extended strike in 1939, but informal understandings and "harmony" clauses did enlist company assistance in dues collection and union membership in some Big Four plants. San Francisco and northwestern plants of the major packers also granted wages and hours gains, responding both to Amalgamated strength and the local labor market on the Pacific Coast.

The Big Four and some major independents held firm in the big plants of the Middle West, and here collective bargaining itself seemed at issue in the early period. The struggle included a number of points—bargaining "in good faith," signed contracts, union security, arbitration of grievances—but the prime issue became master agreements. National contracts seemed the only effective arrangement with a multi-plant company, both for bargaining and organizing purposes. But, beyond that, master agreements would signify the standing and permanence of unionism in the Big Four. For that very reason the packers committed themselves stubbornly to local negotiation. Master contracts were the issue around which crystallized the struggle for collective bargaining. And since the resistance came in the area of CIO strength, the PWOC was the union which had to grasp the initiative in the fight.

Armour & Co. was the first target. On September 25, 1938, a conference of Armour delegates directed Chairman Van Bittner to request a meeting with the company. Armour President Robert H. Cabell summarily dismissed the proposal. "We cannot see at this time that any purpose would be served through a conference with your Committee." [5] Armour thus established its opposition to relations above the plant level. The PWOC was in no position to force the issue. The opening move nevertheless served to define the line of conflict. The issue of a national agreement figured in the PWOC victory in the key Chicago plant, and stimulated organization elsewhere in the next months. By April 1939, five more plants had been added to the eight already certified in September 1938, leaving few important Armour houses outside the PWOC.

The recession of 1938 having passed, the PWOC began in earnest with another request on April 24, 1939 for a joint conference with Armour. Avoiding an earlier tactical error, union officials proposed a master agreement only for plants certified to the PWOC, contending that these "constitute an appropriate bargaining unit." When Cabell again refused to meet, the PWOC began to raise strike sentiment. A National Policy Conference was scheduled for July 16 in the Chicago Coliseum. The PWOC "seems to be developing quite a build up," a Federal mediator reported. "The mass meeting, on that evening . . . may develop into just the push needed to start the real fireworks." [6] That did not quite happen, despite a fiery speech by John L. Lewis. But a national strike was authorized. The packinghouse workers, Lewis warned, "are now serving notice on Armour & Company that their patience is nearing an end, and that if the company continues to refuse collective bargaining, it must accept the consequences of its own actions." [7]

The threat was actually not directed at Armour. Strike sentiment within the PWOC was real enough. But Chairman Bittner, a cool-headed, experienced trade unionist, knew that the PWOC could not hope to mount a successful strike against Armour; the company might, in fact, welcome a walkout as the opportunity to break the union in its plants. Rather, the strategy was, by manufacturing a crisis, to put public pressure on the company to enter negotiations. This was why the PWOC was threatening to strike, not Armour alone, but all of the Big Four. This also explained the assiduous efforts to enlist public support. The Chicago City Council did in fact make an unsuccessful attempt to persuade Armour officials to meet with the union. [8] But the main hope of the PWOC resided with the Federal government.

The Department of Labor had been in touch from the outset. As early as February 18, 1939, national officials had conferred in Des Moines with Labor Secretary Frances Perkins and Dr. John R. Steelman of the Conciliation Service. A Federal mediator was promised if Armour rejected another PWOC request for a conference. That having transpired, Director Don Harris and Assistant Director Henry Johnson hurried to Washington in early May 1939. They went, the mediator accurately surmised, "with the hope of being able to put pressure upon agencies there to bring pressure upon the company to sit down and bargain with them." In the following weeks, PWOC and CIO officers warned several cabinet members, and even the president, of the threatening situation in meat packing. The emphasis was on the magnitude of "the problem confronting the Nation and the New Deal if a National strike takes place." [9]

This strategy bore fruit in August 1939. After being informed by a delegation of the imminence of the strike, Secretary Perkins sent a wire to H. S. Eldred, the Armour general manager. The executive agreed to be in Washington the next day, August 18. The groundwork for a compromise had already been laid by the quiet work of a Federal mediator. Bittner had retreated a few days earlier on the issue of a master agreement, although still

demanding a general conference; for their part, company officials had indicated a willingness to engage in real plant-by-plant negotiations. Nevertheless,
the PWOC needed Federal intervention as an outside guarantee of company
good faith. Secretary Perkins did extract a pledge from Eldred. This, as she
restated it, was "your extension of authority to your local managers to bargain
collectively and reach agreements with representatives of the local unions
(and such officers and advisers as they desire)." But Armour officials refused to confront the union representatives in Washington (separate meetings were held), or to schedule a general meeting afterward. Nor would
they bow to Secretary Perkins' plea for signatures on the plant agreements.
Bittner acceded to Armour's position; he had attained at least the minimum
PWOC requirement of a promise of authentic collective bargaining. The
test would come at the negotiations scheduled to begin at the main Chicago
plant on August 23, 1939.[10]

Amalgamated officers had followed these events with mounting concern.
If the PWOC succeeded, the AFL cause in the packing industry would be
irreparably damaged. Counter-measures were imperative. President Gorman
used the AFL radio station in Chicago in late July 1939 to refute statements
made by John L. Lewis at the PWOC rally. The Chicago City Council was
urged to be neutral. But, above all, the Amalgamated had to make itself
heard in Washington. Gorman complained to Secretary Perkins that the
PWOC "is making a deliberate effort, in our opinion, to use you and the
Department of Labor to save for them the Armour situation." But that fact
did not enable the Amalgamated to inject itself into the controversy. When
Gorman demanded an interview with the secretary, Dr. Steelman replied
that the threatened strike—the basis for public intervention—concerned only
the CIO plants. Both PWOC and Armour representatives had assured the
Commissioner of Conciliation that the Amalgamated, while it represented several company plants, had no dispute with Armour. "I hope this is the situation," was Steelman's ironical conclusion. The Amalgamated thus seemed
shut out of the decisive events in Chicago and Washington.[11]

One opportunity, however, still remained to blunt the PWOC drive against
Armour. Aroused by CIO successes in the Chicago yards, the Amalgamated
and its Chicago allies had launched a rival drive in 1938. The Amalgamated
was claiming in the summer of 1939 a majority in the Armour plant, notwithstanding the certification to the PWOC. Gorman warned the company on
August 23, the first day of negotiations at the Chicago plant, "that any agreement entered into with the C.I.O. would be contrary to the wishes of your
employees and which, if consummated, would force us to bring this whole

matter to the attention of the convention of the American Federation of Labor." [12]

The threat was needless. Armour itself was refusing to accept the NLRB certification of the PWOC at the Chicago plant because the vote had not given the PWOC a majority of the normal 6000 employees in the plant. Despite the *R.C.A.* decision that a majority was to be based on the total number of cast votes, the company was taking the case into the Federal courts. Negotiations at the Chicago plant therefore quickly broke down because "the management . . . would not give us sole collective bargaining." The PWOC again resorted to the threat of a national strike. Rank-and-file pressure was patently real. Outbursts occurred in Chicago and Indianapolis, and a general walkout seemed imminent. Again responding to the union plea, Secretary Perkins telegraphed Armour President R. H. Cabell on September 8, 1939 to attend a joint conference in Washington. The company at first demurred, claiming that the union had not exhausted, or even used, the remedies available through the NLRB. Public pressure finally prevailed, and an Armour delegation came to Washington on September 19, 1939. [13]

The next days saw the decisive compromises involving the three parties. Armour conceded further guarantees for collective bargaining. Federal mediators would be present during plant negotiations, and, more important, responsible executives would participate in the bargaining sessions to avoid what the PWOC called "buck passing" by the local managements. [14] The PWOC, however, had to make an important counterconcession on the key Chicago plant. The Amalgamated had finally presented its case to Secretary Perkins on September 20. Gorman and Joseph Keenan of the Chicago Federation of Labor made a persuasive claim to the Chicago plant. As a result, Van Bittner later told the CIO Executive Board, "the Secretary of Labor argued with me that in her opinion the A.F. of L. had a majority of those workers, and for that reason it was awfully difficult for her to do anything in the situation." [15] The Amalgamated maneuver, plus the company's legal course, resulted in the PWOC consent to a new election at the Chicago plant.

At the critical point, the Amalgamated secured the chance to block the consolidation of the PWOC gains. The high stakes intensified the bitter campaign that ensued in the Chicago yards. Despite its great effort, the Amalgamated was engaged in a losing cause in the Armour plant. Union lawyers sought to delay the election toward which the NLRB was proceeding, as one of them remarked, with "unholy haste." But on November 4 President Gorman called off these legal efforts in order to "get the matter over with." [16]

His pessimism was justified. The PWOC won overwhelmingly on November 21, 1939: 4006 for the CIO, 1047 for the Amalgamated, 1254 for neither. This was the point—insofar as a specific place could be marked—where greater popular appeal guaranteed the success of the PWOC.

Pending negotiations now could move to a rapid conclusion. In January 1940 the PWOC reached agreements at two plants that became the basis for the agreement at the main Chicago plant two weeks later, followed one by one at the other PWOC units in the Armour chain in subsequent months. The plant negotiations finally ended in September 1940 with an accord at Fargo, North Dakota, where an earlier strike had added complications. The agreements improved seniority, vacations, the weekly guarantee, and the grievance procedure. Above all, the PWOC received recognition as the sole bargaining agent and the first substance of collective bargaining in the involved plants. "It marks the acceptance of collective bargaining by the first of the Big Four companies," exulted the union's paper.[17]

The limited nature of the achievement was obvious. The PWOC did not get a national contract, although substantially uniform terms were reached at all the plants. Nor had the company signed the written agreements. No provisions were included for any form of union security, and wage rates, hours, and overtime remained unchanged. Yet Van Bittner, according to a Federal mediator, greeted the first accord at Kansas City with "extreme delight." [18]

The modest victory excited his enthusiasm because he knew the actual weakness of the PWOC. Despite its brave front, the union could not have struck Armour with any hope of success. The company was in fact responding less to PWOC strength than to public censure, Secretary Perkins' influence, and the evident vulnerability of its bargaining policy under the Wagner Act. Bittner had good reason to be pleased with the outcome—that Armour was "ready and willing to do business with our organization." [19] Swift and Cudahy indicated an intent to follow suit. Collective bargaining was, however, far from full attainment.

Time favored the CIO union. The defense prosperity arrived, and packinghouse profits rose: the net return, before taxes, of Armour & Co. went from 4.9 per cent in 1939, to 5.4 per cent in 1940, to 11.2 per cent in 1941.[20] By the same token, the labor market was tightening for the first time in over a decade. Finally, the Federal government began to take a more direct hand in labor relations. Secretary Perkins had played a significant role in the Armour affair with very limited powers. In 1941 and after, the need to maintain full production greatly enhanced the authority of Federal agencies in American labor relations.

The renewal of the Armour drive came early in 1941. Contract negotiations had to begin between local plant and union officials under the Armour agreements. The local unions, therefore, were instructed to make "whatever demands and thoughts our people have in mind in the various plants." The presentation of maximum proposals wold serve, first, to convince the company "that our local people are having a first-hand interest," and, second, to shift the negotiations from the local level. The plant managers responded in the expected negative fashion, making in some places only the barest pretense that they had the authority to negotiate basic issues.[21] The second phase began with the convening of an Armour National Wage Conference in Chicago on March 23, 1941. A plan was devised to undermine Armour's continued insistence on plant-by-plant negotiations. The Armour agreement called for national representatives from both sides to enter plant negotiations if local officials failed to reach settlements. Vice-Chairman John Doherty suggested that each Armour union be represented at the resumed local negotiations. To keep within the technical meaning of the agreements, these delegates might have to be designated "National Officials." But the group would actually be "a national negotiating committee and would carry behind it the full bargaining pressure of all Armour locals." Since the initial accord would be the model for the entire chain, the PWOC would have in effect achieved a national agreement.[22]

This strategem did not fool the Armour officials. They refused to deal with the National Negotiating Committee when the first meetings began at the Omaha plant in early April 1941. The PWOC fared no better on the substantive issues. Then on April 23, while talks were recessed and without even prior notice, Armour in unison with the others of the Big Four announced a voluntary increase of 8 per cent. This unilateral action, beyond being an inadequate raise from the union viewpoint, was a direct affront to the PWOC and a denial of its function as a bargaining agent (and also a violation of the Wagner Act). Negotiations were nevertheless resumed. The hope was that, once differences were reduced to a hard core, the company would recognize that plant-by-plant negotiations "are futile insofar as the changes in the basic fundamentals . . . are concerned." [23] But Armour remained adamant both on a national agreement and on questions of substance. On July 10, 1941 the PWOC finally broke off the negotiations.

The deadlock forced another appeal to Washington. Chairman J. C. Lewis informed Secretary Perkins on July 11, 1941 that a strike was imminent in the Armour chain. The secretary's failure to act prompted Lewis to telegraph her on July 24:

Now in order to protect the interests of the United Packinghouse Workers of America I am recommending to them a general strike throughout the Armour and Company Packinghouse Chain.

That brought a response. Secretary Perkins certified the case to the National Defense Mediation Board two days later, and the parties came before a panel of the Board on August 7, 1941.[24]

Two days sufficed to break the stalemate. Armour surrendered on the key issue. A master agreement would be acceptable to the company (it had already indicated that it would sign agreements). Thus, Chairman Lewis jubilantly reported, the PWOC fulfilled "a standing demand of our Union since its inception" and opened "the only way genuine collective bargaining on all matters can be realized."[25]

Agreement on two other issues at the NDMB hearing eased the subsequent negotiations for a national contract. The minimum male rate at Chicago was set at 72½ cents an hour, an increase of five cents, and other rates were shifted upward accordingly. The grievance procedure permitted an employee, if he desired, to be accompanied by a steward, and as a final step either party could request a Federal mediator (the PWOC had desired arbitration). Direct negotiation then proceeded without a hitch. The PWOC had tacitly dropped the demand for union security, and it also conceded the overtime issue. On September 6, 1941, Armour & Co. and the PWOC signed the first master contract in the annals of the packing industry.[26] The historic achievement was extended to Cudahy on November 1, 1941, and on April 2, 1942, Swift unexpectedly informed the PWOC of its willingness to enter negotiations for a master agreement.

The packers were moving with the tide of events. They were unwilling to jeopardize the highly profitable operations which had at last come to the industry. In addition, defense needs made the interruption of packinghouse production a matter of keen public interest. The resistance to organized labor was crumbling in the open-shop strongholds of American industry (for instance, Ford and Little Steel). In the packing industry itself, the surrender of Armour & Co. exerted a compelling influence on the remaining Big Four.

Positive reasons also entered into the decisions to accept master contracts. Local bargaining proved a costly way of resisting organized labor. Notwithstanding industry claims, labor decisions were not made at the plant level. The union strategy of presenting uniform demands at all plants made local bargaining an increasingly pointless, but consuming, process for the busy executives of the packing companies.[27] Local irresponsibility also argued for master contracts. Cudahy was clearly gratified by the work of Chairman J. C. Lewis against wildcat strikes at several plants during the summer of 1941. "We appreciate very much your attitude in the Kansas City case," the general superintendent told Lewis in the telephone conversation confirming the Cudahy decision to negotiate a national agreement, "and we will do all in our power to do things on a fair basis and know . . . that you will work with us in that way." A master agreement would strengthen Lewis' hand against the local unions by tying in the national office both with collective bargaining and the grievance procedure. Beyond these specific considerations, the PWOC was clearly becoming a permanent institution with which the packers would have to live. The acceptance of national contracts reflected that realization.[28]

The foregoing facts evoked no response only from Wilson. Deeply antipathetic to organized labor, its management adhered rigidly to the original position of the Big Four. By 1942, the PWOC had not achieved either written or signed agreements in any of its five Wilson plants—not to mention a master contract. The company submitted only under the direct compulsion of the Federal government. Over the company's objections, the National War Labor Board ordered Wilson to negotiate a master agreement with the PWOC. A contract was finally signed in March 1943. The architect of the company's labor policies, Judge James D. Cooney, later remarked that "Wilson would have resisted a master agreement if it had not been for the war. Such an agreement is an unsatisfactory arrangement; today or yesterday." [29] This attitude exacerbated company-union relations. But it could not prevent Wilson from being drawn after the other major packers into the pattern of national bargaining.

Four years were thus consumed in the achievement of the basics of collective bargaining with the major firms: authentic negotiation over the terms of employment; written and signed agreements; and, above all, national contracts. This was only the frame of collective bargaining. The processes and substance of collective bargaining had yet to be defined, and the relations among the parties placed on a stable basis. As it happened, the industry was not permitted to plunge directly into this precarious formative period. War intervened.

World War II imposed two vital demands on the domestic economy: uninterrupted production and wage-price stability. The execution of these tasks injected the Federal government into the private sector of labor-management relations. The impact was immense. For better or worse, the war left a permanent stamp on meat-packing unionism.

When hostilities opened, collective bargaining was just starting between the Big Four and the PWOC. The shaping experience came under controlled wartime conditions. Collective bargaining in its full sense could not proceed. The National War Labor Board had to hand down decisions, dated February 8, 1943, and February 20, 1945, that determined the basic terms of employment in the industry. But government intervention did not entirely wipe out collective bargaining. Negotiation between the parties both preceded and followed the board decisions, and, limited although this was, it provided the training and consensus vital to collective bargaining.

Disputes were certified to the War Labor Board only after the exhaustion of collective bargaining. Even then, the preliminary panel could attempt mediation before submitting recommendations to the board. The first round of Big Four negotiations, begun with Armour in March 1942, made negligible progress, but the mediation work of the WLB panel in July and August had substantial success. Armour and Cudahy were able to sign new contracts, leaving for the board only the issues of overtime, wages, union security, and, in the case of Cudahy, vacation pay. Negotiations with Swift advanced even further; the company acceded to the demand for time and a half after eight hours in exchange for union concessions on vacations and guaranteed time. Wilson & Co., as always, was "the hardest to deal with." [30] The panel itself had to draw up the first Wilson contract on the basis of the company-union discussions and the master agreements of the other packers.

The second round of opening negotiations had even less success. There had been a deterioration of CIO relations with the packers, especially with Armour, as a result of unresolved grievances and wildcat strikes. Dissatisfied with the WLB award of February 8, 1943, the PWOC submitted a wide range of demands to the packers at the expiration of the contracts on August 11, 1943. The companies, in turn, submitted tough counterdemands. Relatively fruitful talks with Swift & Co. resulted in a signed contract on December 14, 1943, the major issues being left for board decision. But the union was able to settle only a few matters by direct negotiation with Cudahy, but one with Armour, and none at all with Wilson. The negotiations, the CIO union concluded, "did not truly meet the normal standards of collective bargaining. Armour and Wilson particularly were not interested in exchanging ideas and bargaining out . . . a satisfactory solution to joint problems." [31]

Collective bargaining also followed WLB decisions. The labor board was explicit on most questions, and on a few matters—for instance, union security—even provided the language to be incorporated into the contracts. Other issues, however, required further negotiation. The award of February 8, 1943, for example, approved of premium pay for night work, but left to the parties the definition of such work, the eligibility of employees, and the premium rate. The second decision directed the companies to negotiate the elimination of inter- and intraplant inequalities and the amount of certain "fringe" payments. The consummation of master agreements required several months of negotiation after both board decisions. The closing bargaining proceeded easily to final contracts, for the issues were limited in scope and the general lines already defined by the board. The parties were able to resort to the board when negotiations hit a snag. The WLB in fact assigned one of its staff, Nathan Feinsinger, to assist the parties after the 1943 award.

A limited form of collective bargaining thus did continue in meat packing despite the determination of the central issues by the War Labor Board. Beyond direct negotiations, moreover, labor and management representatives faced each other during the extended hearings before the board and its panels, and in 1945 they worked together amicably in the Meat Packing Commission on the problems of geographical differentials and intraplant inequalities in the industry.[32]

These experiences made collective bargaining a reality by the close of the war. For instance, a Swift representative urged the renewal of discussions, rather than WLB action, to resolve rate inequalities:

> I think this is a practical matter. I don't know how the Union looks at it, but I think they look at it as I do, that it is a practical matter and they want to get all they can, and I will tell you frankly the way we look at it we want to get out of it as cheaply as we can.

This view—the essence of collective bargaining—was reflected also in the Cudahy negotiations on the same issue, in which, each side conceding ground, agreement was finally reached on all but two questions. "After lunch we came back and settled those points, and people shook hands . . . they even had a drink on the proposition," remarked a company official afterward.[33] That happy note was not the invariable conclusion; the bargaining was often unyielding on both sides. The war situation itself, by imposing restraints, engendered hostility and tension; there were mutual accusations of using the emergency unfairly; and, in opening negotiations, the very likelihood of WLB intercession reduced the willingness for mutual concession. Nevertheless, both sides had by 1946 become accustomed to the concept of collective bargaining and practiced in the techniques of negotiation.

Relations with the packers advanced in other ways during the war. A union shop stood high on the list of union requirements. Informal shop action had, in fact, achieved an approximation of union security. First inaugurated at East St. Louis in June 1938, a practice of refusing entry to plants to men without union cards had become common.[34] But such irregular measures, although somewhat successful, could not substitute for contractual union security. Actually, some weakening among the packers had already developed. An oral understanding with Cudahy, in connection with the "harmony clause" in the 1941 contract, was construed by the union to mean "that in due time every plant employee, where the PWOC has jurisdiction, must become a member of our organization." Moreover, according to the WLB panel, both Armour and Cudahy in the 1941 negotiations had given "some indication" of a willingness to consider a union shop the next year.[35]

But, when it was proposed in 1942, the demand was categorically rejected by the Big Four, and it went to the War Labor Board along with the other unresolved issues. Following established policy, the board awarded maintenance-of-membership and, because of the nature of packing-plant operations, the check-off. This considerable union triumph—it was so considered by the PWOC—was resisted by the packers. Armour, for instance, insisted on writing into its contract a statement that the union security provision was "included, over the protest and without the voluntary concurrence of Armour and Company, solely in compliance with [the WLB] Order issued in time of war." Only Swift would apply maintenance-of-membership to newly certified plants without a specific directive from the WLB, and the PWOC accused the packers of trying to undermine the application of the clause.[36] The packers tried to rid themselves of the obnoxious provision in the second round of WLB hearings, in 1943–44, on the grounds that it had not provided the harmony and full production by which it was justified.[37] The War Labor Board rejected the packers' demands (as well as the union request for the elimination of the escape clause); maintenance-of-membership and the check-off were continued with another fifteen-day escape period beginning February 20, 1945.

That decision marked the end of the debate. The packers were already reconciling themselves to the limited form of union security.[38] The moral argument, the cause of so much heat, was no longer much in evidence, and the provisions remained in effect without apparent resistance from the Big Four in the immediate postwar period.

The significance of the achievement went beyond the important benefits to the union. The acceptance of union security by the packers represented a major shift in their view of the CIO union. In 1942, Swift & Co. had opposed a union shop partly on the grounds that the PWOC "has not yet reached the permanency of organization to rightfully demand such security." The support for the union "properly may be considered experimental on the part of the employees." [39] Conversely, the acceptance of maintenance-of-membership by 1945 also signified an acknowledgement by the packers of the permanent status of the union.

The war period inaugurated another kind of advance in the union-management relationship. Industrywide bargaining—that is, multi-company—was a further union objective. Chairman J. C. Lewis remarked during the opening of the first wartime negotiations for master contracts in March 1942:

The time is here when we of the PWOC must function in unity . . . not one group here presenting one of set of demands to the packers, and another group elsewhere presenting an entirely different set of demands. Our demands must be uniform as a whole.

To this end, a policy committee of ten was soon selected, two representing each of the Big Four and two the independents.[40] The packers could not effectively resist this pressure. For one thing, they did in fact act as a unit on basic labor-cost issues. Moreover, the War Labor Board lumped together the Big Four cases, adding in the second round Morrell & Co. The resulting contracts varied in some details, but were identical on the main issues; and, in addition, a uniform expiration date—August 11—became part of all the master contracts.[41] Joint bargaining, although not in a formal sense, was a reality by 1945. The arrangement—whatever its actual strategic advantages—seemed a great advance at the time, for it raised the union status in the industry, and imposed a restraint on the more belligerent of the packers.

The wartime changes did not exclude mutual hostility. Particularly with Wilson and Armour, relations were embittered by company charges of irresponsibility and union countercharges of stalling and "union-busting." In the heat of a dispute in December 1943, Wilson officials assured a Packinghouse Workers' representative of the persistence of the company's prewar attitude. The chilling conclusion of the unionist was "that they are negotiating with the Organization only because they are forced to do so by the Board's Order but at the first opportunity, they will tear up the Agreement and go back to the old days." [42]

But, if the earlier belligerence had not everywhere abated by 1945, the advances nevertheless proved irreversible. Collective bargaining was a fact, and the future of authentic negotiations was assured.

World War II had another kind of effect on the development of labor-management relations in meat packing.* When collective bargaining began in 1941, no clear guidelines defined the areas open for negotiation. What were appropriate issues? What should unions demand, packers concede? The guidelines emerged by the end of the war, making possible fruitful and coherent collective bargaining in the postwar years.

Economic conditions in 1942, the first meat panel of the WLB observed, would assuredly justify substantial wage increases "in ordinary times." Even after taxes, company profits were running very high.[43] The labor shortage was acute. Living costs had risen 10.2 per cent in the eleven months since the last advance in August 1941. And the two increases in 1941, the first in four years, had seemed inadequate even then. Wage demands—20 cents an hour by the PWOC, 15 cents by the Amalgamated, 10 cents by the Swift Brotherhood—constituted the chief issue in the meat-packing cases certified to the War Labor Board in June 1942.

The next month, wage stabilization overtook this move. The WLB enunciated the "Little Steel Formula": no increases where wages had risen 15 per cent or more since January 1, 1941. The Economic Stabilization Act of October 3, 1942, designed to stabilize prices, wages, and salaries at the levels of September 15, 1942, reinforced the Little Steel Formula as the standard for wage adjustments. Since straight-time hourly rates had increased between 17 and 18 per cent since January 1, 1941, the meat unions based their case primarily on the historical connection between steel and packing wages. The WLB panel accepted this contention. Because the WLB had raised steel rates by 5½ cents an hour in July 1942, that increase should also go to meat packing (in accord with exceptions permitted under the Economic Stabilization Act). The War Labor Board unexpectedly rejected the panel recommendation on February 8, 1943. This was, a PWOC official acknowledged, "a severe blow to all of us." There were bitter complaints of changes in "the rules of the game" after the start of the case. Amalgamated Secretary Gorman and President Earl Jimerson warned the WLB about

* The following discussion includes all the meat unions dealing with the major packers. The preceding pages dealt only with the PWOC, for it alone was concerned with the fight for master contracts, and it alone needed the war situation to stabilize its relations with management. The wartime impact on the substantive guidelines of negotiation, however, was uniform, and this section therefore involves all the packinghouse unions. For the reasons the Amalgamated was drawn into the common bargaining pattern, see Ch. 11.

"local unions threatening to strike if the . . . 5½ cent per hour increase in pay is not granted."

> Believe us when we say the unrest that is developing . . . is most alarming . . . Something must be done in this matter because . . . the situation is becoming uncontrollable.[44]

The board did not recede, and it's Directive Order of February 9, 1943 fixed the basic wage level of the industry for the duration of the war. No upward revision was possible even by voluntary agreement under the Stabilization Act and Executive Order of October 3, 1942. The WLB had to approve all wage changes. The Executive Order of April 8, 1943 (the "hold-the-line" order) tightened the application of the Little Steel Formula. A second attempt at a general increase was made in the reopening of the master agreements in the summer of 1943. The unions argued before the board that theirs was "a rare and unusual case." But no one was surprised at the board's denial on February 20, 1945.[45] The basic wage level in meat packing thus remained frozen during the entire course of the war.

The stabilization policy turned the unions to side approaches to more pay. PWOC officers asserted in July 1943 that "full advantage must be taken of what leeway is afforded" so "that the greatest possible gains may be made for the packinghouse workers." And the Amalgamated similarly vowed after the second meat-packing case "that every advantage offered us through this award will be capitalized upon in the interests of our membership."[46]

The revision of wage relationships seemed to offer the chief opportunity. The 1943 decision ordered the parties to eliminate "intra-plant inequalities between wage rates for individuals and between job classifications" and also "those inequalities in wage rates between plants in different localities which represent manifest injustice." For the next two years, however, the unions secured only negligible results in both areas. They were stymied by the confusion over the intention of the WLB, by the lack of data and adequate procedural machinery, and by the resistance of the packers.[47] The WLB finally acknowledged the failure of the piecemeal approach to inequalities. An Interim Order of December 7, 1944 directed the submission of complete rate data and the creation of a procedure to effectuate an equitable wage structure. The Directive Order of February 20, 1945 affirmed this action. The parties were then given 60 days to negotiate a comprehensive solution to inequalities. These matters would be overseen by a Meat Packing Commission, for which there was ample precedent in other industries. The commission, which was headed by Clark Kerr, had authority to decide the issues in the event of stalemate, to review agreements in case of success, and to assist in handling the complex problems. The way was thus opened to a rapid, comprehensive conclusion of the inequalities issues.

The basic approach was suggested by the Amalgamated: namely, to demand a general increase—$5\frac{1}{2}$ cents an hour—to compensate for intraplant inequalities. The packers finally agreed to an "across-the-board" increase of 2 cents an hour with the three unions. The WLB, however, permitted the 2 cents to be applied, not generally, but only to inequalities; that is, the total cost to the packers would be limited to an average 2 cent hourly increase.[48] The commission meanwhile undertook a study of the job-and-rate structure in packing plants. It formulated a "labor grade system" for purposes of simplification and standardization. This provided for 25 job groupings at intervals of $2\frac{1}{2}$ cents. It was then left to the parties to assign the jobs in each plant to the appropriate "concentration points" or grades. This process went forward without much difficulty. The review of voluntary agreements by the commission covered about 85 per cent of the jobs in the 93 master-agreement plants by the end of 1945. At that time the War Labor Board went out of existence, but the packers and unions voted to continue the commission (which had not yet begun the adjudication of the relatively few rate disagreements) with final and binding power until the completion of the task. The resulting job and rate structure was a permanent and important achievement.[49]

Interplant inequalities were likewise resolved through the calculation of the total cost to the packers. The Amalgamated agreement called for an expenditure for inequalities computed as the cost of a $\frac{3}{4}$ cent hourly increase. Modest as a charge on the total labor bill, this action breached the geographical wage pattern of the industry. Over half a century, there had developed four main classifications, based on unskilled male hourly rates, for Big Four plants:

Metropolitan (Chicago, New York, Boston, Pittsburgh, Cleveland, and Milwaukee)	$72\frac{1}{2}$ cents
River	70 cents
West Coast	$77\frac{1}{2}$ cents
Southern: Near south (Dallas, Fort Worth, Oklahoma City)	64 cents
Far south	$52\frac{1}{2}$–60 cents

Despite the WLB ruling against "elimination of long-established differentials," the 1945 negotiations resulted in fundamental changes in the geographical pattern. The Metropolitan classification was enlarged, and the river rate became the minimum for the entire north. Above all, the far south differential was substantially narrowed; no plant in that area received an increase of less than five cents an hour.[50] The future would reveal the competitive disadvantage to the major firms when the industry began a geographic shift away from the old centers.

"Fringe" adjustments offered another opportunity for escaping the wage freeze. The WLB Order of 1945 directed the packers to pay for time spent changing clothes and repairing tools and also to furnish work clothes, tools, and equipment. The pattern-setting Amalgamated clearly was seeking a disguised general increase in the negotiations with Swift. First, the union opposed a narrow application to the actual users of special clothes or tools: it was "necessary that all people be covered in the resolving of these issues . . . and not a limited few." Second, the Amalgamated desired to fix, so far as possible, specific uniform figures for each fringe adjustment. Its opening demand was pared down to 50 cents a week for clothes, twelve minutes' pay per day for clothes-changing time, and tool preparation on company time. Even this settlement, asserted a Swift representative, did not correspond to the actual value of time, clothes, and tools. The WLB nevertheless approved what was in fact a general increase of more than 3 cents an hour.[51]

Premium pay was also in the process of change. In the 1930's, the big firms had generally paid time and a half after ten hours a day or 54 hours a week. The Fair Labor Standards Act of 1938 exempted meat packers from the overtime provisions for fourteen "tolerance" weeks a year. A controversy quickly sprang up over this concession. The view of the American Meat Institute was that it applied to all employees, but the government argued that the exemption, since its justification was the seasonal character of the industry, covered only the slaughtering and dressing departments. The government in 1940 began a test case against Swift for violation of the act. The suit, in which the PWOC and the Brotherhood intervened, was decided against the industry, and litigation was begun by the unions for the recovery of back pay.[52] In late 1941 the packers entirely discarded the exemption; all employees would receive time and a half after forty hours.[53]

The war brought further gains in premium pay. In the first meat-packing case, the WLB ordered overtime rates after eight hours a day as well as forty hours a week. That decision also resulted in the negotiation of a premium rate of five cents an hour for night work. Voluntary agreements or board orders later led to time and a half after five consecutive hours of work without a meal (except in the case of Wilson), a furnished meal and a paid 20-minute meal period for working more than five hours after the first meal, and time and a half for a minimum of four hours for employees recalled to work after leaving the plant.[54]

The war thus improved the terms of employment in the industry. From the standpoint of future collective bargaining, the significant fact was that guidelines had been drawn in the areas of the internal rate structure, geographical differentials, premium pay, and a host of lesser matters. The precedents that would guide negotiations now existed.

The fight for collective bargaining had been won by 1945. It remained to be seen whether stability would follow. The turbulent postwar situation would subject collective bargaining to intense pressures. There was another disturbing circumstance: the rival unionism in meat packing. Collective bargaining could not stabilize so long as it was an arena of conflict between the Packinghouse Workers and the Amalgamated. The unions were entering a time of trial.

CHAPTER 11

RIVAL UNIONISM:
FROM WARFARE TO
INTERDEPENDENCE

In 1941 the AFL and CIO meat unions had a simple relationship: they were rivals and enemies. It was the inescapable logic of the organizing era.

Van Bittner, PWOC Chairman, expressed part of the logic when in June 1939 the CIO Executive Board was discussing peace negotiations with the AFL. "We as a labor organization can never have peace," Bittner insisted. "That would do more to destroy the spirit of the membership of the C.I.O. than anything I know of . . . When anybody asks me in a meeting of our people what the A.F. of L. is trying to do I simply tell them they are trying to destroy our union—and that is exactly what they are trying to do." "If there are going to be peace negotiations carried on," Bittner added the following year, "I doubt whether . . . we could organize the packing industry, because those men swear by all that is holy that they will not belong to the A.F. of L. . . . That union we have built up and the progress we are making would be materially, if not wholly, destroyed."[1]

Amalgamated chieftains returned this animus. The PWOC was an interloper in a field rightfully theirs. The CIO was reaping what they had sowed by years of effort and sacrifice—and this by ruthless tactics and false claims. The resentment intensified with leadership changes within the rival camp. Patrick Gorman believed that, so long as John L. Lewis was in power, there was a chance that the rival unions would be brought into the Amalgamated. Lewis' break with the CIO spiked that hope. And men were rising to prominence within the PWOC for whom Gorman had a thorough antipathy—in particular, Lewis J. Clark, vice-chairman, then secretary-treasurer of the PWOC, and with prospects of even higher office. Clark had been a local business agent who had led his union out of the Amalgamated and had helped found the PWOC. He was the object of that special hatred reserved in the labor movement for disruptionists and secessionists.

The mutual repugnance was bound to relent as the unorganized area narrowed and as the rival unions consolidated their fields of control. This was beginning to happen after 1941. But there was soon a greater imperative toward the normalization of relations.

Collective bargaining, as it developed during World War II, began to force the AFL and CIO rivals into an interdependent relationship. Their hostility persisted, but it could not prevent a common interest in negotiations. The erratic course of their relations reflected the conflicting pressures of rivalry and interdependence. Slowly, the latter became dominant.

Initially, the Amalgamated regarded collective bargaining as an extension of its field of battle. It consequently hindered PWOC efforts insofar as this was possible. There had been vigorous resistance to the CIO attempt to win a master contract from Armour through Federal influence. In November 1940 and later, Gorman privately urged the Big Four packers to announce a general wage increase. In part, his object was to permit a rise in wage levels in Amalgamated plants, particularly Kingan. But Gorman also had in mind the effect on the PWOC, as his correspondence with the packers revealed. He had rejected the possibility of using the wage issue "for the purpose of propaganda in bringing more members into our organization throughout the packing industry." He believed rather "that this is a matter that could best be handled quietly in the same manner as former national increases were handled by the four large packers." Their unilateral action would frustrate the bargaining aims of the PWOC—as indeed it did in April 1941.[2] Similarly, Gorman discouraged the demands of the Amalgamated local for a closed shop in the Armour plant in New York City. The company would then have to "sign it in dozens of other plants," and the CIO would be the chief gainer.[3]

The Amalgamated ultimately could not prevent the gains of its rival. Nor, what was more, could it long divorce itself from the system of collective bargaining being formed by the Big Four, the PWOC, and the War Labor Board. The Amalgamated had ridiculed the first Armour master contract in 1941: "we would have no part of it."[4] Rejection of national bargaining was a feasible course, for the Amalgamated's Big Four plants were mainly outside the labor pattern of the major centers. But that choice was not possible in wartime.

Only one Amalgamated union—Local 78 of the major Swift plant in East St. Louis—participated in the first meat-packing case before the WLB in 1942–43. Early in the proceedings, however, the implications for other Amalgamated units became apparent. Swift & Co. suggested during the panel hearings that the decision on wages be applied to all company plants not represented in the case. Gorman himself was willing to apply the East St. Louis decision to West Coast plants at which controversies were pending:

From a practical standpoint . . . we have nothing else to do with the war upon us. As to whether or not those people would be satisfied if the board would not grant a pay increase, I couldn't vouch for that . . . But all the agencies of our international union would be used to prevent trouble there.

At that time, during the panel hearings in the summer of 1942, an Amalgamated official sounded out Swift representatives about a master agreement. As the result, the directive order of February 8, 1943 became the basis for a national agreement between Swift and the Amalgamated on April 1, 1943. Armour & Co. likewise agreed to a master contract for the union's twelve plants.[5]

This course raised certain difficulties. The primary problem was that objections arose among some Amalgamated locals with an independent tradition and good working agreements. In San Francisco, for instance, Local 508 had to sacrifice its superior overtime rates at Swift. A local official grumbled to Gorman: "Why should they set up the master agreement to create any confusion now?"[6] Later, Sausage Makers' Local 503 received an increase from the independent plants, only to discover that, as a result of inclusion in the master agreements, the rise could not be applied to its Armour and Swift membership. Clearly, national agreements were not an unmitigated blessing to the Amalgamated.

Yet the advantages were weightier. For one thing, master contracts permitted the immediate extension of the benefits of the first WLB order to all the locals in Armour and Swift (excluding branch houses and egg and poultry plants). Moreover, the uncertain and slow processes of the regional boards were avoided. The Amalgamated had direct access to the national board and optimum effectiveness in presenting its case. This also concerned the many Amalgamated locals in the independent plants, for the national meat-packing cases set the pattern for the entire industry. As a result of the union's efforts, the authority of the Meat Packing Commission in 1945 was expanded to pass on voluntary adjustments by independents matching the "fringe" provisions of the second meat-packing award.[7] These considerations sprang from the special circumstances of the wartime regulation of labor relations. Beyond that, of course, the Amalgamated understood the normal advantages accruing from national bargaining with multi-plant firms.

Master contracts with a single employer inevitably raised the question of the relations between the rival unions. Wartime collective bargaining, limited as it was, foreshadowed the inescapable bonds. When in the fall of 1943 Amalgamated negotiations with Swift broke down, President E. W. Jimerson regretfully told the company that "if our organization was the only organization negotiating . . . an understanding could have been had." [8] The extent of the interrelationship was revealed in the bargaining following the board's order of February 20, 1945. The WLB left open the precise terms of the fringe payments and rate inequalities. The Amalgamated reached a quick accord with Swift and Armour. The CIO union then accepted its rival's advantageous fringe agreement but pressed further on the inequalities issue. The companies, however, refused to split the "package." When the Packinghouse Workers pointed out that AFL and CIO plants existed in the same centers, the companies responded that the same logic applied to rate inequalities. "So . . . we have got to treat them all alike . . . The Board should order that the Union accept the A.F. of L. agreement." The Packinghouse Workers received the choice of accepting the package or renegotiating both issues. Although it took the latter course, the union settled in the end on about the same terms as the Amalgamated.[9] The strategy of using the rival organization as leverage in negotiations clearly was a failure. Indeed, the packers' insistence on equal terms reversed the leverage effect. Cooperation would obviously maximize benefits.

This fact slowly became apparent to both unions. Their leaders had made a joint presentation before the War Labor Board in December 1942 in favor of the panel's recommendation of a $5\frac{1}{2}$ cent an hour increase. Board rejection intensified the sense of common purpose. "The need for a cooperative unity on the wage issue is of the utmost importance," proclaimed PWOC Chairman Sam Sponseller. The Amalgamated responded "that on this issue we would fight as one organization. . . . The unity that this issue has created between our International Union and the C.I.O. only makes our army all the more strong—and victory will be ours!" [10] But that sentiment was short-lived.

The imperative of cooperation was not compelling during the war period. Strike action—the point at which unity was most important—was not permissible then, nor, given the role of the WLB, could cooperation have appreciable effect on the basic terms of employment. The fact was that divisive outweighed unifying considerations during the war. The organizational drive was still creating animosity. And the temptation was very great to turn the wartime gains to propaganda purposes. The competitive outlook was evident in Amalgamated opposition to the CIO efforts to eliminate the wage differential between the Big Four plants in Los Angeles, which were CIO, and the independents, which were AFL:

We would hate to see the C.I.O., who have . . . for the most part paper organizations, ride on our back in the splendid wage scales we have established on the West Coast . . . With our wage scale remaining as it is in the Los Angeles area, it will enable us to bring the Big Four plants in Los Angeles into our organization and when this is accomplished we could find other avenues available to bring proper stabilization to these plants within the Big Four.[11]

This attitude reached its apex in the late spring of 1945 in the acrimonious debate between the two unions over the credit for and merits of their master contracts.

Yet the interdependence of the rival unions could not be mistaken. In May 1945 Gorman and Jimerson sent an "Open Letter" to the Packinghouse Workers on the need for cooperation in the approaching postwar period. They made two proposals. First, the Amalgamated would work with the CIO union "in the compilation and presentation of all future demands that will be made upon the big four packers and to stand shoulder to shoulder . . . in using our combined pressure upon the officials of the four large packers and NWLB." That was for purposes of collective bargaining. The second proposal was a defensive alliance in anticipation of a packers' counterattack at the end of the war. The Amalgamated suggested a conference "to outline a plan . . . of organizational protection if the four large packers will again attempt to destroy our unions . . . Sincerity should prompt us to resolve that we shall not permit the packers to attack your union . . . and vice versa."[12] Nothing came of this opening. It was overwhelmed in a burst of personal acrimony and fierce recrimination over the master agreements just then being consummated. The war ended with the two unions more at odds than ever. Nevertheless, the dim recognition of common interest pointed toward cooperation in the uncertain future.

No such eventuality seemed possible in the case of the Independent Brotherhood of Swift Employees (National Brotherhood of Packinghouse

Workers after 1944). The AFL and CIO unions had not reconciled themselves to the permanent existence of the independent, as they were beginning to with one another.[13] Brotherhood plants were subject to unremitting
organizing pressure during the war period. Moreover, the Swift independent
was held in contempt as a "company union." It alone did not reopen its
master agreement in the summer of 1943. When negotiations were drawing
to a close in December 1943, the Packinghouse Workers warned that "if
Swift extends these changes to the Brotherhood agreement, we will consider it an act of support and assistance to the Independent Union of such a
nature as to raise seriously the question of Company's relationship with that
Union." Despite this threat to resort to the NLRB, the same concessions were
granted to the Brotherhood. On December 9, 1943, it also reopened the wage
issue, which, as in the case of the other two unions, was certified to the War
Labor Board. "The Brotherhood thus has with no effort obtained the benefits of the UPWA negotiations."[14] And the independent had the further
benefit, since it was not negotiating a new agreement, of avoiding another
fifteen-day escape period before the continuance of maintenance-of-membership.

The conclusion seemed inescapable that, notwithstanding its denial, Swift
was partial to the Brotherhood; that the independent was content to feed off
the gains won by its rivals; and that it could never be counted as an ally
in a showdown with the packers. "This company union presents a constant
threat to our organization," Packinghouse Workers' leaders judged, "and is
a barrier to complete and satisfactory collective bargaining with the Swift
management."[15] The pressure of events might overcome the enmity between the AFL and CIO unions. But the Brotherhood seemed beyond the
pale of the bona fide labor movement. Here was a weak link in future dealings with the major packers.

Meanwhile, the PWOC was advancing toward organizational maturity.
Interunion amity depended on it. For the Amalgamated would ultimately
accept and work with its rival only when the latter became an autonomous
and viable national union—that is, when the last hope was laid that the
PWOC might disintegrate and the Amalgamated pick up the pieces.

Autonomy had been the primary issue in the internal fight that had culminated in the rebellion of 1941. The dissidence had been crushed, but not
the pressures for autonomy. The CIO slowly responded. Packinghouse representation on the committee was augmented in July 1942, and eminent outsiders—David MacDonald, R. J. Thomas, and Allan Haywood—brought in

for advice and to forestall new charges of "dictatorship" against the chairman. The rank-and-file voice would be increased by national and district conferences every three months. The PWOC would thus develop "to the stage where it will be possible for the convening of an International Constitutional Convention at some future time." [16] These concessions did not lessen the desire for more rapid and certain progress. In September 1942 a joint conference of Districts 3 (Western Iowa) and 10 (Eastern Iowa and Nebraska) adopted a motion for a constitutional convention within four months after the conclusion of the pending War Labor Board case. PWOC officers bowed to this demand two months later. The CIO Executive Board, while accepting a preliminary conference, at first refused to commit itself to an international union, and only reluctantly did so later.[17] Finally meeting July 8–10, 1943, a PWOC conference recommended a constitutional convention between September 30 and October 15, 1943, and elected a committee to handle arrangements and draw up a constitution.

The imminence of autonomy raised the absorbing question: who would grasp the reins of the new international union? That was the basis for the second stage of factionalism within the packinghouse union. And the need to put together a majority would determine the line of conflict. Three identifiable groups were in contention: the 1941 dissidents, the Communists, and the PWOC administration.

The dissident unions of 1941 had the least hope of gaining majority support. While resuming per capita payments, they had remained in opposition. Their identity was sustained, not so much from a reasoned policy, but from the outside by the growing split between John L. Lewis and the CIO. Several insurgent leaders entered the service of the UMWA's District 50. Other local dissidents, while remaining within the PWOC, were on Lewis' payroll or conspired with District 50 representatives.[18] The faction, shifting somewhat as it was in composition, did gain an identity from the connection: it was known at the first constitutional convention in 1943 as the "District 50 group." But this identification inescapably isolated the dissident group from the main body of the movement.

PWOC Chairman J. C. Lewis was soon caught in the crossfire of the CIO-UMWA fight. Unlike other CIO officials, J. C. Lewis had not shifted his allegiance and career from the Mine Workers. When the Armour and Swift Chicago locals passed a resolution attacking John L. Lewis, the PWOC chairman hastily assured Lewis that he had rebuked the district director for the "poor policy." This position soon came under the attack of CIO adherents. Chairman Lewis vehemently denied any attempt to draw the

PWOC out of the CIO or to conspire with the John L. Lewis supporters in the PWOC. But his neutral position ("We are out to organize . . . the packinghouse workers and we are not inclined to become embroiled in fights between any organizations") was not enough once the break between John L. Lewis and the CIO became complete. During the general overhaul of the PWOC in July 1942, J. C. Lewis was replaced as chairman by Sam Sponseller, a district director, a former Mine Workers' official, and a loyal CIO man.[19] The identification with John L. Lewis in the same way—but with greater cause—isolated the dissident group from the majority loyal to the CIO.

The left wing was meanwhile gaining influence. From the start, the PWOC had included a substantial, but never dominant, radical element. The powerful Austin local had been a center of syndicalist and Trotskyite agitation ever since its formation in 1933. Communist infiltration sometimes came from the local parties or, as in Chicago, from remnants of earlier Communist packinghouse unions. Left-wingers—for instance, Don Harris, Henry Johnson, and Frank Alsup—had occupied key posts in the PWOC. The Chicago district director in 1939, Herb March, was an acknowledged Communist who later became a member of the party's national committee. And these officers were able to appoint like-minded organizers. The CIO Arkansas-Oklahoma director later complained "that most of the people who came into that section seem to have interests in common with the 'Comrades.'" This was not in itself disturbing to the PWOC leadership. Van Bittner enunciated the standard CIO position on Communist participation when his Chicago subordinates came under attack in December 1939 from the House Un-American Activities Committee: "As far as politics is concerned, we are for those who are for us and against those who are against us."[20] But that cut both ways. For the left wingers were in the dissident group against Bittner; and they were treated accordingly.

A significant change occurred after the 1941 rebellion. The left-wing elements until then had by no means constituted a united group. The Communist Herb March, who had been demoted and then dismissed separately, had not been closely connected with the Johnson faction. Some time in 1941, March returned to the Chicago stockyards. He shortly helped to reduce the opposition party in Chicago to a minority position. He was on the ticket—running for vice-president—that defeated the dissidents in the crucial local election at the Armour plant in the early summer of 1941. Following this signal contribution, March was restored to his place as director of the Chicago district.[21] In New York and Boston, party members or sympathizers were also assuming important administration posts. The basis for this align-ment was, of course, the Nazi invasion of Russia in June 1941. Thereafter, the Reds could not participate in a dissident movement identified with District 50. The Minnesota district director told J. C. Lewis: "The C.P.s . . . claim, the packinghouse workers should not follow your direction. As you are on the payroll of the Mine Workers and John Lewis is the modern Benedict Arnold who has betrayed the workers." PWOC Communist sym-pathizers attacked the UMWA because "its leaders oppose President Roose-velt and the War Program of the Government." The Russo-American alli-ance made obstructionist policies intolerable to the segment of the PWOC left wing close to or in the Communist Party.[22]

The collaboration with the administration, however, was an uneasy one. For one thing, the PWOC leadership became actively anti-Communist after the departure of Van Bittner. His successor J. C. Lewis, before returning to the PWOC, had led the campaign to break the Communist control of the Washington State Industrial Union. The other administration leaders—Lewis Clark of Cedar Rapids, Roy Franklin of Austin, and others at the district level—had a similar animus. J. C. Lewis recommended the director of the Oklahoma-Texas district, James Dean, as one who once "played a little with the fellow travelers but no man has done more to clean them out than has Dean." The appointment of Sam Sponseller as chairman in mid-1942 did not alter the hostile attitude of the administration.[23]

The second consideration was the weakness of the appointed PWOC leaders. The turnover at the top had been too frequent to permit the de-velopment of broad popular support. Chairman Sponseller did not have a personal following. Nor could the packinghouse workers on the committee, except Lewis Clark of Iowa, deliver the support of their home territories to the administration. And only modest strength would be forthcoming from loyal district directors and representatives. As autonomy drew closer, it became clear that the administration itself lacked sufficient support to carry an election.

On both counts, the left wingers began to take an increasingly independent part in union politics. Obviously concerned, Vice-Chairman Roy Franklin warned a supporter in November 1942 against permitting "an officership that no one wants."

You know the party members will bring in the slate and wherever party dele-gates come from, votes will be cast for their candidate. I admit, party members are far in minority but if they are the only ones at the convention with an organized program and slate the majority is bound to take a licking.

Five months later, he was reporting gloomily that "the comrads [sic] are very busy organizing not for the Revolution, but for the International constitutional convention." [24]

The party group actually was, as Franklin noted, not very numerous. But the Communists were strategically located, especially in Chicago, and could draw on a much wider base of support during the war. Herb March dominated the Chicago district although in 1943 his own Armour union was the only major local there under actual Communist control. This influence extended beyond local limits; Meyer Stern and Herb March—both Communists—received the chairmanships of the two committees charged with preparing for the constitutional convention.

A coalition was meanwhile materializing. The core consisted of: the Chicago and northeastern districts, which were under direct Communist influence; the Minnesota–South Dakota area under Frank Ellis of the powerful Austin local; the Iowa-Kansas-Nebraska territory under PWOC Secretary-Treasurer Lewis Clark, who, with characteristic dexterity, deserted the administration when he saw the main chance. This group, significantly dubbed "Communist" at the convention just as the dissidents were called "District 50," clearly had the strength to capture the new international. [25]

The constitutional convention opened in Chicago on October 13, 1943 in an atmosphere charged with suspicion and division. The spokesmen from the CIO—Allan Haywood, R. J. Thomas, and David MacDonald—worked manfully for unity. Their preference was naturally for the perpetuation of the PWOC incumbents in office, but unity took precedence over experience in the view of the CIO representatives. Nightlong talks on October 16 finally hammered out a compromise solution. The coalition group would have three of the four international officers and would recommend someone from the minority elements for secretary-treasurer. Haywood presented the slate in the morning:

> President: Lewis J. Clark, Cedar Rapids, Iowa
> Vice-President: Frank Ellis, Austin, Minnesota
> Vice-President: Philip Weightman, Chicago, Illinois
> Secretary-Treasurer: Edward F. Roche, East St. Louis, Illinois

Even then, Haywood despaired "that we failed last night in our last sincere efforts to have . . . unity in this delegation on the officers for the coming term." [26] Further efforts during the lunch recess achieved superficial unity. The compromise slate was elected by acclamation, as the CIO representatives had hoped, to lead the new United Packinghouse Workers of America.

The achievement of autonomy did not end factionalism in the packing-house union. The divisive forces generated in the course of unionization took new forms, but they did not disappear. The UPWA failed to achieve the one-party rule characteristic of American trade unions.

The decisive fact was the continuance of a dissident group into the international period. The issues of the 1941 revolt and District 50 of course passed away. But new questions, particularly communism, kept alive anti-administration sentiment. Beyond specific issues or personalities was the binding element of accustomed opposition. An attempt to incorporate opponents into the ruling party through staff appointments succeeded with some of the PWOC administration adherents, but not with the dissidents. For the latter were able to establish independent bases of power within the structure of the UPWA.

Its constitution was shaped by the experience of the PWOC period and by the example of other CIO unions. (Herb March, chairman of the constitutional committee at the first convention, remarked that "consultation was held with all of the outstanding leaders of the C.I.O. that we could get the ear of.") [27] Like other CIO industrial unions, the UPWA was divided into districts. The delegations from these met separately during international conventions to elect directors who would be the chief regional executives and also members of the international executive board together with the four international officers. This arrangement permitted men to hold high places independent of the administration; at least three of the ten district directorships in 1943 were carried by dissidents. Opposition was thus built into the structure of the UPWA. [28]

The second characteristic of UPWA internal affairs, arising partly out of the first, was the continuing alliance of the left wing and the administration. None of the four international officers elected in 1943 or after was Communist by party affiliation or ideology. But March did become director of District 1 (Illinois-Wisconsin) and Meyer Stern of District 6 (northeastern states). More important was the fact that, from the start of autonomy, the left wingers consistently supported the men in power. Ralph Helstein, the UPWA general counsel, assumed the presidency with their backing in 1946. (President Lewis Clark became secretary-treasurer in a compromise maneuver during the turbulent convention.) [29] The broad influence of the Communists diminished with the onset of the Cold War, but there was a compensating expansion of their area of direct control. They captured the Wilson local in Chicago in 1944, as well as some small locals there, and they continued to be strong in the northeast. [30] Factionalism imparted to the Reds a strategic importance beyond their actual numbers. In a close election, they could command the decisive votes.

The accommodation of the Communist minority rested on more than expediency. President Helstein and other administration leaders, for example Vice-President Frank Ellis, were strongly liberal in politics and insistent on the protection of the rights of unpopular minorities. Even after the war the Communists shared with others in the UPWA many points of agreement: militancy in negotiations and on strike action, racial equality, and a variety of domestic political issues such as price control, FEPC, social security, and so on. The left wingers were, moreover, demonstrably good union men; March could point to bullet wounds incurred in the service of the union. Finally, the invocation of "unity" by the Reds always was persuasive within the union.

When the Communist issue became heated in 1947, the executive board formulated a "Statement of Policy." On the one hand, "red baiting" was denounced. "Repression of any sort which attacks civil liberties is the weapon of reaction . . . whose goal is to destroy the trade union movement." On the other hand, the union sought to advance the interests of its own members "through our existing institutions, and we reaffirm our unshakable [sic] faith in its processes." The UPWA would resist Communist domination for party ends.

Those who for whatever reason attempt to influence our actions in other directions will receive no aid and comfort from us. We resent attempts . . . to infiltrate, dictate, meddle or interfere in any way with the functions of our organization. We will make our own decisions free from all outside influence.

This statement, adopted by acclamation at the 1947 convention, was a position which was, in theory, acceptable to the whole of the UPWA.[31] The practice would be another matter.

The CIO packinghouse union thus emerged into relative maturity. By contrast to the Amalgamated, the UPWA still lacked a secure ruling group and internal unity. There were charged issues—communism and negotiating policy—which would wrack the UPWA. But clearly also it was a going concern. When the postwar years opened, the UPWA was a union with which the Amalgamated would have to live.

Pressure on the existing wage level dominated the scene at the close of the war. It was true that income had risen substantially: average hourly earnings in meat packing were up from 78 cents in September 1941 to 95.8

cents in September 1945, and average weekly earnings from $30.63 to $45.81. But the rising cost of living had partially erased the money gains. Overtime would soon be coming to an end, and the industry had gone through the war without a single general increase. The packinghouse workers had scant cause for satisfaction in the midst of booming prosperity. In meat packing, as elsewhere, the demand was imperative for higher wages.

The master agreements, which ran until August 1946, permitted earlier wage negotiations through reopening clauses. Meeting immediately after V–J Day, the UPWA Executive Board decided on a demand of 17½ cents, subsequently hiked to 25 cents the next month. On October 29, 1945, before the actual start of negotiations, the union petitioned the government for a strike vote in accordance with the War Labor Disputes Act. Talks raised the offer from Armour to 7½ cents and lowered the union demand to 17½ cents. There progress ended, and the UPWA set a strike date for January 16, 1946.

The Amalgamated took a separate course. A conference in early October 1945 also placed the demand at a high level: a guaranteed minimum of 36 dollars for a 40-hour week. President Jimerson and Secretary Gorman, however, repudiated the "trigger" tactics of the UPWA. When the CIO union on January 2, 1946 fixed its strike deadline, Gorman announced that the Amalgamated would remain at work. But this position was quickly reversed. On January 4, Gorman and Jimerson telegraphed President Truman that, unless there was a conference within a week to avert a strike, the Amalgamated would have to join the UPWA. "We are faced with the indisputable fact that division among the packing plant workers now might create disaster as far as unions in the industry is [sic] concerned." The next day Gorman publicly pledged financial strike support to the UPWA. Failing to get a substantial offer from the packers by January 11, the Amalgamated took formal action to strike also on January 16.[32] The rival unions were thus drawn into alliance.

The wage issue was tied in with two public questions: price controls and the critical meat shortage. The packers insisted that the OPA price ceilings made it impossible to meet the unions' demands. They were in substance seeking price concessions from the government in exchange for wage concessions to the unions. The meat shortage gave the packers leverage on this point. Secretary of Agriculture Clinton A. Anderson had warned of low meat reserves. On January 4, 1946, he suggested a price adjustment to permit the packers to meet union demands and avoid a ruinous strike. Although two offers made by the Wage Stabilization Board were rejected as inadequate by the packers, it was clear to both labor and management that the government would have the decisive role in the ultimate settlement.[33]

The strike on January 16, 1946 was extremely effective. Everywhere AFL and CIO men came out solidly. There was little or no violence and, on the part of the packers, no attempt to resume operations. Most of the independents, including Hormel and Hygrade, had made terms with the unions or had binding contracts; and the Swift plants of the nonstriking Brotherhood, although picketed, operated with curtailed production. The impact of the strike still was great; the output of Federally inspected meat dropped more than 50 per cent in the first week of the strike. It was a far cry from 1921.

The Amalgamated leadership nevertheless had grave misgivings. Gorman later remarked that he was determined "to guard zealously the progress made and to suffer no setbacks, particularly with so many defeats in strikes glaring at me from out of the past." [34] He wanted a quick end to the controversy. On the first day, without the foreknowledge of the UPWA, the Amalgamated offered to settle for 15 cents. A conference the next day in Washington failed to bring an agreement, but it did result in the appointment of a fact-finding board. While UPWA representatives sat "sullen and dejected," Gorman welcomed this action. The Amalgamated favored government seizure of the packing plants. Such a move, which could be based on the president's war powers and the War Labor Disputes Act, had been widely discussed even before the strike. With no settlement in sight, Gorman urged the measure on the reluctant Secretary of Labor Louis D. Schwellenbach. When they were sounded out at a Washington conference on January 22, Amalgamated representatives assured the government that they would order the men back to seized plants, and did so with alacrity when the government acted two days later.[35]

For its part, the UPWA took an opposite line. A conference refused on January 25 to order a return to the packing houses on January 26 when government seizure would take effect. The union accused President Truman of "a strikebreaking action, the sole effect of which can be to play into the hands of the packers." This resistance, however, carried heavy risks. There were clear signs of hesitation in the ranks. The law, as Secretary of Agriculture Anderson pointed out, provided criminal penalties for impeding the return of strikers to seized plants. And the Amalgamated, UPWA Director Meyer Stern asserted, was "stabbing us in the back" by obeying the order.[36]

In the end the hard line claimed some reward. The UPWA had

charged the administration with a "double-cross" for failing to include in the seizure order provisions for putting into effect the findings of the fact-finding board. Within the past two weeks, General Motors had rejected with impunity the recommendations of a panel, and U.S. Steel had done so with President Truman himself. What assurance did the UPWA have of a better fate? On the night of January 25, Secretary of Agriculture Anderson telegraphed the UPWA that he would seek immediate approval from the Wage Stabilization Board for any increase suggested by the fact-finding board. The order was thereupon given for a resumption of work on Monday, January 28.[37]

The tangled dispute had a satisfactory conclusion for both the unions and the packers. On February 7, the fact-finding board recommended an increase of 16 cents an hour. The Amalgamated certainly and the UPWA probably would have settled for less during the strike. (Shortly after, however, U.S. Steel set the general pattern at 18½ cents.) At the end of February, the Wage Stabilization Board put into effect the recommendation and gave "preapproval" for the increase to all firms which followed the Big Five pattern. The packers, for their part, were amply compensated. The fact-finding board had suggested that 5 cents of the wage hike should be absorbed by the industry—provoking, of course, vehement protests from the American Meat Institute. The OPA subsequently ordered price advances, effective March 11, 1946, to cover in full the wage increase; and funds were made available to repay the packers for the heavier labor costs of the interim period.[38]

Within a few months, collective bargaining resumed on a broader basis. The expiration of the master agreements on August 11, 1946 permitted the unions to press not only for higher wages, but also for major innovations. Although proceeding independently, both the Amalgamated and the UPWA asked for the union shop, health and welfare funds, and, above all, a guaranteed annual wage on the lines of the Hormel plan.[39] The packers rejected all the major demands and, for their own part, submitted counterdemands to end maintenance-of-membership and the check-off, eliminate arbitration from the grievance procedure (except Swift), and to make other sweeping changes. The companies were taking the opportunity, as Armour President George A. Eastwood said, to insist on the "recognition of several fundamental [management] principles which should govern the relationship between employer and employee." [40] The negotiations of late July and August 1946 ended in complete deadlock.

No crisis ensued, for the industry was then experiencing troubles of another sort. After two months of de-control, meat price ceilings were resumed at the end of August 1946. A severe livestock shortage accompanied the restoration of controls. Layoffs mounted to an estimated 40,000 in meat packing by mid-September. It was not a propitious time for a showdown. Despite frantic efforts by the UPWA (the Amalgamated opposed controls) to save price ceilings, President Truman announced the end of controls on meat and livestock prices on October 14, 1946.[41] Production quickly picked up, and bargaining resumed. Then in early December, the Amalgamated accepted a two-year contract with Swift: a 7½ cent hourly increase and improvements in geographical differentials, female rates, night-work premiums, combination and multiple job rates, vacations, and other minor matters. This became the basis for the speedy conclusion of all negotiations. The industry attack on union security and grievance arbitration had been no more than a tactical move. The meat packers surrendered their demands for retrenchment, and the unions for innovations. The best the UPWA could do was to persuade the packers to make a study of the "practicability" of the guaranteed annual wage.[42] The following June another increase of six cents was extracted from the packers through the reopening provision.

Postwar collective bargaining was taking a direction. The question of trade unionism itself seemed no longer at issue. Nor would there be any easy break from the pattern of concessions shaped by World War II. But, as elsewhere, the wage-price push began in meat packing in 1946. In two years the basic rate rose 29½ cents an hour, and the expectation was for repeated increases thereafter.

Rival unionism was responding to the new order of labor relations in the industry. The 1946 controversy, it was true, had revealed that Amalgamated-UPWA relations had another troublesome dimension: conflicting approaches to bargaining strategy and industrial strife. The UPWA had castigated the Amalgamated for lacking militancy, for being "only too willing to sell short at various stages during the negotiations." The Amalgamated, in turn, attacked the "trigger" tactics of the "donkey-like leadership" of the UPWA. The balance then shifted toward amity.

The UPWA hard line began to loosen. Observers remarked on the "unexpected ease in settling wage demands without strikes" in the second bargaining round in 1946.[43] Notwithstanding continued raiding, liaison developed during the lengthy course of those negotiations. The UPWA did not accuse the Amalgamated of "sellout" for settling first with the packers. For its part, the Amalgamated took a happier view of the UPWA after the accession of the lawyer and negotiator Ralph Helstein to the presidency. Lewis Clark was demoted to secretary-treasurer, and his star was clearly in the de-

cline. (In 1950 another setback would end his UPWA career.) The UPWA
Executive Board wanted to establish "the fullest cooperation" with the
Amalgamated in the 1947 negotiations and "to work out the most effective
means possible of coordinating our parallel efforts to obtain a satisfactory
wage increase from the Meat Trust." As a sign of good faith, UPWA District 1 withdrew from a representation election in Chicago. District officials
were hopeful that a precedent was being set for a no-raiding agreement.
When the 1947 round came, the two unions did cooperate with entire success.[44]

Finally, the developing stability had to be tempered by fire. During the
autumn of 1947 signs of renewed militancy multiplied within the CIO
union. A general strike in the Canadian meat industry, where Swift was the
leading firm, started in September 1947 and lasted for seven bitter weeks.
Recalling that the 1921 failure had been preceded by Canadian setbacks,
UPWA officials warned that this was the beginning of the conspiracy to
"smash" the union.

Factionalism was also on the rise inside the CIO union. Notwithstanding
the 1947 resolve, communism was becoming an increasingly heated issue.
This was caused partly by the debate over the Taft-Hartley requirement for
non-Communist affidavits from union officers as a prerequisite for using the
NLRB. (In July 1947 the UPWA Executive Board had decided against
compliance.) It was also a response to the tensions of the Cold War. Differences developed over issues such as the Marshall Plan and the Progressive
Party. Finally, a power struggle in the Chicago area lent substance to the
charge "that the Communist Party is attempting to take over and control
the trade union movement." The left wingers were trying to capture the important Swift Local 28 there. Striking back, the local union and International
Vice-President Philip Weightman, formerly the local president, assumed the
lead of the anti-Communist minority in the district, and also participated in
the successful campaign to eliminate leftist influence from the CIO Chicago
and Illinois central labor bodies.[45] These circumstances broadened the anti-
Communist sentiment and, since President Helstein would not respond to it,
resulted in the gravitation of right wingers to the anti-administration groups
within the UPWA. One reflex of the internal conflict was rising militancy in
external relations.

The UPWA Executive Board in late November 1947 voted to hold a referendum to authorize a special monthly assessment of 50 cents from February
to May 1948. "We called it a defense fund because we knew that in the year

1948 we were coming into a strike period," Vice-President Philip Weight-man later asserted. "A strike was inevitable because we have continually said that all we would get from the packers is what we were able to take." [46] So reasoning, the union subsequently fixed the demand at 29 cents an hour.

The UPWA stand broke the alliance with the Amalgamated. An early meeting failed to establish a basis for cooperation. According to Secretary Gorman, UPWA's Helstein wanted commitments for a strike; he was told that "the Amalgamated was not going into a strike before it went into nego-tiations." In any event, the Meat Cutters proceeded independently to a 9 cent agreement with Armour and Swift on January 29, 1948. The characteristic exchange of accusations followed. But, unlike 1946, the UPWA was now resolved "to continue the fight for a decent wage settlement." President Hel-stein added the pious hope that the UPWA effort "also will bring its benefits to the A.F.L. members, despite the betrayal of their leadership." [47]

With Amalgamated and Brotherhood agreements in hand, the packers were not inclined to make concessions. A last-minute attempt by President Truman to secure a postponement pending the report of a Taft-Hartley board of inquiry failed, and the strike began on schedule at 12:01 A.M., March 16, 1948 at the CIO plants of Swift, Armour, Cudahy, Wilson, Mor-rell, and Rath.[48]

The UPWA labored under heavy disadvantages from the outset. Fore-most was the inconclusiveness of the walkout. The strike was effective at the CIO plants, but the 34 Big Four units not affected by the dispute ran full blast. The strike, moreover, came after the busy season of the industry. The UPWA also lacked strong public support. The President's board of inquiry, reporting on April 8, did not endorse the union's wage demand. Union rivalry received a share of the blame:

> The fact that the companies must negotiate with three competing unions is of inescapable significance . . . The companies might have taken a more flexible at-titude toward the UPWA demands, had there been only one union to consider, or had all three unions been bargaining with the companies jointly, or had not the companies previously settled with the Amalgamated and the NBPW. On the other hand, the UPWA might have accepted nine cents an hour . . . rather than strike, had not the companies previously settled with the other two unions at nine cents.[49]

In addition, internal problems plagued the UPWA. The strike, although hard-fought, lacked the unity and purpose which could keep men out on the streets indefinitely.

"We underestimated the length to which the packers would go in attempting to defeat our strike," Herb March admitted. Relations became increasingly embittered in the course of the strike. Third-round wage increases, then pending in steel and automobiles, created widespread business support for the rigid stand of the packers. A series of Chicago and Washington meetings with Federal mediators failed to evoke from the packers a single concession.[50] More alarming was the decision, in contrast to 1946, to resume operations during the strike. Armour and Wilson, and later the other companies, announced that they were opening their plants to back-to-work movements.

The strike proceeded irrevocably toward defeat. The ranks, firm at the outset, began gradually to weaken. Only the men at the Wilson plant in Los Angeles returned *en masse;* but everywhere growing numbers were reporting for work before the end of April. Injunctions restricted the effectiveness of picketing in practically every center. Violence mounted in the later days. By the end, three strikers were dead, numbers injured or arrested, and the National Guard present in South St. Paul, Waterloo, Iowa, and Albert Lea, Minnesota. There was little response to the union's desperate decision on May 2 to spread the strike to branch houses, independents and stockyards (which could cut production at the source of supply). In fact, the Chicago Stock Handlers' Local 44 accepted a mere 6 cent increase during the second week of May. ("Bingo, the membership balloon just deflated like you put a big sword in it, and not just a pin," a Milwaukee representative later remarked, when the news arrived of the collapse of this "ace in the hole.") [51] By then, the UPWA had exhausted its own resources and outside assistance. At last, the packers unilaterally put into effect the 9 cent increase on May 3, and set May 10 as the deadline after which the jobs of strikers would be forfeited. A return to work, the Strike Committee soon decided, "affords the only possible basis upon which to rebuild our union . . . and to once again become an effective fighting force." [52]

The official end came on May 24, 1948—ten weeks after the start of the strike. The UPWA members not only gained nothing from the dispute; they lost the 9 cent raise for the two months preceding the strike when it was in effect at the AFL and Brotherhood plants. The packers did make a concession, however, on the question of strikers guilty of "unlawful activity." While insisting on their right to discharge such men, the companies agreed to take up these cases through the grievance procedure. The refusal of Wilson & Co. to accept this condition resulted in the futile continuation of the strike at its plants for two additional weeks. The official walkout also dragged on for a short time at the Rath plant because of the company's abrogation of several clauses in the contract and the stubbornness of the strikers.

The UPWA had not yet reached its nadir. A month after the strike—the end of June 1948—the international convention met. The determined anti-Communists organized a "UPWA–CIO Policy Caucus" to fight for the CIO political action program and to oust President Helstein's administration. They had an issue in the disastrous strike. Although no one openly ascribed the defeat to the Communists, the anti-administrationists had been critical of the action. They could consequently charge "here and now that this strike was ill advised, ill handled and misconducted throughout," and demand a change in leadership. The outcome of that turbulent convention ("there were even doubts in many of our minds whether the Packinghouse Workers would have another convention for us to attend," a CIO official later remarked) was the narrowest of victories for the administration.[53] President Helstein and Secretary Clark retained their posts with 56½ and 54 per cent of the vote, and Phillip Weightman—leader of the Policy Caucus—lost his vice-presidency by a tiny margin. But six of ten elected district directors—including A. J. Pittman, who had run against Clark—were in the opposition camp. Nor was that the end of it. Summoned to a meeting in Cedar Rapids on August 15, 1948 by Swift-Chicago Local 28, the Policy Caucus voted to withhold per capita from the international. Factionalism had erupted into open revolt.

Far from reversing events, the humbling of the UPWA ultimately reinforced the developing stable relations in meat packing. The central fact was that the crushing defeat and internal breakdown did not destroy the UPWA; recovery was quick and convincing. Beyond that, all the interested parties—not the UPWA alone—were chastened by the strike of 1948. It was a healthy lesson all around.

With one exception, normal relations soon resumed with the big packers. Despite their hostile strike measures and threatening posture afterward, they shortly sat down with the UPWA to renegotiate the master contracts expiring on August 11, 1948. (The recent fight had been limited to wages under the reopening clause of the contracts.) The only UPWA sacrifice involved union security. The packers seized the chance to eliminate maintenance-of-membership. Actually, the loss was not substantial. The check-off was retained and, moreover, was strengthened by irrevocability during the life of the contract and by automatic renewal except by written notice of cancellation. Since the Taft-Hartley Act permitted dismissals where union membership was a condition of employment only for failure to pay dues, the situation was not materially altered by the removal of maintenance-of-membership. No immediate money gains resulted from the post-strike negotiations, except the introduction of the triple-time rate for holiday work. But an informal understanding was reached with Swift for further talks in October without exhausting the reopening provision in the contract. The outcome was a 4 cent increase which, as usual, became general in the industry.[54] Meanwhile, the problems created by the strike were ironed out. Most of the discharged strikers—541 out of 591—were ultimately restored to their jobs at Armour, Swift, Cudahy, Morrell, and Rath.[55]

Only Wilson & Co., true to its past, long resisted the UPWA. In addition to refusing to submit strike discharge cases to the grievance machinery, the company would not renew collective bargaining with the UPWA. Under intense pressure, the company was forced to reenter collective bargaining in time for the 1949 negotiations. But a stalemate then developed on the issue of "super-seniority," that is, the advance of men who had crossed the picket lines in 1948 to the top of the seniority lists. The union could not accede to this "invitation to suicide." As a result, the company, although agreeing to the other terms and in fact abiding by them, refused to sign a master agreement. The UPWA had to work without a contract for several years until the super-seniority issue was settled during the Korean War. By its intransigence, Wilson as usual revealed what made the other packers more reasonable.

Foremost was the fact that the UPWA did not lose the grip on its share of the industry. The *National Provisioner* had noted that the 1948 defeat was the first of a major union in ten years. "For this reason, some observers anticipate far-reaching effects." The expectation was not fulfilled. Protected by Federal legislation, UPWA representational rights were unimpaired by the strike failure—a vital difference from the 1904 experience. Certifications were of course challenged in many places, but the UPWA—notwithstanding a bankrupt treasury, internal troubles, and a demoralized membership— won all of more than twenty major elections it was forced to undergo in the year after the strike. For the first time, the industry experienced "a 'lost' strike that did not mean a lost union."[56] That was the starting point for the resumption of normal relations.

The second consideration for the packers was the costliness of their victory in 1948. It was among the reasons for reduced earnings in that fiscal year. Armour & Co., the worst hit, fell from net profits of 31 million dollars in 1947 to a 2 million dollar deficit. The shutdown, the retiring president admitted in his stockholders' report, "had a tremendously adverse effect on our results."[57] There was little to show for the expenditure. No further doubt could be nurtured about the permanence of collective bargaining in the industry. It would pay to seek agreements rather than showdowns. Subsequent negotiations reflected that realization.

While defending its field and restoring industry relations, the UPWA was resolving its internal problems. The rump movement had only a short history. As in 1941, external crisis militated against disruption. The UPWA officers surely hit home with their charge that, while they were fighting for the union's life, "a group of disappointed factional leaders have abandoned the slightest pretence of union principle or concern for the welfare of the membership of this organization." Most of the rebel locals resumed per capita payments within a short time. In a few instances, dissident leaders were repudiated in local elections, and on the West Coast a recall election removed the district director from office. In Chicago, an administrator was placed over Locals 28 (Swift) and 100 (Armour 31st St. auxiliary plant). After resisting for a year, Local 28 finally accepted the administrator and a special election in July 1949. The defeat of the dissident slate marked the conclusion of the insurrection.[58]

From this shattering series of events emerged a working relationship between the rival unions. If the Amalgamated had thought of benefiting from UPWA difficulties, it was sadly disappointed. A raiding campaign launched immediately after the packinghouse strike yielded almost no return; the CIO plants could not be captured even when the UPWA was *in extremis*. That was conclusive corroboration of what had become increasingly evident in the recent past. Raiding simply did not work once a union—UPWA, Amalgamated, or Swift Brotherhood—had consolidated its position in a plant. The major source of rivalry had lost its sting.

Neither the Amalgamated nor the UPWA could be driven back by the other. Nor could either advance without the other in collective bargaining. The breakdown of cooperation in 1948 had, of course, unmistakable meaning for the UPWA leadership. The executive board, meeting after the strike, had "a long and important" discussion on the need for joint action. Herb March, always a proponent of unity, asserted in his analysis of the strike that "from all this we have got to learn a lesson of so working that the A.F. of L. cannot do what they are doing." March's District 1 later passed a resolution for a no-raiding agreement and for a joint national wage conference.

The Amalgamated was forced to the same conclusion. For its own bargaining posture was weakened by the UPWA defeat. Swift's position in the 1948 negotiations made it seem as if the Amalgamated was "going to have to pay for the untenable mistakes of the UPWA leadership." The AFL union, like the UPWA a month later, came away in June 1948 without a wage increase or maintenance-of-membership. In submitting to this, the Amalgamated National Negotiating Committee acknowledged "particularly the ill effects of the recent C.I.O. strike." The Meat Cutters, no less than the UPWA, had to recognize their mutual dependence in collective bargaining.[59]

When the difficult negotiations during the recession of 1949 came, therefore, UPWA and Amalgamated chiefs entered into closer cooperation than ever before. In early October, Presidents Jimerson and Helstein issued a statement castigating the packers and announcing a joint conference of local representatives (which does not appear to have been held). The Swift agreement followed on October 14. "Our experience in the prolonged discussions of the past year," the UPWA later concluded, "shows conclusively that where there is a united front of labor . . . the position of both organizations was materially strengthened." The two unions pledged themselves to joint action in future negotiations and to end raiding each other.[60]

Political differences were no material bar to cooperation. The Amalgamated was of course impeccably non-Communist. The UPWA, on the other hand, had an entrenched Red minority that had been decisive in Helstein's reelection in 1948. The executive board decision, reversing an earlier vote, to comply with the Taft-Hartley requirement for non-Communist affidavits from officers had little real effect. Resigning as director of District 1, Herb March—the key man—became an organizer for his Armour Local 347, and from there continued to dominate the Chicago district. The Communist grip on the district tightened in 1949 with the capture of the dissident Swift local. For its part, the UPWA leadership did nothing to oppose the Reds, in fact, connived with them in the Swift-Chicago affair. President Helstein permitted Communists—or, in the 1950's ex-Communists —on the union staff and was notably restrained in his pronouncements on the Cold War—in contrast to his volubility on domestic affairs.[61] Yet the UPWA was not a Communist-dominated union. The Reds, always a small minority even at the height of their influence in the 1948 convention, could not prevent the UPWA from endorsing CIO–PAC policy; and immediately after the convention the executive board came out for the Marshall Plan. The 1952 convention resolved against "all forms of totalitarianism, including fascism, communism, and any other authoritarian group" and "any activity within our organization on the part of communists, fascists or any other groups designed to destroy our democratic process to further their objectives." [62] The CIO had grounds for omitting the UPWA from its expulsion list. Political disagreement between the Amalgamated and UPWA, real although it was, was not great enough to counterbalance the pressures toward interdependence.

The past was, however, not easy to exorcise. A breakdown occurred during the second wage reopener of 1951. The Amalgamated accepted a 6 cent increase, effective December 17, on the grounds that this was adequate to cover the living cost rise and was the most that would be approved by the Wage Stabilization Board. The familiar cry of "sellout" then came from the UPWA, followed by an Amalgamated retort of "double-cross." UPWA President Helstein, it was charged, had indicated during negotiations that his union would go along with an Amalgamated agreement. "The failure of the UPWA officers to keep their word spells the end of cooperation between the two groups in contract negotiations." [63] The next round in 1952 found the Amalgamated accusing the UPWA of sellout for accepting a mere 4 cent rise from Armour. Renewed raiding accompanied the negotiating failure, and the Amalgamated had a few minor victories in the Armour branch houses in New York City and elsewhere. But a major Amalgamated breakthrough was averted by the failure to win a representation election in the Wilson-Chicago plant in early 1953. Amity was then quickly restored.

On June 23, 1953 a formal agreement was signed by the Amalgamated and the UPWA. In addition to a no-raiding clause, the pact contained provisions that went a long way toward negotiating unity:

1. In negotiations where each union represented one or more plants of the same company
   a. every effort to conduct negotiations in joint sessions with committees of both unions present
   b. neither union to execute an agreement without the approval of the other, except in the case of different expiration dates.
2. Joint economic action to the extent practicable; neither union to make a settlement that would weaken the position of the other.
3. Recognition of each other's picket lines to the extent lawfully possible.
4. In Big Four negotiations, joint meetings of negotiating committees, including local representatives, to receive recommendations representing the unanimous agreement of both joint committees and to reach accord on policies. [64]

There were loopholes: the 1952 master agreements of the two unions had different expiration dates, and no reference was made to common agreement on initial demands. But, given mutual good faith, the agreement could serve as the basis for a common front; in fact it did so. Discord would by no means disappear. But, however hard the words and distinct the formal actions, in practice the Amalgamated and UPWA accepted the fact of inter-dependence in collective bargaining. There would be no repetition of 1948.

Rival unionism had been a by-product of the organizing push. Unre-strained, union conflict put a brake on the final stage of unionization. Collec-tive bargaining could not reach maturity while the UPWA and Amalga-mated fought each other in negotiations. Becoming a reality by the early 1950's, interdependence marked the practical completion of the unioniza-tion process among the butcher workmen.

CHAPTER 12

THE NEW EXPANSIONISM

Unionization had run most of its original course by the opening of the 1950's. The bulk of the butcher workmen was organized, internal union order was secured, and collective bargaining was in full effect. The meat unions were increasingly caught up in the representational function that justified their existence. The shifting focus was sharpened by the emergence of issues harder to resolve by collective bargaining, above all, by industry change and technological unemployment. Yet the unionizing process was not at an end. Success generated its own momentum. And there was no lack of targets for the expansionist impulse. No longer the compelling concern that it had earlier been, unionization remained a significant activity to the union able to hold to it.

For the meat unions, as for all organized labor, the 1950's marked the start of a new era. American industry experienced a new burst of technological creativity—with uncertain, often injurious, consequences for trade unionism. The meat industry was no exception; it was, in fact, undergoing an unusually sharp break with past practice. Meat packing had been at the forefront of the mass-production revolution of the late nineteenth century. But the moving assembly line and the division of labor had seemed to exhaust the technological possibilities of packinghouse operation; only modest developments were registered in the twentieth century. Meat packing lagged far behind American industry as a whole. Output per man-hour increased at an annual average of only .5 per cent from 1899 to 1954, compared to 2.2 per cent for manufacturing generally. Then after World War II a second technological revolution hit meat packing. New methods and machinery began to appear: stunners, mechanical knives and hide skinners, power saws, electronic slicing and weighing devices. At the end of 1955, an engineer told the American Meat Institute that automation—that is, electronic control of fully mechanized production—was a technical reality in sausage and bacon operations. Labor productivity rose by nearly 15 per cent from 1954 to 1958. Demand could not keep pace, and employment began to fall. Total man-hours declined by 13 per cent during the four-year period. That represented nearly 18,000 production jobs.

Nor was that the whole story. Technological change speeded locational movement within the industry. Decentralization had been proceeding for many years in response to the dispersion of livestock sources, population changes, and merchandising requirements. From 1925 to 1947, for instance, packinghouse employment in Illinois fell from 26 to 14 per cent of the industry total. Now technological advances, improved plant design, and a less seasonal livestock supply rendered obsolete the old multi-storied plants in Chicago and the river points. Rather than rebuilding on the old sites, the packers preferred the decentralization made possible by motorized transportation; location was no longer tied to the railroad network. The outcome was, beginning with the closing of four Cudahy houses in 1954, the rapid abandonment of the giant plants of the older centers. By 1960, all the major houses in the great Chicago yards were silent or nearly so. An estimated 30,000 employees of the Big Four had been affected by plant closings during the 1950's.[1]

The retail trade was meanwhile undergoing its own transformation. The method of dispensing meat had not changed since the emergence of the modern industry in the late nineteenth century. Even after the chains had entered the field, the retail butcher had remained a custom worker, cutting to the order of the customer. The vital change started in the later 1930's with the appearance of supermarkets. These rested on the concept of self-service. Meat retailing shared in the change. Self-service departments jumped from roughly 300 in 1948 to 7000 in April 1953; by then, an industry expert estimated, 20 per cent of meat sales were from open counters. As the cutting operations moved to the back of the market, the way was opened for economies of scale. A division of labor began: weighing and wrapping became the province of unskilled females. Mechanical devices were introduced: electric saws, wrapping machines, automatic scales, conveyers. And, so the Amalgamated claimed, the meat cutter was being subjected to speed-up and to time-and-motion studies. The logical culmination of this development was the complete withdrawal of meat cutting from the stores. Both packing houses and chain warehouses began to establish cutting facilities. While slowed by spoilage and deterioration problems in fresh meat and consumer resistance to pre-packaged frozen meat, centralized production for retail was clearly under way by the end of the 1950's. Both Safeway and A & P had central cutting plants in operation, and packers, notably Swift, were making greater progress in the frozen field.[2]

Besides these sweeping industry changes, the larger situation altered markedly after World War II. The public scene darkened; organized labor lost some of the favor it had acquired during the New Deal. The consequences of Taft-Hartley, state right-to-work laws, a Republican administration in Washington, and hostile Congressional investigations could only be measured in the specific context of union activity, and the negative effects were exaggerated. But labor leaders did not doubt that the good old days were gone. "The pendulum that once swung favorably toward labor is now swinging in the other direction," intoned Patrick Gorman in 1955.[3] Social and economic changes were also proceeding. Unionization had started under the impetus of the Great Depression; it had advanced during the war and postwar booms. By the 1950's, rank-and-file militancy had dissipated, and the Cold War economy approached an equilibrium interspersed with recessions in 1949, 1954, and 1958. The labor movement, finally, was itself settling down. The AFL–CIO merger of 1955 meant the acceptance of the status quo and the legitimization of jurisdictional dualism. While hardly over, the era was passing when union conflict would generate much of the energy—as well as waste—that went into unionization.

Organized labor consequently lost momentum during the new decade. But, against the waning tide of the general movement, certain unions continued to expand. Among these was the Amalgamated Meat Cutters and Butcher Workmen. The figures speak for themselves: the average dues-paying membership for 1948 was 164,468; for the year ending February 1960 it was 326,676. The UPWA, by contrast, fared badly. After substantial growth during the Korean War, membership leveled off and then entered a perceptible decline. The average paid membership (excluding the Puerto Rican section) for 1959 was slightly over 110,000, 5000 less than 1957, and probably on a par with the level at the start of the decade.[4] Nor was there any promise of a reversal of this downward trend. The future, it seemed clear, rested with the Amalgamated. Why had it maintained, indeed accelerated, its effectiveness as a unionizing vehicle in a period of labor stagnation?

Part of the answer lay in the vigorous international administration. With the death of Secretary Dennis Lane in 1942, Patrick Gorman moved from a tenure of nearly twenty years as president to the secretary-treasurership. From that top post (as it in reality was in the Amalgamated), Gorman dominated the union. His peculiar mixture of tender- and tough-mindedness somehow suited the organization and the 1950's. An Irish wit, a sentimentalist, an idealist of old-time Socialist persuasion, and a confirmed iconoclast, Gorman infused vigor and enthusiasm into his staff. The Amalgamated, observers noted, escaped the lethargy that was spreading through the higher reaches of the labor movement. But Gorman was also a shrewd administrator who strove for an efficient, tight operation. On becoming secretary, he had immediately embarked on a program of streamlining the national office. The staff level was raised by the recruitment of capable younger men, by pay increases, and a retirement plan. An educational program began in 1945 with the selection of the two youngest vice-presidents, Harry Poole and Marvin Hook, to spend a year at Harvard, and with the attendance of the entire organizing staff at a two-week course at the University of Wisconsin.[5] (In 1955, Hook and Poole were transferred from district work to the international office, Hook to head the Retail Department and Poole to serve as executive assistant to President Jimerson and Secretary Gorman.) The international created a research department in 1941, then other departments for the areas of Amalgamated activity, and an insurance company to serve the local unions. In January 1947, the district system was inaugurated, bringing the international authority closer to the local level through the vice-presidents who headed the districts. The symbol of Amalgamated stature was the completion of an imposing building on the north side of Chicago to house the national offices.

The continuing fact of local autonomy contributed at least as much to the Amalgamated success. The comparison with the highly centralized UPWA was revealing. Because the local unions depended on the international, the UPWA could invest little of its income in organizing work. In 1950 the UPWA was expending three-and-a-half times as much on servicing local unions as on organizing activities. Trying to shift responsibilities from field representatives to local officials, the UPWA was able to increase its organizing expenditures from 4.3 per cent of total income in 1953 to 13.1 per cent in 1957.[6] The Amalgamated was much better situated. Local autonomy left the servicing of local unions largely in the hands of paid local officers. The international could use its representatives, numbering eighty in 1956, primarily for organizing purposes. Equally important, the Amalgamated locals engaged in organizing work. In part, they did so because local officials, paid from local dues, were directly interested in increasing the membership, but there was also the initiative fostered by the tradition of local autonomy. In one instance a New Jersey union, Local 56, sent one of its stewards on an organizing errand to Jacksonville, Florida.[7] Mergers of locals and the growth of district and state organizations—both continuing trends during the 1950's—enhanced the means for carrying on organizing work below the national level. The Amalgamated thus put its larger resources to work more efficiently to expand the union's membership.

Changes within meat packing meanwhile tended to favor the Amalgamated. During the great organizing period, the union had failed primarily in the big plants, in the packing centers, and in the major firms. All these sectors of concentration were lessening during the 1950's. The old multi-story factories were being replaced by smaller, modern plants. Production was also dispersing from the packing centers of Chicago and the river points. And the Big Four's share of the market continued to shrink; their portion of the beef trade fell from 38.3 per cent in 1947 to 30.8 per cent in 1955. These shifts permitted the Amalgamated to bring its decentralized power to bear more effectively. Its methods of persuasion and pressure on employers for organizing purposes now had broader scope. The Taft-Hartley Act did restrict secondary boycotts, but in practice circumvention was frequently possible. And the provision was balanced somewhat by the new freedom of employers to express preference for unions. A proved technique in retail and among the local independents in meat packing, organizing through employers continued to be important in the new decade.

The force of the Amalgamated tactics was further augmented by improved relations with the Teamsters. Cooperation with the strategic truckmen had been hindered in the past by jurisdictional troubles and a feud between Dennis Lane and the Teamsters' Dan Tobin. When both men passed from the scene, the national leaders of the two unions put aside old antagonisms. The jurisdictional irritant was finally settled by an agreement signed on September 24, 1954, defining the bounds of the two unions in the food field and establishing a joint committee which, among other things, would "review and settle any jurisdictional disputes which may arise from time to time." The agreement had an important unionizing by-product. The unions would engage in joint organizing followed by immediate division along the agreed lines. Although cooperative organizing was included in subsequent pacts with other unions, large benefits came only with the powerful Teamsters. By December 1958, 26 campaigns had been launched and almost 9000 workers organized, over half of whom went to the Amalgamated, and others were organized through joint drives under entirely local auspices. The achievement was even more impressive because, as Amalgamated Joint Chairman Leon Schachter observed, "we have often tackled the tough ones—the shops which were holding up contract gains." Despite the success of the experiment, the Amalgamated reluctantly terminated the agreement in 1958 as a result of the expulsion of the Teamsters from the AFL–CIO. Local cooperation nevertheless continued, and, with the passage of time, the desire increased to renew the profitable national connection with the Teamsters.[8]

The meat industry itself, however, offered a diminishing field by the end of the 1940's. Unorganized areas did still exist in the retail trade. In 1949, for instance, three of eleven A & P divisions and 400 of 1800 Kroger stores remained nonunion. So were some scattered smaller chains such as Weiss Food Stores in Pennsylvania and the Winn-Dixie chain in the south. The independents, who were becoming increasingly competitive under cooperative buying arrangements such as IGA, suffered some neglect. "There is the feeling with too many of our paid local officers," complained Gorman, "that it is too much trouble to visit so many stores for the purpose of collecting one or two men's dues in each place."[9] Certain areas—the deep south, upper New England, and Canada—were relatively untouched. There was uneven progress during the decade. Slow advances had been registered in the south, particularly in southern Florida, in the face of stiff employer resistance; in the southwest, only San Antonio lacked a meat cutters' local. The greatest single achievement came in the New York metropolitan area where, in 1955, final NLRB certification gave the Amalgamated 12,000 A & P employees. On the other hand, the union failed to capture the A & P New England division. The union, pushing against the hard core of the retail trade, still faced an incompleted task in 1960.

Meat packing had been closer to complete organization at the start of the 1950's. Industry changes, however, required continued organizing activity. For the larger part, geographical shifts did not reduce the extent of organization, although tending to benefit the Amalgamated and, to some extent, contributing to its growth. But one major industry trend raised new organizing difficulties. This was the substantial move to the south. Static as a whole until 1956 and then beginning a decline, packinghouse employment in the southern segment increased numerically by almost half from 1947 to 1958, and as a share of the total, from 9 to 13½ per cent. The southern independents, who, unlike the regional plants of the national companies, had not been organized, encompassed most of this growth. Still not overly important in total production, they posed a competitive danger to the unionized industry.

The southern workers were exceedingly hard to organize. The full-time efforts of six UPWA organizers for three years netted only 620 members. The UPWA won half of its NLRB elections in southern plants compared to two thirds elsewhere, and successful elections frequently led to bitter strikes for a contract (for instance, those fought by the Amalgamated against the Roegelein Provision and Peyton Packing companies in Texas). The difficulties stemmed partly from the Taft-Hartley provisions that permitted employers to influence workers and to initiate representation elections; from a slow-moving NLRB that, as a UPWA organizer complained after futile efforts at the Smithfield Packing company in Virginia, gave "nothing but a run-around . . . when we filed complaints"; and restrictive state laws and hostile public officials.[10] But the greater barrier was the southern pattern of employer and community resistance, small towns, race antagonisms, and violence—for instance, the beating of Amalgamated representatives in Gainsville, Georgia, in March 1951 and the dynamiting of the tourist cabin of an Amalgamated organizer in Center, Texas, in 1954. The southern independents remained a threat in 1960.

The meat industry narrowed as a field for the organizing vitality of the Amalgamated. As it happened, however, the union was expanding the scope of its membership. The process had started years before. The Amalgamated convention of 1936 had opened the union to poultry, egg and creamery, and wholesale and retail fish workers. The 1940 convention added sheep shearers and stockyard workers. Actually, activity spread far beyond these formal limits to canneries, fishing and oystering enterprises, and a miscellany of other food-processing firms. The executive board had initially felt bound to find some connection with meat production, limiting the union's interest in tanneries to departments in packing houses, and cannery workers to products containing meat. But by 1948 Gorman was claiming "that the whole giant canning industry should be organized under the banner of our International Union . . . Certainly no other International Union has more right than our own to organize this field generally." [11] The Amalgamated was far along in its transformation into a diversified food union: thus the inclusion in its constitution of a blanket claim to "any employee engaged in the processing of foods."

The logic was compelling. For one thing, the processing of meat and other foods was interconnected in a variety of ways. Urging the AFL to grant the Amalgamated jurisdiction over fish markets, for instance, Gorman pointed out that often these were on the same premises as meat departments and "the two fields cannot be separated"; and, moreover, that "the filleting and preparation of fish . . . is a slaughtering operation . . . the tools . . . are identically the same as those used by butchers in packing establishments." [12] The new fields were extensive and, in some cases, rapidly expanding. After World War II poultry processing experienced a transformation similar to the development of meat packing more than half a century before. Chickens had been sold either live and fresh-killed or "New York-dressed," that is, by the inferior method of refrigeration without evisceration. The trade, employing 14,506 in 1939, had been local, seasonal, and unimportant. Advances in sharp-freezing, packaging, storage, and transportation permitted production of cleaned chickens at efficient plants. With poultry consumption rising, employment nearly tripled during the 1950's. Finally, the related food-processing fields were largely unorganized and neglected; some had, in fact, not been allocated by the AFL to any national union.

Actuated by these considerations, the Amalgamated had expanded outward on a small scale during earlier growth periods, for instance, during World War I.[13] The impulse intensified during the 1930's and then became irresistible as successes and failures began to set limits on further growth among meat workers. Vice-President T. J. Lloyd expressed the motive clearly for expansionism within his Nebraska territory: "I couldn't see much future in just working on Meat Cutters and Packing House Workers, so I started on a campaign on the Creamery, Poultry and Egg Houses." [14] The processing workers, for their part, were ripe for organization; they were eager to share the union benefits of the more privileged butcher workmen.

Initially, the expansionist impulse came fom the local unions. The completion of organization turned ambitious leaders to new fields for more dues-paying members. Often young and fresh from quick victories, they were not restrained by the older tradition of jurisdictional appropriateness, and they did see that diversification would permit greater resources in a fight in any single area. The addition of provision men and wholesale fish workers, explained an official of Los Angeles Local 421, "have made possible not only a greater financial assistance, but have added strength and unity to the Meat Industry." [15] San Francisco Local 115 enrolled poultry workers and retail fish men in 1936, expanded to wholesale "fish butchers" in October 1937, and then to ice cream wrappers, oyster men, and egg workers. Depending on what was available, other California locals took in tannery workers, wool pullers, egg candlers, apple peelers, crab pickers, and so on. In the northwest, the state branches took the lead. The Washington Federation of

Butcher Workmen in 1935 and 1936 organized the poultry industry and negotiated a statewide agreement with the Washington Cooperative Poultry and Egg Association. Eastward such expansion proceeded on a more limited scale. The outstanding achievement was among the food workers of southern New Jersey. Leon Schachter, a business representative of Philadelphia Local 195, began in early 1940 to work in the Camden vicinity with remarkable success. He organized the extensive mechanized Seabrook Farms at Bridgeton, as well as the related quick-frozen packing plant. Local 56 soon covered parts of the whole spectrum of south Jersey food industries, including seasonal agricultural workers.

The international of course applauded and assisted this progress, but it launched no concerted drives in the new fields until after 1941. During the war period, many small packing plants were shut down by price controls and livestock shortages, and large numbers of meat cutters left for defense jobs or military service. These setbacks prompted the Amalgamated to start a drive on the many midwestern poultry and egg plants in February 1943, and then in the important Delaware industry. The National Poultry Conference, meeting in July 1943 to consider the substandard conditions in the industry, reflected the mounting importance of the poultry workers to the Amalgamated. The international also became active among the fish and oyster workers on the East Coast, particularly in the Chesapeake area, and among cannery workers. By the end of the war, expansionism had become national policy.

The new, diverse groups were easily accommodated by the Amalgamated, representing as it always had a variety of trades. Allocation into locals depended largely on the source of organization. Local unions usually held on to recruits, placing them generally in auxiliary or subordinate bodies. Organizing by the international or state branches resulted in separate unions of poultry workers, fish handlers, and so on. This arrangement was, however, tempered by practical considerations. For better servicing, small groups from different fields were combined into single unions, or, less frequently, a large union was broken up. In the latter case, the business agent usually protected his interests by multiple office-holding. Occasional jurisdictional squabbles did occur between local unions, for instance, over retail fish workers in New York City, but such disputes had little importance.[16] At the international headquarters, departments were eventually created to service the new groups.

Amalgamated expansionism ran into resistance by the AFL. In February 1937, the executive council rejected the Amalgamated claim to fish workers.

The decision was reversed in August after (and probably because of) the open break between the Amalgamated and the CIO. When the Amalgamated sought in 1940 to affiliate the Sheep Shearers, a small international union, the AFL refused its consent. The Amalgamated nevertheless went ahead. "No reason whatever was given why the Executive Council would not officially approve," Gorman protested to President Green on June 7, 1940. "These people will be enrolled . . . Each of them have [sic] already paid their initiation fee." Despite the surrender of the Sheep Shearers' charter, the AFL kept the union on its arrears list for two years before removal—still withholding official recognition of the merger.[17] The Meat Cutters also entered the canneries over the opposition of the federation. The fact was that the Amalgamated could not be limited by AFL rulings. It was responding to basic needs, and it was operating in neglected fields. Perhaps more important, the traditional concept of jurisdiction was loosening as a result of the CIO and the worker's right of choice under the Wagner Act. And, in the final analysis, the power to decide rested, not with the AFL, but with the Amalgamated itself.

The UPWA found diversification a harder undertaking. For one thing, the CIO union started later. It still was occupied with the consolidation of its successes in meat packing when the Amalgamated was capturing allied fields. The industrial-union concept also had a restricting effect, for it made precise the appropriate area of activity of the UPWA. For a time, the union amicably handed over retail, cannery, and seafood workers to other unions. Paradoxically, the CIO stimulus within meat packing was an inhibiting factor outside that industrial scope. The end of meat-packing growth eventually prompted the union to follow the Amalgamated path. (In 1960, in fact, the union changed its name to United Packinghouse, Food and Allied Workers, although retaining the initials UPWA.) But it lacked some Amalgamated advantages: its concentrated organizational strength and centralized organizing procedure were ill-suited to the dispersed food-processing industries; it was weaker financially; it could not exert the boycott pressure from the retail end as could the Amalgamated, for instance, against the Delaware poultry houses; and it lacked the assistance of the Teamsters. Thus limited, the UPWA did expand its scope to include CIO sugar locals in 1947, followed by sugar unions in Puerto Rico; Canadian shoe, flour, and cereal workers; the huge Campbell Soup plant in Camden, New Jersey, as well as a variety of other food-processing plants; agricultural workers, including 2000 California migratory workers in Local 78.

The UPWA could not match its AFL rival. By 1948, Amalgamated enrollment of fish workers reached 4000, cannery workers 10,000. The Amalgamated claimed 22,000 poultry workers in 1953. The shift of the industry to the south raised a block to further poultry expansion. A major drive, inaugurated in 1958, was halted midway with disappointing results. The southern industry remained unfinished business. Yet, on the whole, miscellaneous food workers were a continuing source of membership in the 1950's. In Chicago, for instance, Local 55 was chartered in 1953 to cover grocery clerks of the Jewel Tea Company. The new union began to take in fish workers, poultry workers, egg candlers and breakers, and so on; by 1960 it had over 3000 members. At its fifteenth anniversary in 1956, Food and Allied Workers Local 56 of southern New Jersey had a membership of over 6000 and contracts with 162 companies in fifteen food fields.

The expansionist facility imparted a new dynamism to the Amalgamated. It was not only a matter of numbers. The union could easily have sunk into the conservatism of a craft-oriented meat cutters' organization after its defeat in the major packing centers. But it was invigorated by its entry into the canneries, the seafood regions, and the poultry and egg plants. These fields involved seasonal labor, substandard conditions and pay, Negro and Spanish-speaking workers, and many women. The Amalgamated was liberalized by the needs of the newcomers. Among other things, the union became more aggressive on race questions and in political and legislative affairs. Migrant labor, for example, became a chief concern of Leon Schachter, who, from his start in south Jersey, became an international vice-president and Amalgamated representative in Washington. The Amalgamated took the lead in the campaign for Federal poultry inspection, culminating in a law in 1957, and for the extension of protective labor legislation. In the 1950's, the Amalgamated developed a progressive spirit that had been lacking in the 1930's.

The extension into new fields, at bottom, sprang from enlarged power and resources accruing from successes within the regular Amalgamated field. So did expansionism of a more traditional kind. Unionization of the retail butcher workmen spilled over to other store employees, drawing the Amalgamated into a bitter and prolonged dispute with the Retail Clerks' International Association. Grocery clerks became an important source of growth. But their recruitment brought also the liabilities to the Amalgamated—all the greater for being considered the aggressor—that went with jurisdictional conflict. And, having embarked on retail expansionism, the Amalgamated later found it impossible to extricate itself from the commitment and its accompanying difficulties.

Troubles with the Clerks dated back to the very beginning of the Amalgamated. The Retail Clerks' International Association, not easily reconciled to the AFL confirmation of Amalgamated retail jurisdiction, had continued to organize meat cutters at some points. On the other hand, butchers' unions—for instance, Local 534 of East St. Louis—took in grocery clerks. During World War I, President Gompers tried to bring about an understanding between the two organizations, but with little success.[18] The dispute, however, was of no particular moment, for the infringements generally occurred where the other union was inactive. The conflict could not have much substance while the retail trade was largely unorganized.

This situation changed during the union upsurge of the New Deal era. Initially, neither side paid attention to the jurisdictional niceties. The Retail Clerks, for instance, signed agreements for all retail workers with the Economy Grocery Company of Boston, the American Stores in several eastern cities, the Sanitary Stores in Washington, and the Loblaw Company in Buffalo. For its part, the Amalgamated did the same. The Clerks, Gorman observed, "found out that two could play at this game."[19] This no doubt justified the enrollment of grocery clerks, but the fact was that both unions responded identically to the organizing opportunity.

The contest proved patently unequal. The Clerks' Association was weak and impoverished. In 1937, several important New York locals seceded to become the nucleus of the rival United Retail and Wholesale Employees (CIO). Since it was an organization of unskilled, low-paid workers, the RCIA did not attract the craft-conscious butchers. The RCIA tried to overcome these weaknesses by offering cheap contracts and cheap dues.[20] The Meat Cutters easily overbalanced such measures. Strong on all counts where its rival was weak, the Amalgamated quickly demonstrated its ability to cover ground in the unorganized retail field. The Clerks could not hold even the initial gains among the meat cutters. The Loblaw Stores in Buffalo, for example, had signed an agreement with the RCIA covering all retail employees. Having organized the rest of the trade, the Amalgamated local addressed itself to the Loblaw meat cutters. In December 1937, the meat cutters came out in conjunction with a Teamsters' strike against the chain. The outcome was the enrollment not only of the Loblaw butchers but also the clerks and, in addition, an understanding that the Amalgamated should represent the clerks for the entire Loblaw Company.[21]

Righteousness therefore became profitable to the Retail Clerks. The executive board, meeting in October 1937, "went on record that it was not the purpose of our International to accept for membership any person who comes under the jurisdiction of any other organization . . . Officials of Local Unions are requested to refuse to accept for membership any and all who are not eligible to our Association." And protests were lodged with the AFL against the jurisdictional infringements on the Retail Clerks.[22]

The Amalgamated was meanwhile reaching a different conclusion. The proper place for grocery clerks, the international decided, was with the Meat Cutters and Butcher Workmen. This was a new position. In 1932 the Amalgamated, while urging a campaign among the retail clerks, had not suggested itself for the job. Now, as the clerks came within grasp, the Amalgamated's appetite was whetted. A Wisconsin official thus pointed to "at least 2,500 clerks working in combination stores that are not asking but begging to come into the Amalgamated." And he proceeded to calculate the income that would accrue from this added membership. Moreover, the Amalgamated was accomplishing the job. Why should the Retail Clerks reap the benefit? A New England organizer complained of "the lack of any great amount of work . . . until we had made substantial progress and then they have attempted to chisel in." The Clerks entered his territory, another official charged, "with the sole intent of 'rapeing' [sic] our organization of fifty per cent of our membership." The clerk membership represented too heavy an investment to be lightly given away.[23] Besides that, a single union in food stores would preclude sympathy strikes, petty jurisdictional squabbles, and duplicated effort. "Is it reasonable," Gorman asked, "to insist that in a store employing three people and many cases only two, that two internationals must represent those two or three employees?" [24]

These considerations—as well, no doubt, as the need to respond to the charges already lodged by the Retail Clerks with the AFL—prompted the Amalgamated to apply for an extension of its jurisdiction to cover grocery clerks. Pending a decision, the executive board agreed on March 23, 1938 "that our policy with regard to food handlers be continued—to organize them where there is no organization of clerks in that particular locality." The Amalgamated was proceeding in the established AFL tradition of *realpolitik*.

President Green handed down his decision on November 19, 1938. As was altogether expected, he flatly rejected the Amalgamated bid for broader jurisdiction. He was concerned only with defining the jurisdictional line between the two unions:

Where said meat cutters and clerks who sell meats are employed in selling groceries and food products in connection with their meat selling and meat cutting duties, they come under the jurisdiction of the Amalgamated . . . Those employed in selling groceries, spices and food products, *other than meat,* handled and sold in grocery stores, come under the jurisdiction of the Retail Clerks.

The statement contained enough ambiguity to serve Amalgamated purposes. The award meant, Gorman immediately wrote to Charles Shimmat of A & P, "that not only meat cutters . . . but also the clerks who sell meats in connection with their other duties as a clerk also come under the jurisdiction of our organization." Green "has officially decided that if they handle meat, even in grocery stores, they belong to the Amalgamated." [25] Gorman's reading of the decision, in effect, awarded the Amalgamated the broad retail jurisdiction it had requested.

Such had not, of course, been President Green's intention. He endorsed the interpretation of Secretary C. C. Coulter of the Retail Clerks that the award gave the Amalgamated only those "who may sell other food products in combination with their regular duties in meat departments." But while some uncertainty remained concerning combination markets, President Green drew the line at purely grocery outlets. He told the AFL Executive Council that he was trying "to lay down the rule that the problem is such that if there is no butcher shop then there is no skilled butcher workman." Gorman refused to accept the distinction between fresh and preserved meat. He insisted "that our organization should have jurisdiction over every store where meats were sold and all men employed therein should belong to our organization." Despite Coulter's protests and Green's strictures, the Amalgamated continued to enroll grocery clerks along with meat cutters.[26]

Restraints more compelling than the authority of the AFL, however, were at work. Particularly in less populous regions, mutual interests bound the Meat Cutters and Retail Clerks. Often they negotiated jointly and not uncommonly shared a business agent.[27] Jasper Rose, later an Amalgamated vice-president, acted in this dual capacity in Davenport, Iowa. Ray F. Wentz, also a future vice-president, helped to organize the St. Paul clerks' union in 1933 and was thereafter counted among its closest friends. A network of alliances thus existed at the local level of the rival organizations.

Nor did all meat cutters' unions want the lowlier retail clerks. The Seattle local, for example, refused to accept the delicatessen employees in department stores after they had been organized by the business agent. The international endeavored to overcome this craft opposition by separating the clerks in auxiliary bodies. Secretary Lane explained the reasoning to the reluctant Auburn, New York, local:

We advise this program, because it would not be good to put them right into Local 2 and mix the membership with meat cutters, as the clerks would soon outnumber the meat cutters and dominate the local . . . By organizing the clerks . . . under the charter of local 2, known as branch "B," you would have more control over them and could assist them in negotiating working conditions and a wage scale that would be fair and just for the kind of work they are performing. We have the grocery clerks organized in this manner in lots of cities.[28]

It was possible, especially in the later period, to organize the clerks into separate locals; but since much of the organizing work was done by butchers' unions, their craft consciousness even then constituted an inhibiting factor.

Opposition also came from important employers. Some chains—for instance, Loblaw and First National—were entirely agreeable to dealing solely with the Meat Cutters.[29] But others did not want, as a Safeway official said, "to get in the middle of a Union scrap." By not raiding the Clerks' locals, the Amalgamated had hoped to obviate this possibility, but the RCIA refused to be quiescent. In 1941, C. C. Coulter threatened to picket Kroger in Pittsburgh if the company recognized the Amalgamated as bargaining agent for the clerks. The company was not comforted by Gorman's assurance that Coulter was merely "using his usual well-worn bluff." Moreover, chains were reluctant to jeopardize good relations with the Retail Clerks by siding with the Meat Cutters at some points. "We have over-all commitments to the R.C.I.P.A. just as we do with the Amalgamated Meat Cutters," a Safeway executive pointed out to Gorman.[30] So major chains resisted Amalgamated expansionism. The National Tea Company, for instance, sought an arrangement with the AFL to channel employees into the appropriate unions. More decisively, Safeway instructed its district managers to respect the jurisdictions of national unions: "we will refuse to recognize a Clerks Union's jurisdiction over butchers unless the butchers relinquish jurisdiction, or Amalgamated Meat Cutters' jurisdiction over clerks unless R.C.I.P.A. relinquishes jurisdiction."[31]

Meanwhile, pressure from the AFL mounted. President Green made a certain concession in interpreting his decision of November 1938: Amalgamated jurisdiction could cover all employees in combination stores who handled meat, even if only occasionally; strictly grocery clerks and all employees in stores without fresh meat counters belonged to the Retail Clerks. Simultaneously, the AFL became insistent on acquiescence. On August 29, 1940, Green telegraphed Gorman angrily to "instruct your representatives to refrain from organizing grocery clerks into your meat cutters unions." Finally, C. C. Coulter of the Retail Clerks on December 31, 1941 filed formal charges with the AFL Executive Council against the Amalgamated "on the ground of wilful and intentional infringement against the jurisdiction accorded to our Association under its charter from the American Federation of Labor." [32]

The Amalgamated was ready for an accommodation. By the end of 1941, the retail food trade (except in the south) no longer offered the open field of previous years. Raiding the Retail Clerks had, after all, never been the aim of the Amalgamated. Moreover, the country was at war. It was not a time to become involved in jurisdictional strikes. On April 21, 1942, Gorman and Coulter met with a mediator appointed by President Green, Reuben G. Soderstrom of the Illinois Federation of Labor. At first, Soderstrom reported, "both sides were a little testy." But they agreed "that the AFL decision was workable and adequate." Further meetings culminated in an accord on October 12, 1942 on the basis of the jurisdictional decision of 1938. To avoid further trouble, representatives of the unions would meet quarterly to discuss jurisdictional problems. [33]

But changing retailing methods prevented a permanent peace. Supermarkets and self-service of course had an unsettling impact on jurisdictions. Who should claim female weighers and wrappers? pre-packaged meat, poultry, and fish? sausage in delicatessen counters and other meat products sold far from the meat departments in modern supermarkets? Jurisdictional strife soon intensified. In February 1949, Amalgamated leaders reappeared before the AFL Executive Board to answer the charges of the Retail Clerks. The outcome was another agreement, dated July 7, 1949, which attempted to deal with the new retail practices. The Meat Cutters were awarded the handling of pre-packaged meat in self-service counters and also frozen meats "where similar or like methods . . . may apply." The Retail Clerks received canned meat products normally sold in grocery sections, as well as checkers and cashiers when meat and groceries were paid for together. The knotty delicatessen question was to be resolved by "common sense . . . where the preparation or sale of meat is a major or substantial factor." The parent bodies, when such problems could not be settled, would select a neutral party to make a binding decision. [34]

Yet the issue could not be laid to rest. Neither side felt that its jurisdiction, as each chose to define it in the agreement, was being properly protected. And there were also gross infringements. The climax came in California, where the Clerks were strongly organized. A controversy there, accompanied by injunctions, law suits, and resort to the NLRB, forced a new accommodation. An agreement, signed December 12, 1955, called for a joint committee to handle disputes. President George Meany of the AFL–CIO, who was a party to the agreement, would be final arbiter of unresolved disputes. Soon the Clerks raised a question over the definition of the meat department area in a retail store. The issue went to George Meany for a decision. His ruling against the Amalgamated angered Gorman, who vowed "that if we had a thousand disputes with the Retail Clerks, we were going to keep away from arbitration and going to President Meany." [35] Ultimately, a memorandum, dated October 7, 1958, established a definition more in accord with the Amalgamated's broad concept of a meat department. No sooner had the territorial question been settled than wrangling began over the meaning of "meat and sausage products." Precision seemed impossible to achieve while retailing methods were in flux.

There were larger obstacles to a resolution of the differences with Retail Clerks. The dispute involved exclusive jurisdictions, not the dual jurisdictions of AFL and CIO unions. Even if exact lines could be drawn and be punctiliously honored by both unions, there would remain the problem of transferring existing memberships. For exclusive jurisdiction did not permit the acceptance of the status quo, as, for example, was possible between the Amalgamated and the UPWA. The agreements of 1942, 1949, and 1955 thus spoke of the need to correct "present inconsistencies" or "jurisdictional irregularities." No actual moves followed until the 1955 agreement, which did result in the transfer of a "substantial number of members" to the Clerks and of "a small number of members" to the Amalgamated. The 1960 Amalgamated convention revealed widespread resentment against such exchanges. Clearly, here was a dilemma of major proportions.[36]

The other fact was that modern retailing methods whetted the Amalgamated appetite for expansion in the food stores. Since the meat-cutting tasks were shrinking, the surest defense of its retail interest was to stake out the largest possible claim to food-store workers. Whatever remained of the craft concept of jurisdiction, moreover, was breaking down. The journeyman meat cutter, in the old sense, was disappearing in the face of self-service and central cutting, and with him went the inhibiting idea of jurisdiction based on the kind of work. Now the Amalgamated defined its retail coverage by the product or work area.

It was ultimately logical for the Amalgamated to desire all food retail employees. In 1949 Gorman reiterated the arguments for a transfer of jurisdiction to his union. He also denied any intent of acting unilaterally and, failing to win his case, abandoned it.[37] But it was not so easy to abandon the clerks. Gorman never relinquished his conviction that they should be granted to the Amalgamated. (He repeatedly pointed to the fact that, even without food stores, the Retail Clerks would still have the largest jurisdiction of any union.) This view doubtless hindered a permanent accommodation with the Retail Clerks. When the 1955 agreement expired on August 1, 1960, it was not renewed.

Whatever strains the dispute created within the labor movement, there was no question that it contributed to Amalgamated growth in the 1950's. The constant pressure for redefining jurisdictions did enlarge the Amalgamated territory in the supermarkets. And, beyond that, there were repeated breaches of established lines, most notably perhaps in the capture of the A & P clerks in New York City in 1952. The continuing expansion of the Amalgamated rested partly on its ability to tap the larger retail food field.

Amalgamated expansionism took one form that involved, not the recruitment of new members, but of other unions. Four internationals, three of them small ones, merged with the Amalgamated, accounting for probably a seventh of its size in 1960. Besides the members they brought in, the mergers added to the diversified character of the Amalgamated. And, what ultimately might be of larger significance, they encouraged the prospects of ending the split that had accompanied the organization of meat packing.

From an institutional standpoint, within organized labor no act was as momentous and as hard to consummate, as the merger of one national union into another. That move, involving as it did the surrender of autonomy, would be undertaken only for compelling reasons. Even then, the act required a host union of a special kind. One such was the Amalgamated; it was, as its journal put it, "a merging union." For one thing, the Amalgamated was so situated that it touched a number of lesser unions. Second, it had a capacious interior. Another union could enter, become part of the Amalgamated in membership and local affiliation, and still retain a degree of its identity, autonomy and leadership. Third, the Amalgamated had financial stability. A union could be sure, at the least, that it was not adding to its burdens by a merger with the Amalgamated. Finally, Patrick Gorman and some of his associates actively sought mergers, devoting themselves to the ambitious and large objective of a unified food workers' union.

The merging procession began back in 1940 with the Sheep Shearers' International Union (AFL), a small organization in the northwest. Despite an existence of many years, the union was weak and lacked recognition from the National Wool Growers' Association. The union expected the merger to secure the backing of the strategic Amalgamated against the hostile employers. The Sheep Shearers became a "division" of the Amalgamated, retaining A. A. Evans as president, its own laws, and "the right . . . to carry on the affairs of their organization in the same manner as though they were still an International Union of their own," but also, as Secretary Lane assured the executive board, subject to "the laws of the International the same as all other affiliated locals." [38] In 1951, the United Leather Workers' International Union (AFL) entered the Amalgamated. The Leather Workers covered only a small part of the tanning industry and, lacking in resources and militancy, faced an unpromising future. Because of its meat jurisdiction and interest in leather workers in packing houses, the Amalgamated seemed a logical haven. The Leather Workers' Union, bringing in roughly 5000 members, received essentially the same semi-autonomous arrangement as had the Sheep Shearers. But the scond merger had an added difficulty: the small staff in the Leather Workers. The Amalgamated agreed to employ four or five as organizers. It was probably decisive, however, that the president, Bernard Quinn, was over seventy and ready for retirement. A third merger occurred in 1955 with the affiliation of the Stockyard Workers' Association of America. This was an organization of four livestock handlers' locals that had recently seceded from the UPWA. Never viable, the union was seeking another affiliation. The Amalgamated, through the quick work of Marvin Hook, beat out the Swift Brotherhood and District 50. These accessions, themselves minor, were accompanied by a fourth merger of great significance.

The International Fur and Leather Workers' Union of the United States and Canada was the major union in its fields.[39] For many years, the IFLWU had been a vigorous union with forceful leadership, a loyal membership that numbered 40,000, and a record of achievement in collective bargaining. The union, still strong in 1954, confronted mounting difficulties. Both fur and leather were sick industries with declining employment and, in leather, increasing resistance to the IFLWU. The primary trouble, however, was political. The IFLWU had long been a Communist-dominated union; its president, Ben Gold, and his assistant, Irving Potash, were prominent party members, and Communists were entrenched at the local level. This became an insupportable liability in the Cold War era. The IFLWU was one of the left-wing unions forced out of the CIO. It then became fair game for other unions and, with the Amalgamated now in the leather field, faced a new and potent rival. There were also legal liabilities. The IFLWU could not use the NLRB until its officers signed non-Communist affadavits, and after Gold did so, a Federal grand jury found that his resignation from the party had been a ruse. The IFLWU then again faced the possibility of losing access to the NLRB. Gold himself was found guilty of perjury in April 1954. These pressures forced a leadership change. Gold resigned in October 1954 after the passage of the Communist Control Act destroyed any possibility that the IFLWU might retain its NLRB privileges while he was president. Gold's replacement was Abe Feinglass, a non-Communist official. This was an opportune time, Amalgamated strategists surmised, to broach the idea of merger. Feinglass' situation was probably still insecure, and the Amalgamated had recently been cutting into IFLWU territory. The Amalgamated made contact with Feinglass. After clearing away initial suspicions, negotiations proceeded with unexpected swiftness. A Merger Agreement was signed on December 28, 1954.[40]

The Fur and Leather Workers, Feinglass asserted, had "pride in our own past accomplishments as trade unionists." [41] Merger would have to permit some continuity with that past. The Merger Agreement made the Fur and Leather Workers part of the Amalgamated: the membership would receive Amalgamated cards; finances would be combined; the locals would be under the supervision of the district vice-presidents; and the locals would be subject to the provisions of the international constitution. But it was also possible to retain a measure of the older identity. While receiving Amalgamated charters, the IFLWU locals, district councils, and joint boards "shall continue their respective identities." [42] They would retain their existing constitutions and governing practices, except when these conflicted with the Amalgamated constitution, and their certifications and contracts. A Fur and Leather Department would be created in the international. The fur and leather locals would hold biennial conferences which would choose a departmental council. Two fur and leather representatives would become international executive board members, one of whom (Feinglass) would be director of the department. The Amalgamated would also appoint as organizers and staff members a group of IFLWU employees agreed upon during the negotiations. While surrendering its independent base of power, the Fur and Leather Workers' Union did not expect to be entirely absorbed by its entry into the Amalgamated.

The Communist issue, however, raised unforeseen difficulties. The Amalgamated of course wanted safeguards against "further subversive or anti-American activities" within the Fur and Leather Workers. Under the Merger Agreement, every official down to the lowliest would have to sign the non-Communist affidavit established under the Taft-Hartley Act, those not required to do so by the law to deposit the document with the Amalgamated. It was also understood that neither Ben Gold nor Irving Potash would ever be permitted to hold any union office. These safeguards satisfied the Amalgamated and the IFLWU, but not the executive council of the AFL. George Meany, who had at first seemed favorable, raised sharp opposition to the Merger Agreement. At its meeting in February 1955, the executive council accused the Amalgamated of providing a "temporary haven" for the Communist-dominated union. (The merger, which went into effect when a special IFLWU convention approved it at the end of January 1955, would not be final until the 20th Amalgamated convention in 1960; until then, both sides had the right to terminate the merger.) Acknowledging that it could not prevent an autonomous affiliate from entering a merger, the council refused its approval. It publicly rebuked the Amalgamated for acting in the face of its opposition, and raised the possibility of sterner measures, namely, expulsion.[43]

The Amalgamated refused to reverse its course. But it did seek to answer the objections of the federation leaders by a thorough purge of the Fur and Leather Workers. The international brought to bear its full constitutional powers. The key point was the Fur Workers' Joint Council of New York City. The international demanded the resignation of thirteen officers, and when three refused, suspended them. Elections scheduled for June 22, 1955 were postponed, and Abe Feinglass was appointed receiver of the council. He embarked on a campaign for a "Unity" slate that would be acceptable to all shades of opinion and simultaneously be anti-Communist and pro-merger. His were standard tactics for such an occasion. For the sake of "unity," he urged men to "place their Union above everything else, above all differences." The fierce contest ended in the election of the administration candidates on September 1, 1955. Elsewhere the older leadership of the fur locals was similarly dislodged. The harsh process, Feinglass admitted, meant "some sacrifices which might well be described as ruthless." [44] But the main purpose was served.

These were hard months for Amalgamated leaders. They had invoked the wrath of the tribunes of the labor movement. Afterward, Gorman recalled the hostile reception at his appearances before the AFL Executive Council to argue the Amalgamated case, "feeling somewhat of an outcast in having, yes, many of the Board members pass you by without giving you the time of day." There was also the suspicion that some of the Amalgamated critics were acting from ulterior motives as well as from anti-Communist fears. Still, as the purge proceeded, federation opposition became untenable. Finally, George Meany judged that the "effective steps" by the Amalgamated to take "the fur and leader workers out of the hands of the Communist elements" made them "worthy of admission to the ranks of decent trade unions." Meeting in October 1955, the executive council lifted its objections to the merger.[45]

An array of other problems were meanwhile being encountered. In eastern Massachusetts a group of tannery locals with roughly 5000 members broke away. Except in Lynn, most of the seceding locals were lost in NLRB elections. Another secession began in Canada when the Amalgamated ousted some local leaders, but the unions were quickly clamped into receiverships and the movement was broken. Legal complications also developed in Canada when the Ontario Labour Relations Board refused to transfer certifications from the IFLWU to the Amalgamated. Other sources of internal friction developed. There was, for example, the problem of the IFLWU's hand bag workers. These were claimed by the Hand Bag and Leather Workers' Union. Gorman had assured the AFL Executive Council that the Amalgamated would honor the Hand Bag Workers' jurisdiction. Although Gorman denied having promised to turn any workers over, that was the expectation of the Hand Bag Workers' Union. When the IFLWU Chicago Joint Board officials discovered this, they immediately objected. Gorman refused to force the transfers and thereby raised new criticism within the AFL.[46] On the whole, nevertheless, these troubles were outweighed by the merger gains in collective bargaining, in organizing, and in an end to raiding.

The Merger Agreement of December 1954 had created the formal frame by which the Fur and Leather Workers would live within the Amalgamated. But the implementation was the critical question, and that depended on the good faith and the intentions of the Butcher Workmen. At the 20th convention in Atlantic City in June 1960, fur and leather delegates met in caucus. They were satisfied. The Amalgamated connection had been beneficial. But what was equally important to the delegates, they did not think they had sacrificed the identity of the Fur and Leather Workers. Aside from the Communist issue, they had run their own internal and bargaining affairs, and their department under Abe Feinglass had sustained national ties.[47] The Fur and Leather Workers took their places as one of the diverse groups making up the Amalgamated. The convening of the 20th convention, automatically making the merger permanent, marked the end of the hard term of trial.

One further merger would have climaxed the unifying process. With the UPWA joined to the Amalgamated, the remaining de-stabilizing result of meat-packing unionization would be removed. Men on both sides saw the benefits of a merged organization. "In recent years," UPWA President Helstein told the Amalgamated convention of 1956, "we have been able to overcome many of the disadvantages that grew out of the divisions, but what we have been able to do is nothing compared to what the future holds for us when in a united organization we move forward and together." Merger, the two unions acknowledged, "would be to the best interests of both organizations." [48]

The idea of merger had never been entirely absent. Patrick Gorman, more than anyone else, clung to his ambition for a single organization in the meat industry. Before the break between John L. Lewis and the CIO, Gorman had hoped for a transfer of packinghouse locals to the Amalgamated. After the UPWA was formed in 1943, he began to turn over the thought of a merger. The possibility took hold with emerging interdependence of the two unions in collective bargaining. "More could be accomplished," remarked Gorman, "through joint efforts and, better still, single effort through a single union." [49] In 1947 and again in 1950 there were exploratory talks. On both occasions, a breakdown on the bargaining front intervened. But in 1953 the cooperative relationship reached a firmer basis. The Amalgamated-UPWA agreement of June 23, 1953 not only assured joint effort in collective bargaining. It also promised organizational peace: no raiding, no interference in representation elections, and, in attacking Swift Brotherhood plants, precedence to the union with the greater strength in the area. This *rapprochement,* real as it was, renewed the prospects of a merger. On April 27, 1953, the executive boards of the two unions met together—unprecedented event—to discuss the unity question, and committees were appointed to examine the matter further. But, as Gorman observed on the occasion of the 1954 UPWA convention, the merger effort "was apparently lost in the shuffle" and no real progress was made. [50]

The immediate obstacle was the division of the labor movement. Neither the Amalgamated nor the UPWA was willing to shift its affiliation. In 1950, there had been an ingenious suggestion to surmount the dilemma: the merged union would pay per capita to the AFL for the Amalgamated membership, and to the CIO for the UPWA share, the ratio to be maintained as the combined organization grew. The novel idea was, however, never implemented. [51] By 1955, such ingenuity was no longer needed. The imminent merger of the AFL and CIO disposed of the problem and simultaneously rekindled the desire for unity in the meat industry. Begun in the spring of 1955, talks reached a turning point during the AFL–CIO merger convention in December 1955. During the afternoon session on December 7, George Meany read from the platform a telegram from the Amalgamated and UPWA advising "that we have reached an accord which we are certain represents the basis for an early merger of our two organizations."

This merger posed problems immensely larger than the preceding ones. The UPWA was a major union, so that there was the need to figure a sharing of power; the UPWA covered an identical category of workers, so that there had to be a real integration in the packinghouse branch while preserving something of the UPWA identity; and there was a history of rivalry to overcome. Earlier, these would have been insurmountable obstacles.

By the mid-1950's, however, the UPWA had cause to want a merger. The union was losing momentum: the original militancy was on the wane; industry changes were working against it; membership was beginning to decline. And the problems were mounting: new issues in collective bargaining raised by automation and plant relocation; inadequate financing; and employer resistance that was being toughened by recent applications of labor legislation. (After a bitter strike in 1954-55, for example, the UPWA had been de-certified at the important Colonial Provision Company in Boston.) But, amenable as majority opinion was becoming to the idea of merger, there was no inclination to do so on any terms. At UPWA conferences, resolutions favoring merger often included the proviso that, as District 2 asserted, such an undertaking had to "protect the traditions and democratic principles that have been a part of the UPWA since its inception." And the National Policy Conference that decided to go ahead with the talks in May 1955 made clear that a merged union should have "an internal structure which makes for UPWA integration—but not absorption." [52]

The negotiations during 1955, culminating in a lengthy night session on December 6 during the AFL–CIO merger convention, reached a point that satisfied the UPWA. It would have twelve seats on the thirty-three-man executive board (originally the Amalgamated had offered five, the UPWA had wanted its entire board of fourteen). There would be three executive officers, a president and secretary-treasurer chosen by the Amalgamated and a general vice-president, "ranking with" the first two, chosen by the UPWA (it had originally wanted one of two executive posts). Six of twelve districts would be directed by vice-presidents from the UPWA. It thus received a reasonable share of the control of the merged union, and it would dominate the packinghouse section. An industrial department, covering meat packing, would be under a UPWA appointee. Finally, UPWA views on public affairs would not be submerged in the new union. A UPWA program department handled such matters as civil rights, politics, education, farmer-labor relations, women's and community activities. The merged union also would have a program department, directed by a UPWA man, which would be "charged with carrying out convention program directives in all sections of the country." [53] On the face of it, the UPWA had secured an arrangement that would protect its interests as a minority element in a combined organization.

The Amalgamated would of course be the dominant party in the union. The enlarged union would have its name (with the understanding that this would be altered afterward to indicate the broader scope of the union). The Amalgamated would have a majority of executive offices and board memberships, and the directorships of all but two of the departments. And the internal character of the new union, as it was hammered out in many meetings and eventually incorporated in a draft constitution, followed Amalgamated rather than UPWA lines. The executive officers had broad powers to intervene in the affairs of the local unions and to direct the work of the executive board members. The latter would be chosen, not by the districts they would lead (as in the UPWA), but by general convention vote. The division of powers and functions tended to follow the Amalgamated pattern of locally autonomous operation with strong powers of supervision invested in the international executive officers. The Amalgamated had not expected merger to mean a surrender of control nor a radical change of accustomed ways. Nor did the proposed plan do so.[54]

Yet, as the merger approached realization, Amalgamated doubts began to emerge. For one thing, there was a money question. President Jimerson's initial hesitation arose from "worrying more about finances than anything else." [55] The difficulty was eased somewhat by the decision that UPWA members, who did not have death benefits, should be treated as new members for purposes of computing benefits in the merged union. Still, there were financial sacrifices facing the Amalgamated. For its resources (over three millions, excluding death and retirement funds, in 1956) were larger than the UPWA in relation to membership; and, what was more important, the UPWA policy of centralized servicing of local unions meant a relatively heavier drain on the treasury of the merged union.[56]

The Communist question was also troubling. Left wingers had played a significant role in the internal life of the UPWA. By 1956, the Communists were no longer visible, but many in the Amalgamated did not doubt that they were still entrenched "underground." The Amalgamated had just merged with a union under blatant Communist influence, but the difference—as on many of the problems—was that the Fur and Leather Workers constituted no threat to the internal order of the Amalgamated. The UPWA, reluctantly, accepted a clause in the proposed constitution that prohibited members from organizations advocating or seeking to overthrow the government by unconstitutional means. The uneasiness within the Amalgamated nevertheless persisted.[57]

Finally, a question lurked about the future control of the merged organizations. Age was a factor here. Gorman was sixty-three and Jimerson sixty-seven, whereas Helstein was only forty-seven. Vacancies among executive officers, it was true, were to be filled by vote of the executive board rather than, as was the case with the presidency in the Amalgamated, by seniority. But Helstein was able and ambitious; he could be expected to work hard to establish a claim to primacy in the new union. What was more, the industrial department, comprising the entire UPWA plus the Amalgamated packing-house locals, would constitute the largest division in the merger. Slated to be led by an aggressive UPWA man, A. T. Stephens, the department might well become a solid bloc that would swing behind UPWA candidates.

Some of these fears were articulated a month after the Amalgamated Executive Board had ratified the merger at the end of March 1956. Additional demands were made: that the Amalgamated slots on the executive board be raised from twenty-one to twenty-seven; that the Amalgamated vice-president in the industrial department be designated "first-assistant" (thus making him senior to the UPWA assistant director) and that, should the program department require an assistant, he would be chosen by the Amalgamated; and that all union employees be required to sign non-Communist affidavits annually.[58] The UPWA resisted these changes, finally agreeing only to two additional Amalgamated vice-presidents. By the time this was accomplished, it was too late for the merger to be concluded on schedule in June 1956. Instead, the two unions met separately for regular conventions in Cincinnati to consider, among other business, the Merger Agreement.

When the Amalgamated convention opened on June 11, misgivings had not been entirely stilled. One powerful vice-president, Max Block of New York, had steadfastly refused to make the board decision unanimous. A poll of local unions revealed 41 of 270 opposed to a merger. There was also a coolness among functionaries in the retail area, for which the merger seemed of little value, and among staff men who, while their jobs were assured, might be placed in subordinate positions in the enlarged organization. "Do the delegates know," asked Gorman, "that this was supposed to be a Convention that, by a whispering campaign of those who do not believe in unity, that might have become the most disgraceful Convention ever held in the labor movement? There were some high leaders in labor who actually thought that there were going to be fist fights in this Convention." [59] Nothing of the sort happened. The resolution favoring completion of the merger passed without a recorded dissenting voice following a prolonged demonstration, and UPWA President Helstein was warmly received when he came to address the convention. So long as the top men—above all, Patrick Gorman—wanted a merger, the Amalgamated would go ahead.

The next months were devoted to the two tasks of drafting a constitution and to negotiating new master contracts in meat packing. For the first time, the packers agreed to bargain with a joint Amalgamated-UPWA team. On September 20, 1956, the two unions went out on strike against Swift. Five days later, they reached an agreement with Armour that provided an hourly increase of 25 cents spread over three years, a modified union shop, and other important improvements. The Swift strike, strongly waged because of UPWA-Amalgamated unity and support by the retail locals and other unions, ended a few days later substantially on the Armour terms, but without the union shop. On October 7, delegates from packinghouse locals of both unions gathered in a body—this had never happened before—to review the new master agreements. The happy outcome of the difficult negotiations, men on both sides felt, demonstrated "the practical meaning of cooperation between the Amalgamated and the UPWA." [60] The link, drawn tighter by the joint bargaining, the Swift strike, and the evidence of effectiveness of unity, now only required one formal step to merge the two unions. The Joint Unity Committee, having finally hammered out the details of the proposed constitution, fixed the date of October 24, 1956 for separate conventions to ratify the constitution and the merger. On October 26, the delegates would convene together and the union would be consummated.

That day never came. One very minor issue had remained after the rest of the constitution had been finished, approved by the executive boards, printed and sent out to the local unions. This was the complicated matter of a combined retirement plan for the employees of the two unions. This, in turn, required a precise agreement on salaries in order to calculate the cost of consolidating the UPWA plan with the more generous Amalgamated plan. A week before the scheduled merger conventions, on Monday, October 15, 1956, talks on this matter began and immediately raised hackles on both sides. The Amalgamated objected to the increases granted by the UPWA convention in June. UPWA representatives retorted that their new scale was still below Amalgamated salaries and, more than that, that the

Amalgamated was breaching the understanding that no salary cuts would result from the merger. UPWA officials also were deeply annoyed that Amalgamated representatives came to the union headquarters on Wednesday morning to check the accuracy of UPWA statements on per diem practice against the records.

The minor discord over staff salaries (involving only a dozen or so people directly attached to the UPA national office) led immediately into the larger questions, as Helstein afterward put them, "of what the structure of this new union would be, how its departments would function . . . on staff allocation, on responsibilities and on functions." The Amalgamated position had been to postpone administrative details until after the merger, and the UPWA had reluctantly agreed. But now, as tempers flared and as the salary dispute drew attention to the unsettled questions of staff integration, the UPWA Unity Committee began to press its views on administrative structure and functions during that fateful meeting of Wednesday afternoon, October 17, 1956. With little or no prior consultation among themselves, the UPWA men put forward a number of demands: that the Amalgamated's education department be made part of the program department, which a UPWA vice-president would head; that the Amalgamated's industrial engineering department, or at least its activities in meat packing, be placed under the industrial department, also to be directed by a UPWA vice-president; that all departments be headed by vice-presidents (four in the Amalgamated were not); and that the executive officers should not direct the work of the department heads.

After caucusing, the Amalgamated delegation announced that the new proposals, if the UPWA adhered to them, meant an end to the merger. It was finally agreed to hold another meeting the next day. Instead, the UPWA received a telegram from the Amalgamated in the morning:

At this late date new and impossible interpretations have been applied to the merger agreement by your committee which has [sic] resulted in very serious misunderstandings. Therefore, in the interests of both organizations our Executive Board and Unity Committee has [sic] postponed the special and merger conventions until some later date.

On Saturday, October 20, an emergency session of the Amalgamated Executive Board decided to withdraw from the merger. The UPWA urged a resumption of talks; the Amalgamated could not unilaterally postpone conventions that had been called by the joint unity committee. The UPWA went ahead with its special convention on the appointed date of October 24 and ratified the draft constitution, but this was only in the nature of a gesture. The fact was that the merger was dead.[61]

What had happened? It seemed clear enough what was motivating the UPWA. As the lesser organization, it was vitally concerned—as it had said from the start—that merger should not mean "absorption." What this meant at bottom was that UPWA services and activities would not be impaired. "We had said to you, time and again and again and again," Helstein told the UPWA special convention subsequently, "that no matter what else happened in this merger, you could rest assured that the way we conducted negotiations . . . grievances . . . wage rate problems would continue, that our program activities would continue." That, in turn, required the proper personnel and functions in the two departments—industrial and program— under UPWA leadership. "We expected to have an industrial department . . . that would provide our membership with the same kind of services that they had been accustomed to. We expected to have a program department . . . that would provide functions such as education, our farmer-labor program, our political action program . . . discrimination . . . women's activities." When the salary dispute revealed an apparent intention to include some of this work in existing Amalgamated departments, the UPWA representatives made demands that they felt would protect their established practices.[62] The administrative issue, secondary as it seemed, went to vitals of the UPWA concern in the merger.

The drastic Amalgamated response had less evident sources. This explanation was given to the membership:

The UPWA *NEW* interpretations were of such serious consequence that the entire internal structure of the Amalgamated would have been wrecked—so much so that the three top executive officers of the International Union would have been reduced to figure-head status.[63]

The radical demand was that the departments not be supervised by the executive officers. The UPWA could hardly have hoped to win this point. The proposed constitution certainly did not lend itself to such an "interpretation." Afterward, Helstein even denied that he or the UPWA favored the idea. It had apparently been thrown into the meeting for one of two reasons. Either it had been thought to have some bargaining value, or it had reflected UPWA rancor over the wage issue. In neither case could departmental autonomy have seemed an issue on which the UPWA would commit the fate of the merger. The other UPWA demands, had they been fully met, would have resulted in structural changes of only secondary importance, for the education and industrial engineering departments were recent creations and hardly centers of internal power. The four department heads threatened by the UPWA stand were staff men rather than elected vice-presidents and hence without independent bases of power. They were not important enough to have counted critically in the ultimate decision. (Probably the most influential of the four, David Dolnick of the research department, left the Amalgamated the following year.) Afterward, Helstein remarked that many harder hurdles had been passed earlier on the long merger road. Now, instead of seeking to bargain out a compromise (as the UPWA evidently had expected), the Amalgamated took the injection of the administrative issue as cause to break off the merger.

Clearly, a fundamental reevaluation had occurred at the last minute. Two doubts, in particular, seemed to have deepened as a result of the late UPWA move. The specific one touched the militancy of the Packinghouse Workers. As the merger had approached, Patrick Gorman had warned the UPWA that it was time to leave "its period of adolescence and growing pains" when "it believed in militancy and more militancy to accomplish its ends." "Those who lead the merged union must remember that neither the employers nor the members can be pushed around because of our additional strength." These remarks Gorman read on October 7 to the joint UPWA-Amalgamated conference that was reviewing the new master contracts, to the obvious annoyance of UPWA chieftains. Gorman's reminder of past differences may have helped to trigger the UPWA attempt to pin down an administrative arrangement that would guarantee its existing policies and methods. By the same token, the UPWA move (T. J. Lloyd later remarked that it was typical of the "hard-nosed" bargaining tactics of the UPWA) heightened Gorman's fears about being able to moderate the militant packinghouse group after it was part of the Amalgamated. That was the main thrust of his editorial "This Old Union" following the breakdown of the merger.[64]

The other effect of the dispute was more pervasive and, ultimately, more important. Merger had to rest on mutual confidence. This was a fragile thing even during the final weeks. On both sides, men could not entirely efface the scars of past rivalry and hostility. After the collapse of the joint negotiating effort in the winter of 1952, Gorman had asserted that "we cannot consistently trust any longer most of the present officers of this group [UPWA], and until a change is made we cannot consider them because they are not sincere."[65] That distrust persisted in the background even after bargaining harmony was restored and merger discussions brought leaders on both sides into sustained contact. Gorman himself remained aloof from the detailed talks, thereby passing up the chance to establish closer personal ties with Ralph Helstein.

The UPWA move revived Gorman's suspicions. "Could it possibly be that in the last moment before the merger," Gorman and Jimerson asked in their explanatory letter to the locals, "there might have been a plan to wreck entirely the internal structure of the Amalgamated? We don't know, but it certainly did not appear to us to be sincere." [66] In the face of this thought, all the earlier fears about the merger—the Communist influence, the succession question, finances—assumed a darker cast. Secretary Gorman concluded that the merger became too risky a venture. And, so far as the Amalgamated was concerned, Gorman's conclusion was the ruling one.

The merger foundered on these proximate circumstances. But underlying was another fact of first importance: that the alternative of continued separation was not unacceptable. What, after all, would be the price of the collapse of the merger? "For the time being at least, everything that was hoped for," Gorman and Jimerson immediately assured the membership, "must now be accomplished through cooperation alone. We pledge this cooperation at all times to the workers in the meat packing industry, even though actual organizational unity temporarily has been postponed." [67] Helstein likewise emphasized this same conciliatory view at the special UPWA convention. In the midst of angry recrimination, no one threatened to renew the rivalry of earlier years; the destructive practices of raiding and competitive collective bargaining would not return. Ironically, therefore, durable cooperation worked counter to the logical culmination in merger: the marginal return did not seem to justify the merger risks to either side.

It remained to be seen whether later events might revise this calculation. In 1957, ailing President Jimerson died and was replaced by First Vice-President T. J. Lloyd, lessening somewhat the danger of a succession plot by UPWA elements in a merged union. Max Block of New York City, the anti-merger voice on the executive board, in May 1958 was exposed by the McClellan Committee for misconduct in the affairs of his New York unions, and his association with the Amalgamated came to an immediate end. Within the UPWA, a silent power struggle between Helstein and A. T. Stephens, which in obscure ways certainly inhibited merger decisions, closed with the stormy departure of Stephens. Besides this leadership turnover, the unions' situation in some ways tended to sustain the idea of merger. Within the Amalgamated, concern mounted about industry changes. If the centralization of retail meat cutting became established, most of the work of the retail unions would be lost to the packing houses. The meat cutters' branch, merger advocate John Jurkanin argued, had no reason to be indifferent toward unity with the Packinghouse Workers.[68] For its part, the UPWA was failing to arrest the decline that had started earlier in the decade. And the Swift Brotherhood, a continuing thorn in the side of both unions, could be dislodged, if at all, only in the wake of a merger.

These developments did not in fact rekindle the merger. Both unions still officially favored it, and periodically there were flurries of merger talk (coinciding, men privately claimed, with conventions and with rounds of master contract negotiations). But the fact was that the momentum lost in the 1956 debacle was not regained. The final step toward consolidating the organization of the butcher workmen still had to be taken. Only the future would disclose whether it ever would be taken.

So the unionization process remained incomplete at the opening of the 1960's. The last scar of the organizing push that had begun in the 1930's—rival unionism—had not been entirely effaced. Nor was there an end to the need for unionizing effort. For the meat trade was in process of rapid change. And beyond the industry proper were extensive bodies of unorganized food workers. The unfinished character of unionization in some part reflected union failings, but in greater part it was the consequence of progress and change. Unionization remained a challenge and, to the union responsive to it, a powerful dynamic in an era of labor inertia.

NOTES

ABBREVIATIONS

AMC Files.	Files, Amalgamated Meat Cutters and Butcher Workmen of North America.
AMC Journal.	Amalgamated Meat Cutters and Butcher Workmen of North America, *Official Journal.*
BW.	*Butcher Workman.*
FMCS Files.	Files, Federal Mediation and Conciliation Service, National Archives.
NM.	*New Majority.*
NP.	*National Provisioner.*
NRA Files.	Files, National Recovery Administration, National Archives.
PH News.	*CIO News. Packinghouse Edition.*
PW.	*Packinghouse Worker.*
UPWA Files.	Files, United Packinghouse Workers of America.

NOTES

CHAPTER 1. THE MODERNIZATION OF THE MEAT TRADE

1. U.S. Bureau of Animal Husbandry, *Report*, 1884, pp. 266–7; Bessie L. Pierce, *A History of Chicago* (3 vols., New York, 1937–57), III, 116–7; O. G. Anderson, *Refrigeration in America: History of a New Technology* (Princeton, 1953).

2. Ferdinand Sulzberger, in *Butchers' Advocate*, May 8, 1895, p. 11. On resistance to the western packers, see U.S. Congress, Senate, Select Committee on Transportation and Sale of Meat Products, *Report and Testimony*, 51st Cong., 2nd sess. (1890), *passim*.

3. U.S. Bureau of Corporations, *Report on the Beef Industry*, 1905, pp. 57–8; U.S. Federal Trade Commission, *Food Investigation. Report on the Meat Packing Industry*, 1918–20, II, ch. 1; Lewis Corey, *Meat and Man: A Study of Monopoly, Unionism and Food Policy* (New York, 1950), pp. 49–51.

4. On Swift debts and inventory for 1904, for example, see Bureau of Corps., *Report*, p. 210.

5. For the Sulzberger failure, see FTC, *Food Investigation*, II, ch. 4.

6. Senate Select Committee on Meat Products, *Report and Testimony*, p. 480.

7. Bureau of Corps., *Report*, pp. 253–7.

8. *National Provisioner*, Jan. 27, 1900, p. 25, cited hereafter as *NP;* Rudolf A. Clemen, *The American Meat and Livestock Industry* (New York, 1923), pp. 124–6, 228–31; Illinois Bureau of Labor Statistics, *Report*, 1882, pp. 224–5; Siegfried Giedion, *Mechanization Takes Command* (New York, 1948), pp. 214–8.

9. *NP*, Jan. 27, 1900, p. 25; Harper Leech and John C. Carroll, *Armour and His Times* (New York, 1938), p. 46; Bureau of Corps., *Report*, p. 268, also, pp. 211–48; Clemen, *The American Meat and Livestock Industry*, ch. 16.

10. Institute of American Meat Packers, *The Packing Industry* (Chicago, 1924), pp. 108–13; *NP*, Dec. 24, 1904, pp. 16–7; Jan. 28, 1905, pp. 16–7; Charles Winan, "The Evolution of a Vast Industry, VII." *Harper's Weekly*, Dec. 23, 1905, p. 1879.

11. Clemens, *The American Meat and Livestock Industry*, pp. 122–3; John R. Commons, "Labor Conditions in Slaughtering and Meat Packing," *Quarterly Journal of Economics*, XIX, 3 (Nov. 1904); Bureau of Corps., *Report*, pp. 17–8; *NP*, Nov. 21, 1903, p. 27; Oct. 17, 1908, p. 85.

12. Louis F. Swift, *Yankee of the Yards* (Chicago, 1927), p. 164; Commons, *Quarterly Journal of Economics*, XIX, 7; *NP*, Nov. 17, 1900, p. 17; Amalgamated Meat Cutters and Butcher Workmen of North America, *Official Journal*, Feb. 1903, p. 23, cited hereafter as *AMC Journal;* A. M. Simons, *Packingtown* (Chicago, 1899), p. 24.

13. *NP*, March 10, 1894, p. 16; "Knights of Labor Local Assembly 2802 Minutes, May 27, June 9, 23, 1886," printed in *Butcher Workman*, Oct. 1932, p. 2, cited hereafter as *BW*.

14. *NP*, Jan. 30, 1897, p. 31; *AMC Journal*, March 1903, p. 3; Commons, *Quarterly Journal of Economics*, XIX, 12; C. W. Thompson, "Labor in the Packing Industry," *Journal of Political Economy*, XV, 92 (Feb. 1907); John C. Kennedy, *Wages and Family Budgets in the Chicago Stockyards District* (Chicago, 1914), p. 14; U.S. Commission on Industrial Relations, *Final Report and Testimony*, 1916, IV, pp. 3504–5.

15. U.S. Bureau of Labor, *Bulletin* 77 (1908), pp. 58, 118–20; Illinois Bureau of

Labor Statistics, *Report*, 1886, p. 357; Commons, *Quarterly Journal of Economics*, XIX, 12.

16. U.S. Immigration Commission, *Report*, XIV, 31 (1910); Kennedy, *Wages and Family Budgets*, p. 5; U.S. Commission on Industrial Relations, *Report*, IV, 3470–2; E. Abbott and S. P. Breckenridge, "Women in Industry: The Chicago Stockyards," *Journal of Political Economy*, XIX, 632–54 (Oct. 1911).

17. *Charities and the Commons*, June 6, 1906, pp. 755–6; *NP*, July 14, 1900, p. 30; Sept. 2, 1905, p. 43; Simons, *Packingtown*, p. 12; Illinois Bureau of Labor Statistics, *Reports on Industrial Accidents*, 1907–1910. See also, for example, the graphic descriptions in Upton Sinclair, *The Jungle* (New York, 1905).

18. *Butchers' Advocate*, July 3, 1895, p. 13; *NP*, April 24, 1897, p. 32; Aug. 25, 1900, p. 40; Jan. 4, 1902, p. 34; Thompson, *Journal of Political Economy*, XV, 107–8.

19. Bureau of Corps., *Report*, p. 205.

20. In 1916 there were 525 intrastate slaughterers who accounted for 9.4 per cent of the cattle, 9.2 per cent of the sheep, and 7.1 per cent of the hogs used in the wholesale trade. FTC, *Food Investigation*, III, Exhibit 6, 311–4.

21. Senate Select Committee on Meat Products, *Report and Testimony*, pp. 169, 262. See also *Census of Massachusetts*, II, 659–69 (1875); Mass. Bureau of Labor Statistics, *Report*, 1871, p. 299; Arnold C. Schueren, *Meat Retailing* (Chicago, 1927), pp. 5–12.

22. *Butchers' Advocate*, Nov. 14, 1894, p. 11; Senate Select Committee on Meat Products, *Report and Testimony*, p. 404.

23. See *NP*, Aug. 4, 1894, p. 24; March 28, 1896, p. 37; Jan. 14, 1899, p. 38; Oct. 27, 1909, p. 45; *Butchers' Advocate*, Nov. 21, 1894, p. 18; July 17, 1895, p. 9.

24. *NP*, Jan. 20, 1894, p. 19; July 1, 1893, p. 19; Jan. 7, 1899, p. 41; Jan. 28, 1899, p. 39; Senate Select Committee on Meat Products, *Report and Testimony*, p. 262.

25. U.S. Dept. of Agriculture, *Bulletin 1317* (1920).

26. *NP*, April 25, 1896, p. 37; March 6, 1897, p. 7; June 12, 1897, p. 36; March 11, 1899, p. 31; Dec. 1, 1900, p. 19; Jan. 4, 1902, p. 35; Nov. 1, 1902, p. 31.

27. Dept. of Agriculture, *Bulletin 1317*, pp. 9, 52; *NP*, July 7, 1900, p. 39; Roy C. Lindquist, *Efficient Methods of Retailing Meat* (Washington, 1925); Horace Secrist, *Expenses, Profits and Losses in Retail Meat Stores* (Chicago, 1924).

28. *NP*, Oct. 29, 1910, p. 40; *AMC Journal*, March 1902, pp. 21–2.

29. *NP*, Sept. 14, 1907, p. 41; Sept. 12, 1908, p. 40; Dec. 31, 1910, p. 40.

30. Lindquist, *Efficient Methods*, p. 15; *AMC Journal*, Aug. 1904, p. 24; March 1905, p. 42.

31. *NP*, Jan. 20, 1900, p. 41. For similar agreements, see *NP*, March 18, 1899, p. 41; *Butchers' Advocate*, Nov. 14, 1894, p. 12; Nov. 21, 1894, p. 10; Jan. 27, 1897, p. 18; *Butchers and Packers Gazette*, Oct. 22, 1904, p. 1.

CHAPTER 2. THE UNION IMPULSE

1. Edna L. Clark, "History of the Controversy between Labor and Management in the Slaughtering and Meat Packing Industries in Chicago," unpub. diss., University of Chicago, 1922, p. 63; Illinois Bureau of Labor Statistics, *Report*, 1888, pp. 313–7; *Knights of Labor* (Chicago), Nov. 20, 1886; "Knights of Labor Local Assembly 2802 Minutes, May 27, June 9, June 23, 1886," in *BW*, June 1932, p. 2, and the autobiographical account of J. T. Joyce in *BW*, July 1920, pp. 1, 5. Accounts of the strike and Powderly's disputed role may be found in Norman J. Ware, *The Labor Movement in the United States, 1860–1895* (Gloucester, Mass., 1959), pp. 152–4; John R. Commons et al., *History of Labor in the United States* (4 vols., New York, 1918–35), II, 418–20;

T. V. Powderly, *The Path I Trod* (New York, 1940), ch. 12; Philip S. Foner, *History of Labor in the United States* (2 vols., New York, 1947–55), II, 86–8; Joseph R. Buchanan, *Story of a Labor Agitator* (New York, 1903), pp. 319–21.

2. *NP*, July 14, 21, 28, Aug. 4, 11, 1894, *passim;* account of J. T. Joyce, *BW*, Feb. 1933, p. 2; March 1933, p. 2; April 1933, p. 2; Clark, "Controversy Between Labor and Management in Meat Packing," pp. 87–8; Alma Herbst, *The Negro in the Slaughtering and Meat Packing Industry in Chicago* (Boston, 1932), pp. 17–9.

3. *Butchers' Advocate*, Nov. 11, 1896, p. 13; *NP*, Nov. 7, 1896, p. 31; *BW*, Dec. 1932, p. 2.

4. *Knights of Labor*, Nov. 20, 1886; Clark, "History of the Controversy," p. 87; *BW*, June 1920, p. 5; Feb. 1932, p. 2.

5. Clark, "History of the Controversy," pp. 56, 63–6.

6. *NP*, Aug. 18, 1894, p. 17; Sept. 8, 1894, p. 18; *Butchers' Advocate*, March 13, 1896, p. 15; Clark, "Controversy Between Labor and Management in Meat Packing," p. 87.

7. *By-Laws*, Sheep Butchers Protective Union of New York City, No. 6716, AFL, in Wisconsin State Historical Society; *AMC Journal*, April 1901, p. 12.

8. For instances of the organization of unskilled packinghouse workers in Chicago and Kansas City, see Illinois Bureau of Labor Statistics, *Report*, 1886, p. 278; *American Federationist*, Oct. 1895, p. 149; Feb. 1896, p. 232; March 1896, pp. 21–2; Clark, "History of the Controversy," p. 65.

9. *NP*, Sept. 30, 1893, p. 16; Jan. 27, 1894, p. 18; Aug. 15, 1896, p. 27; Oct. 31, 1896, p. 33; *American Federationist*, Oct. 1895, p. 149; Swift, *Yankee of the Yards*, pp. 10–11.

10. *Knights of Labor*, Nov. 13, 1886; Nov. 20, 1886; Herbst, *Negro in Meat Packing*, p. 16; Clark, "History of the Controversy," p. 77; Leech and Carroll, *Armour*, p. 217.

11. Ware, *The Labor Movement*, p. 152; Clark, "History of the Controversy," pp. 48–9; *Butchers' Advocate*, March 25, 1896, p. 18; April 22, 1896, p. 18; Aug. 12, 1896, p. 12; *American Federationist*, March 1895, p. 17; AFL, *Proceedings*, 1895, p. 27.

12. *NP*, May 16, 1896, pp. 5, 21, 29; May 23, 1896, p. 5; Aug. 24, 1896, p. 27; *Butchers' Advocate*, May 13, 1896, p. 16; *American Federationist*, June 1896, p. 77; Sept. 1896, p. 149; Oct. 1896, p. 174; AFL, *Proceedings*, 1896, pp. 64–5.

13. Chicago *Daily News*, Aug. 29, 1904.

14. Financial Reports, *American Federationist*, Dec. 1896–March 1897, *passim*.

15. *NP*, Nov. 24, 1900, p. 36. See also, Joseph Belsky, *I, the Union* (New York, 1952), chs. 1, 2.

16. *AMC Journal*, Oct. 1903, p. 34; "The Sum of $500," *NP*, May 10, 1903, p. 35; also, *NP*, Sept. 14, 1907, p. 41.

17. *NP*, Jan. 18, 1902, p. 34; Nov. 12, 1892, p. 18; April 4, 1896, p. 31; *Butchers' Advocate*, Nov. 14, 1894, p. 12; July 3, 1895, p. 12; Aug. 5, 1896, p. 10; Grace H. Stimson, *The Rise of the Labor Movement in Los Angeles* (Berkeley, 1955), pp. 91, 138; Illinois Bureau of Labor Statistics, *Report*, 1886, pp. 430–1.

18. *NP*, Oct. 21, 1893, p. 17; July 29, 1893, p. 19; March 10, 1900, p. 40; Sept. 8, 1900, p. 41.

19. Illinois Bureau of Labor Statistics, *Report*, 1886, p. 378; Newark *Advertiser*, March 17, 1902, p. 34, reprinted in *NP*, March 22, 1902, p. 34.

20. This was said by W. C. Wellman, who had headed the Benchmen, after he joined the Amalgamated Meat Cutters. *AMC Journal*, May 1902, p. 21.

21. *Knights of Labor*, Dec. 30, 1886; Illinois Bureau of Labor Statistics, *Report*, 1886, p. 235; *Butchers' Advocate*, Dec. 4, 1895, p. 15; *NP*, May 18, 1895, p. 29; Belsky, *I, the Union*, pp. 13–7; Financial Reports in AFL, *Proceedings*, 1888–92.

22. Senate Select Committee on Meat Products, *Report and Testimony*, p. 151. On the importance of the central bodies at Utica and Hartford, see *Butchers' Advocate*, Nov. 20, 1895, pp. 9–10; Dec. 25, 1895, p. 17.

23. Samuel Gompers to John H. Schofield, National Retail Butchers Protective Association, n.d., Gompers Letterbooks, AFL–CIO Headquarters.

24. *Butchers' Advocate*, Nov. 20, 1895, pp. 9–10; Dec. 4, 1895, p. 15; April 1, 1896, p. 17; Nov. 18, 1896, p. 17; *San Francisco Butchers and Stock Growers Journal*, quoted in *Butchers and Packers Gazette*, Jan. 14, 1905, p. 4; *NP*, May 20, 1893, p. 17; July 8, 1893, p. 22; July 29, 1893, p. 19; April 21, 1894, p. 21.

25. Homer Call, letter to editor, April 26, 1897, *American Federationist*, May 1897, p. 59.

26. Art. 6, Sec. 2, AFL Constitution (1886); AFL, *Proceedings*, 1890, p. 14; 1892, p. 16; Samuel Gompers, *Seventy Years of Life and Labor* (2 vols., New York, 1927), I, 342.

27. *NP*, May 16, 1896, p. 5; *AMC Journal*, March 1903, p. 18.

28. *NP*, Aug. 25, 1894, p. 25; *BW*, July 1920, p. 5, and issues Dec. 1932–July 1933, p. 2.

29. AFL, *Proceedings*, 1891, p. 44; *NP*, June 10, 1893, p. 16; July 8, 1893, p. 22.

30. *AMC Journal*, May 1904, p. 43; AMC, *Proceedings*, 1899, p. 14.

31. AMC, *Proceedings*, 1899, pp. 14–5; *NP*, June 10, 1893, p. 16; June 24, 1893, p. 18.

32. *AMC Journal*, July–Aug. 1905, p. 12. This was in response to a statement by Theodore W. Glocker, "The Unit of Government in the Meat Cutters' and Butcher Workmen's Union," *Johns Hopkins University Circular*, New Series, no. 1 (1905), 21–5, that the reason for amalgamation was that the packinghouse workers wanted the meat cutters "in case of strike." Call denied this, but he was writing in 1905 when experience had shown the weakness of the secondary boycott and, moreover, when it was necessary to placate the retail branch.

33. *AMC Journal*, May 1904, p. 2.

34. AMC, *Proceedings*, 1897, pp. 3–4; 1899, p. 31; *AMC Journal*, May 1904, pp. 1–2; AFL, *Proceedings*, 1906, pp. 8, 36, 78; Gompers to Homer Call, Jan. 2, 1897, Gompers Letterbooks.

35. Art. 2, Sec. 1, AMC Constitution, 1897.

36. Gompers to Call, July 12, 1897, Gompers Letterbooks.

37. AMC, *Proceedings*, 1899, pp. 9–10; *AMC Journal*, May 1904, pp. 8, 10.

38. Gompers to Call, Feb. 18, June 17, 1897, Gompers Letterbooks; AMC, *Proceedings*, 1897, pp. 4–5; 1899, p. 16.

39. *American Federationist*, May 1897, p. 59; Feb. 1898, p. 283; and organizers' reports, 1897–1900, *passim*; *AMC Journal*, Sept. 1902, p. 24; AMC, *Proceedings*, 1899, p. 15.

40. AMC, *Proceedings*, 1899, pp. 10, 12, 26–8; 1900, p. 12; *AMC Journal*, June 1904, p. 29; Aug. 1904, p. 18; *American Federationist*, Dec. 1897, p. 242.

41. AMC, *Proceedings*, 1899, pp. 10–11.

42. AMC, *Proceedings*, 1897, pp. 6–7; 1899, p. 6; 1900, pp. 9–10, 14; Gompers to Call, June 30, July 23, 1897, Gompers Letterbooks.

43. AMC, *Proceedings*, 1899, pp. 24–5; *American Federationist*, Oct. 1897, p. 199; AFL, *Proceedings*, 1897, p. 110; Gompers to Armour Packing Co., March 25, 1898, Gompers to Call, Jan. 10, 25, 1898, Gompers to H. L. Palmer, Dec. 5, 1897, Gompers Letterbooks.

44. AFL, *Proceedings*, 1898, pp. 69, 101; 1899, p. 116; 1900, pp. 74, 162; 1901,

pp. 169–70; AMC, *Proceedings,* 1900, pp. 18, 26–7; U.S. Industrial Commission, *Reports,* 1901, XVII, 313; Gompers to AFL Executive Council, Aug. 29, 1898, Gompers Letterbooks.

45. AMC, *Proceedings,* 1897, pp. 7–11; 1899, pp. 19–22, 25; 1902, pp. 28–9; Gompers to Call, Aug. 28, 1897, Sept. 25, 1897, to Max Morris, Aug. 28, 1897, Gompers Letterbooks; *AMC Journal,* Nov. 1902, p. 12; May 1904, p. 4; AFL, *Proceedings,* 1897, p. 116; 1901, pp. 70, 177; *American Federationist,* July 1900, pp. 221, 223.

46. Gompers to Call, Aug. 5, 1897, Gompers Letterbooks.

47. Unless otherwise indicated, the information in this section is drawn from AMC Constitutions, 1896, 1897, 1899, 1900.

48. AMC, *Proceedings,* 1897, p. 15.

49. Gompers to Call, May 24, Aug. 3, 1897, Gompers to J. A. Cable, Jan. 3, 1898, Gompers Letterbooks; AFL, *Proceedings,* 1899, p. 118; 1901, p. 207.

50. AMC, *Proceedings,* 1900, p. 7. For Gompers' advice to use "promptness, earnestness and aggressiveness" against "hasty action," see Gompers to Call, July 18, 1899, Gompers Letterbooks.

51. *AMC Journal,* March 1900, p. 10.

CHAPTER 3. MEAT PACKING: THE FIRST CYCLE OF UNIONIZATION

1. AMC, *Proceedings,* 1900, pp. 5–6; *AMC Journal,* Aug. 1902, p. 71; Oct. 1903, p. 24.

2. AMC, *Proceedings,* 1900, p. 6; *AMC Journal,* April 1903, p. 40.

3. *AMC Journal,* Dec. 1901, p. 13; Feb. 1902, p. 18; Sept. 1902, p. 30.

4. AMC, *Proceedings,* 1902, pp. 12–3; *AMC Journal,* March 1903, pp. 4, 6.

5. Homer Call to Mary McDowell, Nov. 22, 1902, Mary McDowell Papers, Chicago Historical Society.

6. *AMC Journal,* May 1902, pp. 13–7; AMC, *Proceedings,* 1902, p. 27; *NP,* May 17, 1902, p. 34.

7. AMC, *Proceedings,* 1899, pp. 10–11; 1900, p. 5. The agreement that ended the S & S strike, requiring the men to abide by contracts made by the union and not to strike without official sanction, is reprinted in *NP,* July 29, 1899, p. 18.

8. *AMC Journal,* Feb. 1904, pp. 24–6.

9. *NP,* Nov. 15, 1902, p. 33; *AMC Journal,* May 1902, p. 13; May 1904, p. 105.

10. U.S. Industrial Commission, *Report,* 1901, XVII, 313.

11. Printed in *NP,* Dec. 5, 1903, p. 19.

12. *American Federationist,* June 1897, p. 81; Sept. 1903, p. 909; AFL, *Proceedings,* 1901, p. 226; AMC, *Proceedings,* 1900, p. 4; *BW,* Jan. 1920, p. 2.

13. Howard E. Wilson, *Mary McDowell, Neighbor* (Chicago, 1928), p. 93; *AMC Journal,* March 1902, p. 14; Sept. 1902, p. 23; Immigration Commission, *Report,* XIV, 369.

14. *AMC Journal,* April 1901, p. 12; Sept. 1902, p. 23; Nov. 1904, p. 11; *American Federationist,* Sept. 1903, p. 909.

15. Immigration Commission, *Report,* XIV, 47.

16. *AMC Journal,* March 1901, p. 1; May 1903, p. 42; AMC, *Proceedings,* 1900, p. 4; Gompers to Call, Jan. 27, 1898, Gompers Letterbooks. On federal labor union efforts to boycott Kingan, and AMC opposition, see AMC, *Proceedings,* 1902, p. 24; AFL, *Proceedings,* 1901, p. 226.

17. Call to Mary McDowell, Nov. 22, 1902, McDowell Papers; Foner, *History of Labor,* II, 362.

18. *AMC Journal,* Nov. 1902, p. 20; March 1903, p. 9; Sept. 1904, pp. 26–30; AMC, *Proceedings,* 1902, pp. 12, 52–3; 1904, pp. 52, 72; Immigration Commission, *Report,* XIV, 45–60; Commons, *Quarterly Journal of Economics,* XIX, 28–9.

19. *NP,* July 21, 1894, p. 17; Herbst, *Negro in Meat Packing,* pp. 16–20; S. D. Spero and A. L. Harris, *The Black Worker* (New York, 1931), pp. 264–5; Clark, "Controversy between Labor and Management in Meat Packing," pp. 87–8.

20. "Brother Simms' Funeral," *AMC Journal,* May 1903, pp. 26–7.

21. Wilson, *Mary McDowell, Neighbor,* pp. 86–90, 95–100; Alice Henry, *The Trade Union Woman* (New York, 1915), pp. 53–7; *AMC Journal,* Oct. 1902, p. 29; Nov. 1903, pp. 1–4; May 1904, p. 66; Molly Daly to Mary McDowell, Aug. 1, 1902, McDowell Papers.

22. *NP,* March 1, 1902, p. 37; AMC, *Proceedings,* 1904, pp. 34, 56, 62–3, 71–2, 83; *Sausage Makers' Scales, 1904; American Federationist,* June 1906, pp. 382–3.

23. *American Federationist,* Feb. 1900, p. 44; Gompers to Donnelly, June 23, 1904, Gompers Letterbooks; AMC, *Proceedings,* 1902, p. 14; 1904, pp. 27, 91–2.

24. C. D. Wright, "Influence of Trade-Unionism on Immigrants," U.S. Bureau of Labor, *Bulletin 56* (1905), pp. 1–8.

25. AMC, *Proceedings,* 1900, p. 7; 1902, pp. 14, 52; 1904, pp. 27, 37, 52, 69, 86, 87.

26. AMC, *Proceedings,* 1904, pp. 66, 86–7.

27. *AMC Journal,* March 1902, p. 26; Aug. 1902, p. 63; Nov. 1902, p. 16; Feb. 1903, p. 29; Feb. 1904, p. 23; *NP,* Feb. 14, 1903, p. 40; Chicago Packing Trades Council Constitution, copy in Wisconsin State Historical Society.

28. *AMC Journal,* Aug. 1902, p. 68; July 1903, p. 2; AMC, *Proceedings,* 1900, p. 7; 1902, p. 25. One problem was to determine tax assessments on the auxiliary locals, for these included men of the same trade working outside the stockyards. See Gompers to Call, Sept. 1, 1903, Gompers Letterbooks.

29. *AMC Journal,* July 1901, pp. 13, 95; March 1903, pp. 2, 3, 4, 9, 11; Clark, "Controversy between Labor and Management in Meat Packing," p. 102.

30. *AMC Journal,* Aug. 1902, p. 78; Jan. 1903, p. 45; March 1903, p. 4; Commons, *Quarterly Journal of Economics,* XIX, 7–8; U.S. Commissioner of Labor, *Regulation and Restriction of Output,* 1904, pp. 714–5, and ch. 13.

31. *AMC Journal,* March 1901, p. 22; Aug. 1902, p. 71; Dec. 1902, p. 17; March 1903, p. 2.

32. *AMC Journal,* April 1903, p. 21; also Nov. 1902, p. 20; Feb. 1903, p. 30; May 1903, p. 31.

33. *AMC Journal,* Nov. 1901, p. 6.

34. *AMC Journal,* March 1901, p. 5; Jan. 1902, p. 31; March 1903, pp. 1–12; April 1903, pp. 20–2; May 1903, p. 42. On convention resolutions on national scales, see AMC, *Proceedings,* 1899, p. 41; 1900, p. 32; 1902, p. 26; *AMC Journal,* Sept. 1902, p. 17.

35. *AMC Journal,* Jan. 1902, p. 16; AMC, *Proceedings,* 1902, pp. 13, 25.

36. AMC, *Proceedings,* 1902, pp. 58–9; 1904, pp. 37–8; *AMC Journal,* Feb. 1903, p. 40; Oct. 1903, pp. 23–4; *NP,* Sept. 19, 1903, pp. 32–3; Oct. 3, 1903, p. 33; Oct. 10, 1903, pp. 16, 28, 40; Commons, *Quarterly Journal of Economics,* XIX, 4–16; *Output Regulation,* pp. 713–5.

37. *AMC Journal,* Jan. 1904, p. 21; Call to Mary McDowell, Oct. 16, 1902, McDowell Papers; AMC, *Proceedings,* 1904, pp. 10–11.

38. U.S. Bureau of Labor Statistics, *Bulletin 77,* pp. 118–20.

39. *AMC Journal,* June 1903, p. 14; July 1903, p. 19; Sept. 1903, p. 24.

40. *NP,* Nov. 28, 1903, p. 14.

41. *AMC Journal,* Sept. 1903, p. 22; May 1904, p. 95; July 1904, p. 15; AMC, *Pro-*

ceedings, 1904, pp. 12, 69, 72–3; Chicago *Tribune,* June 13, 1904; *NP,* June 18, 1904, p. 43; June 25, 1904, pp. 13, 26.

42. AMC, *Proceedings,* 1904, pp. 11, 13, 16, 26, 88; Donnelly to Locals, December 12, 1903, in *AMC Journal,* Dec. 1903, p. 38.

43. Gompers to Donnelly, June 15, 1904, Gompers Letterbooks; *AMC Journal,* July 1904, p. 14; AMC, *Proceedings,* 1904, p. 11.

44. AMC, *Proceedings,* 1904, pp. 33, 193. Only two executive board members, Stephen Vail and C. E. Schmidt, favored national scales.

45. AMC, *Proceedings,* 1904, pp. 92–3; 1906, p. 12.

46. AMC, *Proceedings,* 1906, p. 12; *NP,* Aug. 6, 1904, p. 14; Chicago *Tribune,* July 18, 1904; Commons, *Quarterly Journal of Economics,* XIX, 5. There is a printed copy of the projected Amalgamated scale in the Wisconsin State Historical Society.

47. Donnelly to Packers, June 24, 1904, in AMC, *Proceedings,* 1906, pp. 7–8; Donnelly, in *Harpers' Weekly,* Aug. 27, 1904, p. 1321; Chicago *Inter-Ocean,* July 11, 1904.

48. Donnelly, in *Harpers' Weekly,* Aug. 27, 1904, p. 1320; Chicago *Tribune,* July 12, 14, 1904; *NP,* July 16, 1904, p. 14; *AMC Journal,* Dec. 1904, p. 2.

49. *Independent,* July 28, 1904, p. 184; Chicago *Inter-Ocean,* July 24, 1904; Omaha *Bee,* July 13, 1904; Chicago *Tribune,* July 13, 1904.

50. U.S. Bureau of Corps., *Report on the Beef Industry,* p. 84; *NP,* July 23, 1904, pp. 13–4.

51. Statements and letters printed in the Chicago *Tribune,* July 13–18, 1904; *NP,* July 23, 1904, p. 15; *American Federationist,* Sept. 1904, p. 773.

52. Omaha *Bee,* July 18, 1904.

53. AMC, *Proceedings,* 1906, pp. 8–9; *NP,* July 23, 1904, pp. 14–5; Samuel J. Naylor, "History of Labor Organization in the Slaughtering and Meat Packing Industry," unpub. diss., University of Illinois, 1936, p. 57.

54. Gompers to Donnelly, July 21, 1904, Gompers Letterbooks; *NP,* July 23, 1904, p. 15.

55. Chicago *Tribune,* July 21, 1904; Chicago *Inter-Ocean,* July 22, 1904; Omaha *Bee,* July 22, 1904.

56. AMC, *Proceedings,* 1906, p. 12. The details of the start of the second strike are drawn from the Chicago *Tribune,* July 23, 1904, and *Inter-Ocean,* July 23, 1904. See also, *Harper's Weekly,* Aug. 27, 1904, p. 1320; *NP,* July 30, 1904, pp. 13–4; *AMC Journal,* Dec. 1904, pp. 3, 8.

57. *NP,* July 30, 1904, p. 14 (my italics supplied). It seems clear that, from the first week of the strike, Donnelly had begun to see the strategic uses of a sympathetic strike and was deliberately employing the threat. See Omaha *Bee,* July 18, 1904; Chicago *Inter-Ocean,* July 17, 18, 1904; Chicago *Tribune,* July 20, 1904.

58. See statistics in *Butchers and Packers Gazette,* Aug. 27, 1904, p. 4; also, weekly slaughtering reports in *NP.*

59. See Arthur Meeker's statement in *NP,* July 23, 1904, p. 14, and Edward Tilden's in *NP,* Aug. 6, 1904, p. 14.

60. *Harpers' Weekly,* Aug. 27, 1904, p. 1320; *NP,* Sept. 12, 1903, p. 33, Sept. 19, 1903, pp. 19, 33; *AMC Journal,* Oct. 1904, p. 23; Chicago *Tribune,* July 13, 18, 1904.

61. Quoted in W. Hard and E. Poole, "The Stock Yards Strike," *Outlook,* Aug. 13, 1904, p. 886; *AMC Journal,* March 1903, p. 25; *Output Regulation,* pp. 711–13, 716. Union officials argued that output regulations were the result of the excessive speed-up in the nonunion period. President Donnelly himself opposed formal restrictions: "A fool pace setter could be controlled by the union without resorting to a general or uniform working list."

62. Bureau of Corps., *Report,* pp. 163, 260–8.

63. For a chronicle of violence during the strike, see Selig Perlman and Philip Taft, *History of Labor in the U.S., 1896–1932,* vol. 4 of Commons, *History of Labor,* pp. 120–2.

64. Gompers to Donnelly, Aug. 31, 1904, Gompers Letterbooks.

65. On the various peace efforts, see *NP,* Aug. 20, 1904, p. 36; Aug. 27, 1904, p. 16; *Butchers and Packers Gazette,* August 20, 1904, p. 4; Sept. 17, 1904, p. 1; *American Meat Trade and Retail Butchers Journal,* Aug. 25, 1904, p. 5.

66. Wilson, *Mary McDowell, Neighbor,* pp. 113–4; Chicago *Inter-Ocean,* Sept. 6, 1904; AMC, *Proceedings,* 1906, p. 26.

CHAPTER 4.   QUIESCENCE, 1904–1917

1. *Butchers and Packers Gazette,* Sept. 17, 1904, p. 5; *AMC Journal,* Oct. 1904, p. 15.

2. *NP,* Sept. 17, 1904, pp. 15, 36; Oct. 8, 1904, p. 37; *Butchers and Packers Gazette,* Sept. 17, 1904, p. 5. On the failure of strike leaders to be rehired, see letters in *AMC Journal,* Oct. 1905, p. 32; April 1906, p. 26; April 1908, p. 19.

3. *AMC Journal,* Oct. 1904, pp. 28, 39; Jan. 1905, p. 10; AMC, *Proceedings,* 1906, pp. 25, 36.

4. AMC, *Proceedings,* 1906, p. 27. Resolution reprinted on pp. 26–7.

5. "AMC Executive Board Minutes, Sept. 26–8, 1904," in *AMC Journal,* Oct. 1904, p. 36.

6. *AMC Journal,* Sept. 1905, p. 21. On wage movements in meat packing, see U.S. Bureau of Labor Statistics, *Bulletin 77,* pp. 118–20.

7. *NP,* July 11, 1908, p. 34; Aug. 15, 1908, p. 34; Jan. 16, 1909, p. 18; Feb. 13, 1909, p. 18; Nov. 11, 1911, p. 20; U.S. Commission on Industrial Relations, *Report,* IV, 3467, 3476, 3487, 3493; U.S. Bureau of Labor Statistics, *Bulletin 252* (1919), pp. 62–3.

8. Call to Mary McDowell, Jan. 30, 1908, Mary McDowell Papers.

9. *AMC Journal,* April 1906, p. 29.

10. Clark, "Controversy between Labor and Management in Meat Packing," pp. 140–2; *AMC Journal,* June 1905, p. 29; April 1906, p. 29; *NP,* Nov. 2, 1907, p. 15; Homer Call to Mary McDowell, Feb. 5, 1908; McDowell to Call, Feb. 21, 1908, McDowell Papers.

11. *AMC Journal,* Feb. 1907, p. 2; Jan. 1908, p. 1.

12. *AMC Journal,* April 1906, p. 27.

13. *AMC Journal,* Nov. 1904, pp. 33–4.

14. *AMC Journal,* May 1902, p. 21; December 1902, p. 20; March 1903, p. 45; June 1906, pp. 41–2; AMC, *Proceedings,* 1902, pp. 21, 22, 27, 33, 72; *NP,* Feb. 8, 1902, p. 34.

15. AMC, *Proceedings,* 1902, pp. 17, 25, 63; 1904, pp. 20, 33, 41.

16. AMC, *Proceedings,* 1914, pp. 10, 94.

17. *NP,* Oct. 10, 1903, p. 40.

18. *AMC Journal,* March 1903, p. 18; May 1903, p. 42; AMC, *Proceedings,* 1904, pp. 42, 89–90; Thompson, *Journal of Political Economy,* XV, 97.

19. *AMC Journal,* Sept. 1902, p. 16; July 1903, p. 19; AMC, *Proceedings,* 1904, p. 26; 1914, p. 53.

20. *AMC Journal,* March 1903, pp. 21, 45–6; April 1903, pp. 18, 25; July 1903, pp. 12, 20–1; May 1904, p. 94; AMC, *Proceedings,* 1904, pp. 21–4, 28, 42–3; 1914, p. 49; *NP,* May 16, 1903, p. 40.

21. AMC, *Proceedings,* 1906, p. 19.

22. New York *Herald,* Sept. 11, 1904; *American Meat Trade and Retail Butchers Journal,* Sept. 8, 1904, p. 5; Sept. 15, 1904, p. 6; AMC, *Proceedings,* 1906, pp. 29–40.

23. Gompers to Ernest Bohm, Sec., New York Central Federated Union, March 14, 1905; Gompers to Homer Call, March 29, 1905; Gompers to J. J. Keegan, March 21, 1905, Gompers Letterbooks; Call to Mary McDowell, Feb. 5, 1908, McDowell Papers; *AMC Journal,* Jan. 1905, pp. 23–4; Sept.–Oct. 1906, p. 2; Dec.–Jan. 1907, p. 2; July 1908, pp. 14–5, 18.

24. AMC, *Proceedings,* 1906, p. 68; 1914, pp. 131–2; *AMC Journal,* Oct. 1904, p. 34; April 1906, p. 34; June 1906, p. 30; Call to Mary McDowell, Jan. 30, 1908, McDowell Papers.

25. *NP,* Dec. 12, 1908, p. 36; AMC, *Proceedings,* 1914, pp. 62–3, 132; 1917, pp. 58–9.

26. AMC, *Proceedings,* 1904, p. 14; 1906, p. 12; 1914, p. 29; Quarterly Report to July 31, 1905, in *AMC Journal,* Sept. 1905, p. 21. On AFL assistance after the strike, see Gompers to Donnelly, Oct. 14, 1904; Gompers to Call, Jan. 5, 1905, Gompers Letterbooks; *American Federationist,* Nov. 1904, p. 1001; AMC, *Proceedings,* 1906, pp. 31–2.

27. "Executive Board Minutes, May 25–7, 1908," in *AMC Journal,* July 1908, pp. 15–6.

28. AMC, *Proceedings,* 1914, p. 10.

29. Correspondence in AMC, *Proceedings,* 1914, pp. 61–2.

30. *AFL Executive Council Minutes,* Feb. 21–6, 1916, pp. 5–6; Joseph Menhart to Gompers, July 18, 1917, copy in Gompers Letterbooks; *BW,* July 1916, p. 1; March 1916, p. 2; AMC, *Proceedings,* 1917, p. 82.

31. AMC, *Proceedings,* 1914, pp. 86 ff.; 1917, pp. 19–26, 58, 66, 107; *BW,* Jan. 1917, p. 7; Feb. 1917, p. 3.

32. AMC, *Proceedings,* 1917, pp. 32, 56, 97, 125; *BW,* March 1916, p. 2; Feb. 1917, p. 3; *AFL Executive Council Minutes,* Feb. 21–6, 1916, p. 6.

33. AMC, *Proceedings,* 1917, pp. 13–6, 28, 75–6, 133; Lewis Corey, *Meat and Man,* p. 284.

34. AMC, *Proceedings,* 1914, pp. 13–4; Call to McDowell, Jan. 30, 1908, McDowell Papers.

35. Mary McDowell to Call, Feb. 1, 1912; Call to McDowell, May 27, June 10, 1912, McDowell Papers; AMC, *Proceedings,* 1914, pp. 12–4, 46–8, 93–4.

36. *NP,* Oct. 14, 1916, p. 107.

37. *AFL Executive Council Minutes,* Nov. 25, 1916, pp. 3–4; Jan. 20–7, 1917, p. 17.

38. *AFL Executive Council Minutes,* Feb. 21–6, 1916, p. 5; AMC, *Proceedings,* 1917, pp. 13–4, 32, 34; *BW,* June 1916, p. 5; Dec. 1916, p. 2.

39. AMC, *Proceedings,* 1917, pp. 34, 41; *BW,* March 1916, p. 1; May 1916, p. 3; Sept. 1916, p. 6; *NP,* Oct. 14, 1916, p. 131.

CHAPTER 5. THE SECOND CYCLE IN MEAT PACKING

1. Omaha *Bee,* Sept. 5–12, 1917; *BW,* Dec. 1916, p. 1; *NP,* Sept. 22, 1917, p. 31.

2. Gompers to John Fitzpatrick, Aug. 22, 1917, Gompers Letterbooks; Spero and Harris, *Black Worker,* p. 270; O. M. Sullivan, "The Women's Part in the Stockyards Organizing Work," *Life and Labor,* May 1918, pp. 102, 104.

3. Chicago *Tribune,* Sept. 10, 17, 1917; William Z. Foster, *From Bryan to Stalin* (New York, 1937), pp. 94–5; *BW,* Jan. 1918, p. 5; April 1922, p. 1.

4. *Report of the President's Mediation Commission to the President* (Washington, 1918), p. 16; *BW,* Feb. 1918, p. 8.

5. Swift & Co., *Yearbook,* 1917, pp. 36–43; 1918, pp. 10, 55; "AMPA, Proceedings,

Oct. 15–17, 1917," in *NP*, Oct. 20, 1917, p. 115; "Proceedings, Oct. 14–16, 1918," in *NP*, Oct. 19, 1918, p. 131. On the various reforms, see *NP*, 1915–1917, *passim*.

6. John F. Hart to Gompers, Oct. 16, 1917; Gompers to William B. Wilson, Oct. 19, 1917, Gompers Letterbooks.

7. U.S. Commission on Industrial Relations, *Report*, IV, p. 3494; *AFL Executive Council Minutes*, Feb. 21–6, 1916, p. 6.

8. *BW*, Nov. 1917, p. 1; Dec. 1919, pp. 1–2; *NP*, Nov. 17, 1917, p. 36; W. Z. Foster, in *Life and Labor*, April 1918, pp. 65–6.

9. Gompers to Herbert Hoover, May 25, 1917; Gompers to Homer Call, May 31, 1917, Gompers Letterbooks; Chicago *Tribune*, Nov. 2, 1917.

10. Baker to W. B. Wilson, Dec. 19, 1917, Labor Dept. Files, National Archives.

11. Union agreement printed in *BW*, Jan. 1918, p. 2; packers' agreement in Clemens, *American Meat Industry*, pp. 717–8.

12. Philadelphia *North American*, Jan. 19, 1918, clipping in Scrapbook, Frank P. Walsh Papers, New York Public Library; New York *Times*, Jan. 19, 1918; *BW*, Jan. 1918, p. 1; *American Federationist*, June 1919, p. 516.

13. *NP*, Jan. 26, 1918, p. 15.

14. *Labor Herald*, Feb. 1, 1918 and New York *Sun*, Jan. 25, 1918, clippings in Scrapbook, Walsh Papers; New York *Times*, Jan. 25, 28, 1918; *NP*, Feb. 2, 1918, p. 15; *BW*, March 1918, p. 4.

15. Walker to Max Lowenthal, February 21, 1918, Labor Dept. Files.

16. Walsh to John Fitzpatrick, June 14, 1918; Walsh to R. G. Knutson, March 19, 1918, Walsh Papers.

17. Philadelphia *Press*, Feb. 14, 1918, clipping in Scrapbook, Walsh Papers, also, many other clippings on the Hearings. Selections from the Hearings are in *BW*, March 1918, pp. 1–4, and a copy of the complete Transcript is in Labor Dept. Files.

18. Fitzpatrick and Lane to Walsh, March 26, 1918; Walsh to Samuel Alschuler, March 27, 1918; Fitzpatrick to Walsh, March 29, 1918, Walsh Papers.

19. *BW*, April 1918, pp. 1, 5; W. L. Chenery, "Packingtown Steps Forward," *Survey*, April 13, 1918, pp. 35–8; M. E. McDowell, "Easter Day After the Decision," *Survey*, April 13, 1918, p. 38. Among the many congratulatory messages to Walsh was this one from a South Omaha local:

Please except [sic] our heartfelt graditude [sic] and if there is any way in which we can repay you we will gladly do it. The decision . . . will go down in history as one of the most far reaching in the last decayed [sic]. And your name will be on the tongue of every packinghouse worker for years to come as the champion of humanity.

R. L. Rice, Sec., AMC Local 602 to Walsh, April 4, 1918, Walsh Papers.

20. See National War Labor Board, *Dockets*, 80, 81, 177–177d, 234, 235.

21. *AFL Executive Council Minutes*, Feb. 21–6, 1916, p. 66; *AFL Secretary's Report*, December 15, 1919, p. 10; *BW*, Feb. 1918, p. 8; Foster to Walsh, July 6, 1918, Walsh Papers.

22. Samuel Alschuler to Independent Packing Company, Chicago, July 27, 1921, copy in AMC Files; *BW*, Jan. 1919, p. 1; Feb. 1919, p. 3. See also correspondence between Dennis Lane and Armour & Co. during 1920, AMC Files.

23. *BW*, Nov. 1918, p. 1.

24. Award printed in *BW*, Feb. 1919, p. 1.

25. Dennis Lane *et al.* to Affiliated Unions, June 5, 1919; Cudahy, Morris, Swift, Wilson and Armour Companies to W. B. Wilson, April 12, 1919, copy in AMC Files.

26. *AFL Executive Council Minutes,* May 9–19, 1919, pp. 3–4, 49; Oscar F. Nelson to Hugh L. Kerwin, U.S. Conciliator, May 9, 1919; Kerwin to W. B. Wilson, May 21, 1919; Kerwin to E. B. Marsh, June 2, 1919, Labor Dept. Files; *BW,* June 1919, p. 3; Dec. 1919, p. 2; *New Majority,* May 10, 1919, p. 1, cited hereafter as *NM; AMC, Proceedings,* 1920, p. 112; Lane *et al.* to Affiliated Unions, June 5, 1919, AMC Files.

27. Lane *et al.* to Affiliated Unions, June 5, 1919, AMC Files.

28. Immigration Commission, *Report,* XIV, 204, 213, 271, 344; W. A. Crossland, *Industrial Conditions Among Negroes in St. Louis* (St. Louis, 1914), pp. 82–3; Clark, "Controversy between Labor and Management in Meat Packing," p. 152.

29. U.S. Dept. of Labor, *Negro Migration, 1916–1917* (Washington, 1919), pp. 117, 130; Herbst, *The Negro in Meat Packing,* pp. 35–7; Spero and Harris, *Black Worker,* pp. 272–3; *BW,* Oct. 1918, p. 5; Dec. 1918, p. 5; Chicago *Tribune,* Aug. 1, 1919; *NM,* Aug. 9, 1919, p. 1.

30. *Crisis,* Oct. 1919, pp. 294–5.

31. Foster, quoted in Chicago Commission on Race Relations, *The Negro in Chicago* (Chicago, 1922), p. 429; *BW,* Aug. 1920, p. 12.

32. *NM,* June 14–July 26, 1919, *passim; BW,* June 1919, p. 1; July 1919, pp. 1, 3; *NP,* July 26, 1919, p. 25.

33. *NM,* Aug. 2, 1919, p. 1; Aug. 9, 1919, p. 12; "Chicago Federation of Labor Minutes, Aug. 3, 1919," in *NM,* Aug. 9, 1919, p. 14; *Negro in Chicago,* ch. 1; Chicago *Tribune,* July 28–Aug. 3, 1919.

34. *NM,* Aug. 9, 1919, p. 2; Aug. 16, 1919, pp. 1, 2, 9, 16; Chicago *Tribune,* Aug. 5, 6, 10, 11, 1919; *BW,* Aug. 1919, p. 3; New York *Times,* Aug. 7–10, 1919; John Fitzpatrick and J. W. Johnstone to W. B. Wilson, Aug. 23, 1919, Labor Dept. Files.

35. *BW,* Nov. 1919, pp. 1, 2, 5.

36. *BW,* Dec. 1919, p. 2; June 1919, p. 4; Oct. 1919, p. 2.

37. *BW,* Aug. 1919, pp. 1, 4; Sept. 1919, p. 1; *NM,* Dec. 6, 1919, p. 3; Dec. 13, 1919, p. 9; *NP,* Dec. 13, 1919, p. 42; AMC, *Proceedings,* 1920, pp. 85–6.

38. *BW,* Nov. 1919, pp. 2, 5; "CFL Minutes, Jan. 18, 1920," in *NM,* Jan. 24, 1920, pp. 12–3; Clark, "Controversy between Labor and Management in Meat Packing," p. 186.

39. Foster, *From Bryan to Stalin,* p. 97; *AFL Executive Council Minutes,* Oct. 5–22, 1919, pp. 54–5; Dec. 11–8, 1919, pp. 4, 50; Feb. 24–March 3, 1920, p. 28; "CFL Minutes, Oct. 5, 1919," *NM,* Oct. 11, 1919, p. 13; "Minutes, Oct. 19, 1919," *NM,* Oct. 25, 1919, p. 13; "Minutes, Jan. 18, 1920," *NM,* Jan. 24, 1920, pp. 12–3; *BW,* Nov. 1919, p. 1; Jan. 1920, p. 1; Clark, "Controversy between Labor and Management in Meat Packing," p. 186.

40. "CFL Minutes, Dec. 18, 1919," *NM,* Jan. 24, 1920, pp. 13–4; "Minutes, Feb. 15, 1920," *NM,* Feb. 21, 1920, p. 12; "Minutes, March 7, 1920," *NM,* March 13, 1920, p. 12; "Minutes, April 18, 1920," *NM,* April 24, 1920, p. 13; *NM,* Feb. 28, 1920, p. 8; April 3, 1920, p. 2; April 10, 1920, p. 6. One clue to Kikulski's motives can be found in the fact that the Amalgamated was able to have him dismissed from the AFL payroll, and thus perhaps to force him over to its side. *AFL Executive Council Minutes,* Dec. 11–8, 1919, p. 50.

41. AMC, *Proceedings,* 1920, pp. 147–8; *BW,* Jan. 1920, p. 6; June 1920, p. 12; July 1921, p. 1; *AMC Executive Board Minutes,* Aug. 28, 1920.

42. *BW,* Aug. 1920, p. 6; Art. 2, Sec. 10, AMC Constitution, 1920.

43. See *BW,* Aug. 1918, p. 3; April 1920, p. 1.

44. Lane to Hart, Oct. 5, 1920, *AMC Executive Board Minutes,* Oct. 5, 1920; also, Sept. 8, 1920; Oct. 27, 1920; *BW,* July 1921, p. 1.

45. *AMC Executive Board Minutes,* April 27, 1921.

46. The relevant sections of Article 2 read:

Section 2. In case of the dismissal, removal or death of the General International President, the General International Vice President shall succeed to all the powers and privileges of the General International President, and this, and all other vacancies in the International Executive Board shall be filled in accordance with Section 16 of Article 2.

. . . .

Section 16. . . . In case of any vacancy in the International offices, the Executive Board shall have the power to fill such vacancy.

47. *BW,* July 1921, p. 1; Nov. 1921, p. 3.

48. AMC, *Proceedings,* 1920, pp. 11, 37, 139–40; *BW,* Sept. 1918, p. 3; Aug. 1919, p. 2; Nov. 1919, p. 4; Jan. 1920, pp. 5, 7; April 1920, p. 7; May 1920, p. 1; *NP,* June 30, 1917, p. 42; Aug. 2, 1919, p. 26; Oct. 18, 1919, pp. 22, 43; Dec. 13, 1919, p. 48.

49. *AMC Executive Board Minutes,* April 19, 25, 1921; Interview, P. J. Guest, June 28, 1960.

50. Sec. Reports in *AFL Executive Board Minutes,* Dec. 11, 1919; Nov. 14–9, 1921.

51. "IAMP, Proceedings, Sept. 13–15, 1920," in *NP,* Sept. 18, 1920, p. 114.

52. Paul H. Douglas, *Real Wages in the United States, 1890–1926* (Boston, 1930), pp. 101, 108.

53. J. G. Condon and Carl Meyer to W. B. Wilson, Feb. 21, 1921, copy in AMC Files.

54. *BW,* Feb. 1921, pp. 1, 5, 6; H. L. Kerwin to Lane, Feb. 9, 1921, Labor Dept. Files.

55. See Lane to Don Cargill, Feb. 7, 1921, AMC Files; O. F. Nelson to H. L. Kerwin, Jan. 26, 1921; Kerwin to J. E. Spangler, Feb. 1, 1921, Labor Dept. Files.

56. *BW,* March 1921, pp. 1–4; April 1921, p. 3; *NM,* March 26, 1921, pp. 3, 12; New York *Times,* March 9–14, 17, 19, 20, 1921; Gompers to Lane, March 6, 10, 1921, Gompers Letterbooks.

57. *AMC Executive Board Minutes,* April 19, 1921; Gompers to John F. Hart, March 23, 26, 1921, Gompers Letterbooks; New York *Times,* March 22–5, 1921; *BW,* April 1921, p. 1. Agreement printed in *NP,* March 26, 1921, p. 20.

58. *AMC Executive Board Minutes,* April 19, 25–7, 1921; *BW,* May 1921, p. 1.

59. "IAMP Proceedings, Sept. 13–15, 1920," in *NP,* Sept. 25, 1920, pp. 43 ff. M. D. Harding expressed the shifting view on industrial relations:

The main trouble has been doing things for people rather than doing things with people. That is the idea. You want to bring them in and make them a part of the whole. You don't want to take the attitude of giving them something. Make them think they are part of the organization.

60. A. H. Carver, *Personnel and Labor Problems in the Packing Industry* (Chicago, 1928), pp. 149–50; Clemens, *American Meat Industry,* p. 723. On industry thinking, see discussion at Industrial Relations Committee meeting, "IAMP Proceedings, Sept. 13–5, 1920," *NP,* Sept. 25, 1920, pp. 43 ff.

61. L. D. H. Weld, Letter to Editor, *Survey,* Jan. 14, 1922, p. 604; Swift & Co., "Memorandum on the Strike Situation in the Packing Industry," in Clark, "Controversy between Labor and Management in Meat Packing," p. 188.

62. *Armour Magazine,* April 1921, p. 9; May 1921, p. 9; Sept. 1921, p. 11; *NP,* July 9, 1921, p. 22; Clemens, *American Meat Industry,* pp. 723–31.

63. "IAMP Proceedings, Aug. 8–10, 1921," in *NP,* July 30, 1921, p. 20; Clemens, *American Meat Industry,* pp. 730–6; Wilson & Co., *Joint Representation Committee Plan* (May 1921); Swift & Co., *Employee Representation Plan* (Nov. 1, 1921).

64. *Armour Magazine,* June 1921, p. 15; Sept. 1921, p. 11; *NP,* Aug. 13, 1921, p. 131;

BW, June 1921, p. 1; July 1921, pp. 1, 2, 4, 8; "Application for Change in Rate and Working Conditions," June 4, 1921, in Swift & Co. Library, Chicago; Samuel Alschuler to Lane, June 7, 1921, AMC Files.

65. *South Omaha Conference Minutes*, Aug. 15-7, 1921, AMC Files; *Joint Conference Minutes*, Sept. 6, 1921, AMC Files; *BW*, Aug. 1921, p. 1; Sept. 1921, p. 1; Jan. 1922, p. 6; AMC, *Proceedings*, 1922, p. 28.

66. *Armour Magazine*, Dec. 1921, pp. 3, 5; *NP*, Nov. 12, 1921, p. 22; Nov. 19, 1921, p. 40; Nov. 26, 1921, pp. 19-20; *BW*, Feb. 1922, p. 6.

67. *Joint Conference and Executive Board Minutes*, Nov. 17, 21, 30, Dec. 1, 1921, AMC Files.

68. *BW*, Dec. 1921, *passim; NM*, Dec. 10, 1921, p. 1; Dec. 17, 1921, pp. 1, 2; *NP*, Dec. 10, 1921, pp. 17, 33, 35, 40; Dec. 17, 1921, pp. 9, 33, 35, 42; D. J. Saposs, Letter to Editor, *Survey*, Jan. 14, 1922, pp. 605-6; Oscar Nelson to H. L. Kerwin, Dec. 6, 13, 1921, including Dept. of Agriculture reports, Labor Dept. Files.

69. *BW*, Dec. 1921, p. 3; AMC, *Proceedings*, 1922, pp. 36, 106; *NP*, Dec. 17, 1921, p. 47; Sherman Rogers, "The 'Sympathetic Strike' in the New York Packing Industry," *Outlook*, Jan. 18, 1922, pp. 96-7.

70. O. F. Nelson to H. L. Kerwin, Dec. 22, 1921, Labor Dept. Files.

71. *AFL Executive Council Minutes*, Feb. 21-6, 1916, p. 5; Nov. 11-20, 1918, pp. 75-6; May 9-10, 1919, p. 34; Daniel J. Tobin to AFL Executive Council, Nov. 25, 1917, Frank Duffy Papers, Harvard Labor-Management History Center; *AMC Executive Board Minutes*, Jan. 11, 1921.

72. Gompers to Lane, March 6, 10, Aug. 24, 1921, Gompers Letterbooks.

73. *Joint Conference Minutes*, Nov. 17, 21, 30, 1921, AMC Files.

74. "CFL Minutes, Dec. 18, 1921," in *NM*, Dec. 24, 1921, p. 11; *BW*, Dec. 1921, pp. 7, 12.

75. Benjamin Stolberg, "The Stock Yards Strike," *Nation*, Jan. 25, 1922, pp. 91-2. On strike problems, see reports in *BW*, Dec. 1921; Jan. 1922, *passim*.

76. On use of strikebreakers, see Clark, "Controversy between Labor and Management in Meat Packing," p. 202; Herbst, *The Negro in Meat Packing*, pp. 59-65; Spero and Harris, *Black Worker*, pp. 281-2; Paul S. Taylor, *Mexican Labor in the U.S.: Chicago and the Calumet Region* (Berkeley, Calif., 1932), p. 46. It is significant that the percentage of Negroes at the Swift and Armour Chicago plants, despite a decline of 1500 in their labor forces, jumped from 20.3 per cent in 1921 to 32.7 per cent in 1922.

77. Lane to Gompers, Dec. 22, 1921; AFL Executive Council, Doc. 47, Dec. 29, 1921; Doc. 57, Jan. 16, 1922, Duffy Papers; Gompers to Lane, Dec. 29, 1921, Jan. 13, 1922, Gompers Letterbooks.

78. *BW*, Feb. 1922, p. 2; AMC, *Proceedings*, 1922, p. 29.

79. O. F. Nelson to H. L. Kerwin, Jan. 4, 5, 1922, Labor Dept. Files; *Joint Conference Minutes*, Jan. 20, 1922, AMC Files; *BW*, Jan. 1922, pp. 1, 2, 4, 5.

80. *Joint Conference Minutes*, Jan. 23, 1922, AMC Files; *BW*, Feb. 1922, p. 2; *NP*, Jan. 21, 1922, p. 46; *NM*, Feb. 4, 1922, p. 6.

CHAPTER 6. THE MEAT CUTTERS: THE PROCESS OF LOCALIZED UNIONIZATION

1. *AMC Executive Board Minutes*, Nov. 30, 1920.

2. National War Labor Board, *Docket* 829; *BW*, May 1918, p. 5. On the rise of Chicago Local 546, see *BW*, Oct. 1915, p. 5; Dec. 1918, p. 3; Jan. 1919, p. 3; AMC, *Proceedings*, 1917, p. 78; *American Federationist*, April 1917, p. 302.

3. The spread between wholesale and retail beef prices increased by 65 per cent; operating expenses increased 85 per cent. *Monthly Labor Review,* April 1922, p. 77.

4. AMC, *Proceedings,* 1922, p. 13.

5. Entry, Oct. 1, 1935, *Minute Book,* Local 358, in Wisconsin State Historical Society; *AMC Journal,* June 1902, p. 15; W. R. Satterlee, Local 1, to Gorman, Jan. 5, 1938, AMC Files.

6. AMC, *Proceedings,* 1899, pp. 46–7; 1902, p. 27; 1904, p. 17; 1906, p. 25; 1914, pp. 62, 78, 81; 1926, p. 18; 1930, pp. 15, 20; Art. lv, Sec. 2, Local Constitution, 1900.

7. *AMC Journal,* June 1902, pp. 15–6.

8. AMC, *Proceedings,* 1922, pp. 91–2; 1930, pp. 54, 75–7.

9. AMC, *Proceedings,* 1926, pp. 79–80; *BW,* Nov. 1926, p. 7; April 1930, p. 3.

10. *AMC Journal,* April 1903, p. 39; Nov. 1903, p. 39.

11. AMC, *Proceedings,* 1899, p. 26; 1902, pp. 31–2; 1904, pp. 16–16a; 1914, pp. 24, 57; *AMC Journal,* Feb. 1903, p. 2.

12. *AMC Journal,* Feb. 1906, p. 28.

13. *Monthly Labor Review,* June 1927, pp. 201–2; *BW,* March 1927, p. 8.

14. AMC, *Proceedings,* 1926, p. 79; *Monthly Labor Review,* May 1926, p. 208.

15. AMC, *Proceedings,* 1926, p. 80.

16. *AMC Journal,* May 1901, pp. 4–5; Feb. 1902, pp. 13–4; Dec. 1902, pp. 24–5; *American Federationist,* Sept. 1903, p. 909; *NP,* Jan. 23, 1904, p. 43.

17. *NP,* May 16, 1903, p. 40; July 25, 1903, p. 40.

18. *AMC Journal,* Oct. 1902, p. 36; May 1904, p. 93.

19. *AMC Journal,* Jan. 1904, p. 46.

20. *NP,* May 2, 1903, p. 40; Feb. 20, 1904, p. 43; *Butchers and Packers Gazette,* Aug. 20, 1904, p. 4.

21. *NP,* Aug. 29, 1903, p. 39.

22. *AMC Journal,* Nov. 1900, p. 8; July 1902, pp. 33–4; April 1903, p. 31; June 1906, p. 37; *NP,* Feb. 21, 1903, p. 40.

23. *AMC Journal,* March 1903, p. 32; March 1905, p. 12; *NP,* March 7, 1903, p. 39.

24. *AMC Journal,* Nov. 1905, p. 37.

25. AMC, *Proceedings,* 1904, p. 19.

26. In Rochester the district organizer, V. J. Schneider, became sales representative for the supplier, and in Oakland both Vice-President C. E. Schmidt and organizer Herman May resigned to handle the affairs of the California Meat Company.

27. On the Chicago retail strike, see *NM,* Nov. 1, 8, 22, Dec. 6, 1919; *NP,* Nov. 8, 1919, p. 40; Nov. 22, 1919, p. 45; *BW,* Oct. 1919, p. 5; Nov. 1919, p. 3.

28. Belsky, *I, the Union,* pp. 120–1; *BW,* April 1922, p. 3; April 1926, p. 1.

29. Ringer to John Bettendorf, St. Louis RMDA, June 14, 1938, AMC Files. Also, on union-retailer cooperation, see Dennis Lane to Executive Board, May 8, 1929, AMC Files.

30. AMC, *Proceedings,* 1930, p. 15; *BW,* Jan. 1923, p. 5; Feb. 1923, p. 1; Dec. 1925, p. 2. On West Coast "kick-back" cases, see AMC, *Proceedings,* 1940, p. 60; Gorman to William Ingram, Safeway Stores, Jan. 18, 1939, AMC Files.

31. *BW,* March 1930, p. 3; Jan. 1933, p. 5; AMC, *Proceedings,* 1926, pp. 49–50, 75–6, 79.

32. Gompers to Lane, March 31, 1921, to Max Pine, March 26, 1921, Gompers Letterbooks; *AFL Executive Council Minutes,* Nov. 1928, pp. 18–9; *AMC Executive Board Minutes,* Nov. 30, 1920; AMC, *Proceedings,* 1920, pp. 80, 163–4; *Butcher Worker* (Local 234), July 1938, p. 29; Belsky, *I, the Union, passim.*

33. *NP,* Sept. 8, 1906, p. 43; *Federated News,* Jan. 10, 1931; *BW,* Feb. 1931, p. 3;

New York *Times*, Oct. 31, 1933; AMC, *Proceedings*, 1930, p. 6; *AMC Executive Board Minutes*, Jan. 23, 1936.

34. New York *Times*, Aug. 6, 14, 28, Sept. 19, 24, Oct. 11, Dec. 22, 1935; Jan. 28, 30, July 9, Dec. 29, 1936; Jan. 8, April 21, 1938; *Butcher Worker*, Nov. 1938, p. 38; Gorman to W. C. Lilly, Aug. 15, 1935, AMC Files; *AMC Executive Board Minutes*, Jan. 23, 1936; Harold Seidman, *Labor Czars: A History of Labor Racketeering* (New York, 1938), pp. 187–95.

35. Belsky, *I, the Union*, chs. 4, 5; New York *Times*, Sept. 11, 1928; June 1, 10, 14, 1935; *Butcher Worker*, Nov. 1938, p. 37; Dec. 1938, p. 27; M. Epstein, *Jewish Labor in the U.S., 1914–1952* (New York, 1953), ch. 9; Will Herberg, "Jewish Labor Movement in the United States," *Industrial and Labor Relations Review*, VI, 44–66 (Oct. 1952).

36. See *BW*, Dec. 1932, p. 1; June 1, 1936, p. 1; *Butcher Worker*, Nov. 1938, p. 37.

37. AMC, *Proceedings*, 1930, p. 6.

38. See various entries in *Minute Book*, Local 358, 1932–7, Wisconsin State Historical Society.

39. *BW*, Feb. 1931, p. 1; Oct. 1931, p. 5.

40. Telegrams of Dennis Lane, F. C. Yoerk and J. Y. Henderson, June 9–13, 1932, AMC Files.

41. *AFL Executive Council Minutes*, Feb. 18–25, 1929, p. 100; April 20–May 2, 1933, p. 92.

42. *AMC Executive Board Minutes*, Feb. 1, 1922; AMC, *Proceedings*, 1926, p. 9.

43. Quoted, I. W. Ringer to Gorman, April 29, 1938, AMC Files; *BW*, Feb. 1925, p. 3.

44. *Western Butcher Craftsman*, Oct. 1931, p. 5.

45. The vice-presidents in Dec. 1933 were: M. J. Kelly, E. W. Jimerson, J. P. McCoy, J. J. Walsh, John Malone, J. S. Hofmann, M. S. Maxwell, T. J. Lloyd and A. M. Provo (who was replaced in 1936 by Joseph Belsky).

46. *AMC Executive Board Minutes*, Feb. 1, 1922; April 15, 1925; AMC, *Proceedings*, 1922, p. 34; 1926, p. 27; 1930, pp. 6, 13.

47. *AMC Executive Board Minutes*, June 18, 1926; AMC, *Proceedings*, 1926, pp. 69–70, 79–82; 1930, pp. 17–21; *BW*, Feb. 1925, p. 1; Aug. 1926, p. 1; Oct. 1926, p. 8; Dec. 1926, pp. 1, 2.

48. AMC, *Proceedings*, 1926, p. 5; *AMC Executive Board Minutes*, Sept. 14, 1920; Jan. 23, 1931; Jan. 24, 1936; Jan. 21, 1937; March 25, 1938.

49. AMC, *Proceedings*, 1930, pp. 7, 53, 68–71; 1936, p. 102; Lane to the Executive Board, Aug. 23, Oct. 1, 1930, Jan. 29, 1932, AMC Files; *BW*, July 1930, p. 1; March 1931, p. 1.

50. AMC, *Proceedings*, 1926, pp. 39, 66; 1930, p. 39. For examples of trusteeships before 1930, see *Proceedings*, 1922, p. 52; 1930, p. 17; *BW*, March 1926, pp. 1, 8; *AMC Executive Board Minutes*, Dec. 21, 1920.

51. AMC, *Proceedings*, 1926, p. 3.

52. *AMC Executive Board Minutes*, April 16, 1925; Gompers to Lane, Jan. 27, 1922, Gompers Letterbooks; *BW*, Nov. 1923, p. 3; Dec. 1923, p. 8; March 1924, p. 1; Sept. 1924, p. 8; AMC, *Proceedings*, 1926, pp. 3, 22, 28.

53. Sanford to Lane, Dec. 10, 1921, and Lane's remark, in *BW*, March 1925, p. 2.

54. *BW*, Aug. 1924, p. 2; Feb. 1925, p. 1; March 1925, p. 1; April 1925, p. 1; May 1925, p. 1; *Western Butcher Craftsman*, Oct. 1931, p. 5; *AMC Executive Board Minutes*, April 15, 16, 1925; Jan. 23, 1931; Lane to Executive Board, Aug. 23, Oct. 1, 1930, AMC Files. One dimension of the problem was the geographical situation in California. The north was thoroughly organized and was the seat of the state branch. The

unions of the south were ineffectual and, for obscure reasons, generally opposed to the rule extending down from San Francisco. Maxwell deeply resented enemies within the state—specifically, Max R. Grunhof—working for the international in Los Angeles, pocketing a salary that might otherwise have gone to do constructive work in the north —Maxwell considered southern California hopeless for unionism—and at best building up a movement that would be hostile to him.

55. *Western Butcher Craftsman,* Jan. 1932, pp. 1, 11; California State Federation of Labor, *Proceedings,* 1931, pp. 4–5; *BW,* Dec. 1931, p. 2; Aug. 1932, p. 8; Oct. 1932, p. 1.

56. Maxwell to Gorman, Jan. 30, Feb. 9, 1932, including Contract and "Interpretation of Articles Agreed Upon," AMC Files.

CHAPTER 7. NEW FORCES IN THE RETAIL TRADE

1. C. F. Phillips, "The Chain Store: A Study in Retail Development," unpub. diss., Harvard University, 1934, *passim.*

2. See *NP,* March 22, 1902, p. 24; Nov. 5, 1904, p. 43; *Butchers and Packers Gazette,* Feb. 25, 1905, p. 8; *BW,* Oct. 1916, p. 3.

3. *Chain Store Age. Grocery Products Section,* Jan. 1930, p. 3; May 1931, p. 18; Phillips, "The Chain Store," p. 123; New York *Times,* Sept. 23, 1929; U.S. Bureau of Census, *Census of Business,* 1940, I, 11, 66–7, 819; Lewis Corey, *Meat and Man,* pp. 166–9.

4. AMC, *Proceedings,* 1936, p. 35.

5. "Loyalty Begets Loyalty, American Stores Find," *Chain Store Age. Grocery Products Section,* June 1931, pp. 38, 42, 44.

6. *BW,* Feb. 1930, p. 3; Sept. 1930, p. 1; Nov. 1930, p. 1; *AMC Executive Board Minutes,* Jan. 22, 1931.

7. AMC, *Proceedings,* 1930, pp. 20, 28, 67; *BW,* Jan. 1930, p. 6; April 1931, p. 3; May 1933, p. 1; Gorman to Executive Board, Nov. 29, 1930, AMC Files.

8. AMC, *Proceedings,* 1930, p. 67; 1936, p. 9; *AMC Executive Board Minutes,* Jan. 23, 1931; Gorman to Morrell, March 1, 1934, AMC Files; Interview, Patrick Gorman, Nov. 17, 1960. There was, of course, a danger implicit in this tactic. Would the union always be able to resist the temptation of trading on the contract terms in order to get the first contract from the employers? This happened in New York City between New York leader Max Block and A & P in 1952. The McClellan Committee in 1958 found that the chain, in exchange for recognizing the Amalgamated as bargaining agent for the A & P clerks, received an extension of the 45-hour week for five years. New York *Times,* May 16, 17, 1958.

9. The agreement is printed in *BW,* May 1935, p. 1.

10. Gorman to Warren, May 15, 1935, Aug. 27, 1936, AMC Files.

11. See Gorman to Joseph Bappert, Kroger Grocery Co., June 30, 1936, Sept. 22, 1938; Bappert to Gorman, Oct. 5, 1938, AMC Files.

12. Paul Mooney, Kroger Grocery Co., to A. V. Landes, March 19, 1934; Gorman to J. G. Brown, National Tea Co., Aug. 15, 1934; F. H. Massmann, National Tea Co., to Gorman, Sept. 5, 6, 1934; Gorman to Massmann, Sept. 5, 1934; Lane to J. Bappert, May 11, 1934, AMC Files.

13. Gorman to Hy Gregerson, Safeway Stores, Aug. 3, 1939, AMC Files.

14. Gorman to Robert Call, June 6, 1934, quoted in telegram, Gorman to Paul Mooney, Kroger Grocery Co., June 6, 1934, AMC Files.

15. In San Diego an attempt was made to implement the provision through the accounting firm of Price, Waterhouse & Co., which was hired to make a survey. The

result was repeated delays and eventually a serious strike in the winter of 1936–37.

16. Gorman to Joseph Bappert, Kroger Grocery Co., June 18, 1935, AMC Files.

17. Warren to Gorman, July 2, 1937, AMC Files. For an interesting parallel analysis of the impact of chains on retail unionism in England, see Robert E. L. Knight, "Unionism Among Retail Clerks in Postwar Britain," *Industrial and Labor Relation Review*, XIV, 515–27 (July 1961).

18. New York *Times*, Nov. 1–6, 1934; *BW*, Dec. 1934, p. 8; Jan. 1935, p. 1; AMC, *Proceedings*, 1936, p. 6.

19. *BW*, Sept. 1934, p. 3. The wage minima ranged from $13 to $15 in the north and $12 to $14 in the south.

20. *BW*, June 1935, p. 4.

21. On the jurisdiction of the NLRB, see Joseph Rosenfarb, *National Labor Policy* (New York, 1940), ch. 14, particularly pp. 439–40; and J. Ackerman, "The Problem of Jurisdiction of National and State Labor Relations Boards," *Industrial and Labor Relations Review*, II, 360–71 (April 1949). On state laws, see C. C. Killingsworth, *State Labor-Relations Acts* (Chicago, 1948), ch. 12.

22. *BW*, Jan. 1, 1938, p. 6; June 1, 1937, p. 6; July 1, 1937, p. 4; Aug. 1, 1937, p. 8; *AFL Executive Board Minutes*, Aug. 21–Sept. 2, 1937, p. 193; Jan. 24–Feb. 8, 1938, p. 240.

23. Gorman to R. S. Brennan, March 8, 1938, AMC Files. On the use of injunctions involving boycotts before the Norris-LaGuardia Act, see Felix Frankfurter and Nathan Greene, *The Labor Injunction* (New York, 1930), pp. 30–1, 42–6.

24. H. J. Mullally, Local 2, to Gorman, Oct. 5, 1938; Gorman to Gus Kliphon, Local 1, Oct. 8, 1938, AMC Files; *AMC Executive Board Minutes*, Oct. 11, 1941; *BW*, Dec. 1934, p. 8; April 1, 1937, p. 3; Sept. 1, 1937, p. 2; Feb. 1, 1938, p. 7; Nov. 1, 1940, p. 1. More a more general discussion of the Teamsters' role in retail organization, see Martin S. Estey, "The Strategic Alliance as a Factor in Union Growth," *Industrial and Labor Relations Review*, IX, 41–53 (Oct. 1955).

25. *BW*, Sept. 1, 1937, p. 2.

26. *BW*, Nov. 1, 1936, p. 1; April 1, 1937, p. 3; Cincinnati official Michael Schuld quoted, Gorman to Charles Schimmat, A & P, May 4, 1939, AMC Files; *AMC Executive Board Minutes*, Jan. 21, 1937.

27. Gorman to C. W. Calkins, National Tea Co., April 1, 1939; Calkins to Gorman, Feb. 14, March 29, 1938; Gorman and Lane to Executive Board, Oct. 30, 1936, AMC Files.

28. See account of chain opposition and Patman Bill in G. M. Lebhar, *Chain Stores in America* (New York, 1952), chs. 6, 7, 12.

29. The announcement is printed in *Chain Store Age. Grocery Edition*, Oct. 1938, p. 58.

30. Gorman to Joseph Bappert, Kroger Grocery Co., April 12, 1938, AMC Files; AFL, *Proceedings*, 1939, pp. 464–7.

31. Gorman's testimony is reprinted in a pamphlet, *What Organized Labor Did to Defeat the Patman Bill and Why* (n.d.).

2. Gorman to Charles Schimmat, Oct. 1, Nov. 3, 1938; to Carl Byoir, Nov. 30, 1938; Byoir to Gorman, Dec. 3, 1938, AMC Files.

33. Gorman to I. M. Ornburn, AFL, April 18, 1939, AMC Files.

34. AMC, *Proceedings*, 1940, pp. 89, 122–3. The A & P agreements are also listed in these *Proceedings*, p. 86.

35. The agreement is printed in *BW*, Sept. 1, 1937, p. 7.

36. Gorman to I. W. Ringer, Sept. 15, 23, Oct. 29, Nov. 12, 1937; E. Priebe to

Ringer, Nov. 22, 1937, copy; Charles Glatz to Ringer, Jan. 28, 1938; Ringer to Glatz, Feb. 17, 1938, copies, AMC Files.

37. Gorman to V. J. Svetavsky, Nov. 6, 1939; to I. W. Ringer, Jan. 9, 1945; Ringer to Gorman, Sept. 9, 1940, AMC Files.

38. *BW,* March 1, 1940, p. 1.

39. Gorman to H. E. Carlson, Dec. 19, 1939, AMC Files.

40. *AMC Executive Board Minutes,* March 23, 1938.

41. Telegram quoted, Gorman to Joseph Bappert, Kroger Grocery Co., April 30, 1937; Bappert to Gorman, May 14, 1937, AMC Files.

42. *AMC Executive Board Minutes,* March 23, 1938; Lloyd quoted, Gorman to Charles Schimmat, A & P, April 13, 1939; Cincinnati official Michael Schuld quoted, Gorman to Schimmat, May 4, 1939, AMC Files.

43. Gorman to L. A. Warren, Safeway Stores, Oct. 8, 1936, AMC Files.

44. For instance, when the Auburn, New York, local refused to permit any apprentices in a new chain store, Gorman threatened to transfer the store to the jurisdiction of another union. Gorman to H. J. Mullally, Local 2, March 21, 1939, AMC Files.

45. I. W. Ringer to Gorman, March 3, 1938; Gorman to E. J. McCarthy, Local 2, March 28, 1939, AMC Files.

46. Gorman to Marvin Hook, Local 88, Dec. 6, 1940, AMC Files; *BW,* May 1935, p. 1.

47. Warren to Gorman, June 4, 18, 1935, AMC Files.

48. *AMC Executive Board Minutes,* March 24, 1938.

49. *BW,* Sept. 1, 1937, pp. 1, 8; Gorman to I. W. Ringer, July 17, 1937, Jan. 29, 1938; to J. A. Kotal, Nov. 9, 1937; to Marvin Hook, Dec. 6, 1940; Ringer to Gorman, Sept. 14, 1937; Hook to Gorman, Dec. 9, 1940, AMC Files.

50. Gorman to Charles Schimmat, May 4, 9, 1939; to I. W. Ringer, Sept. 14, 1938, AMC Files; *AMC Executive Board Minutes,* Oct. 23, 1938; Nov. 24, 1940; AMC, *Proceedings,* 1940, p. 222.

51. Warren to Gorman, June 18, 1935, AMC Files. It was significant that Michael Schuld, an outspoken critic of the A & P policy, had a very precarious hold on his own Cincinnati union. *AMC Executive Board Minutes,* March 24, 1938. But even secure leaders, such as J. S. Hofmann of Seattle, sometimes acted in negotiations under rank-and-file pressure. See I. W. Ringer to Gorman, Dec. 14, 1937, March 9, 1938, AMC Files.

52. For example, George A. Roach of Los Angeles, U. G. Rich of Cleveland, and Natale Masi of Philadelphia. For Roach's activities, see *AMC Executive Board Minutes,* March 25, Oct. 6, 1938; June 19, 1939; Oct. 10, 1941; *BW,* June 1, 1937, p. 3; July 1, 1938, p. 5; Sept. 1, 1938, p. 8.

53. Sometimes this could be blatant. When the New York locals after a lengthy fight finally won a contract from the Retail Meat Dealers Association in Sept. 1936, one clause called for the "Deliverance of 100 shops to each Local Union by March 14, 1937." New York *Times,* Sept. 15, 26, 1936.

54. Gorman to R. S. Brennan, March 5, 1938; Brennan to Gorman, March 11, 1938, AMC Files.

55. Gorman to Joseph Bappert, April 17, 1938; to Charles Schimmat, April 28, 1939, AMC Files.

56. Gorman to I. W. Ringer, June 6, 1938; to Joseph Bappert, April 12, 1938, AMC Files; Interview, Gorman, Nov. 17, 1960.

57. W. A. Ingram, Safeway, to Gorman, Feb. 9, 1939, Dec. 19, 1940; Gorman to Charles Schimmat, A & P, May 14, 1945, AMC Files.

58. Interview, Philip Guest, June 28, 1960; AMC, *Proceedings,* 1940, p. 51; 26 NLRB 117 (1940); Carl Byoir to Gorman, Nov. 22, 1938; Gorman to Byoir, Nov. 30, 1938, AMC Files.

59. W. S. Brennan to Gorman, March 11, 1938, AMC Files.

60. D. F. Hurley, Report to Director of Conciliation, Oct. 28, Nov. 2, 1939; T. J. Williams to J. R. Steelman, March 19, 26, 1938, Federal Mediation and Conciliation Service Files, National Archives.

61. Gorman to John A. Hartford, A & P, Nov. 1, 15, 1940; Hartford to Gorman, Nov. 13, 1940, Jan. 13, 1942; Gorman to Charles Schimmat, Jan. 14, 1941; Schimmat to Gorman, April 1, 1941, AMC Files.

62. 4 New York State Labor Relations Board 758 (1941); 26 NLRB 1275 (1940), 27 NLRB 518 (1941), 55 NLRB 56 (1944); AMC, *Proceedings,* 1944, pp. 41, 61.

63. Gorman to Charles Schimmat, Dec. 8, 1938, Feb. 28, 1940, AMC Files; *AMC Executive Board Minutes,* Oct. 4, 1939; Nov. 20, 1940; New York *Times,* Dec. 24, 1940; *BW,* July 1, 1939, p. 4; Jan. 1, 1940, p. 6; April 1940, p. 1; 3 NYLRB 933 (1940); 8 NYLRB 256 (1945); 9 NYLRB 166 (1946).

CHAPTER 8. PROLOGUE TO PACKINGHOUSE UNIONIZATION, 1933–1937

1. *AFL Executive Council Minutes,* Sept. 29–Oct. 13, 1923, p. 22; *BW,* March 1924, p. 1.

2. AMC, *Proceedings,* 1926, p. 83; Dennis Lane to Executive Board, May 28, 1929, AMC Files.

3. *NP,* Aug. 5, 1933, pp. 15, 40; Oct. 28, 1933, pp. 84–6; New York *Times,* Aug. 6, 7, 12, 1933. Agreement and interpretation printed in *NP,* Aug. 12, 1933, pp. 17–8, 26, 46.

4. *BW,* July 1933, p. 1; Feb. 1934, p. 1; Jan. 1936, p. 1; *AMC Executive Board Minutes,* Dec. 4, 1933.

5. Circular in Local 28 Correspondence, AMC Files; *BW,* Sept. 1933, p. 1; Jan. 1934, p. 1.

6. Gorman to Wesley Moravec, Local 28, Dec. 13, 1933; Gorman and Lane to Swift & Co., Jan. 6, 1934, AMC Files.

7. *AMC Executive Board Minutes,* Dec. 4–5, 1933; Gorman to W. Moravec, Local 28, Nov. 15, 1933; Gorman and Lane to Executive Board, Feb. 22, 1934, AMC Files. The Memorandum is printed in *BW,* Jan. 1934, p. 2.

8. J. D. Cooney to B. M. Marshman, March 23, 1934, National Mediation and Conciliation Service Files.

9. F. I. Badgley, "IAMP Proceedings, Oct. 23, 1933," in *NP,* Oct. 28, 1933, pp. 82–4.

10. *NP,* Dec. 9, 1933, p. 44; July 28, 1934, p. 45; Sept. 29, 1934, p. 35; Carver, *Personnel Problems in Packing,* p. 149; J. E. Bullmaster, Local 23, to Lane, March 24, 31, 1934, AMC Files; *BW,* April 1934, pp. 3, 8; Aug. 1934, p. 1.

11. Cases 218, 238, 408, *Decisions of the National Labor Relations Board* (Washington, 1935), II, 112–20, 479–80. On these problems, see Twentieth Century Fund, *Labor and the Government* (New York, 1935), pp. 83–6, and chs. 9, 10.

12. W. Moravec, Local 28, to Lane, Feb. 12, March 19, 1934; *Organizer* (S. Omaha), June 1934, in Local 28 Correspondence, AMC Files.

13. Gorman and Lane to Executive Board and Organizers, Feb. 22, 1934, AMC Files, printed in *BW* (March 1934), p. 1.

14. Gorman and R. S. Brennan to Robert F. Wagner, March 12, 1934; Gorman and

Brennan to E. F. McGrady, March 12, 1934, NMCS Files; Brennan to William Green, Nov. 8, 1934, National Recovery Administration Files, National Archives; *BW,* April 1934, p. 1; June 1934, p. 1; AMC, *Proceedings, 1936,* pp. 6–7.

15. *BW,* June 1934, pp. 2, 8. Accounts of the subsequent strikes are in *BW,* NP, and the New York *Times.*

16. The following agreement between Local 304 and Morrell was the typical outcome of negotiations in the NRA period:

May 17, 1934.

The following is a statement of our policy with respect to employment and working conditions:

1. The Company restates its plan to bargain collectively with any of its employees or representatives of their own choosing without discrimination and not to interfere with the employee's right to join any organization as stated in the National Recovery Act.

2. The Company recognizes the principle of seniority rights and will continue this policy in major departments in accordance with the N.R.A. interpretations.

3. A minimum week of thirty-two hours is guaranteed to all regularly employed.

4. The Company will continue to pay a scale of wages equal to the average paid by all packing plants similarly located.

5. The Company will continue to pay time and a half for all time worked in excess of ten hours in any one day, or if men are called back for night work.

6. The Company agrees not to make any change in present working conditions without notifying our employees of such proposed change thirty days in advance. The employees likewise agree not to propose any change in present working conditions without submitting such proposed change to the management thirty days in advance.

JOHN MORRELL & COMPANY
W. H. T. Foster,
Vice President

AMC Files.

17. J. E. Bullmaster, Local 23, to Lane, June 2, 1934; Lane to Bullmaster, June 4, 1934, AMC Files.

18. W. W. Wood to George Carlson, Aug. 17, 1934, NRA Files; L. E. Kline, "History of Negotiations between the Institute of American Meat Packers and the NRA," Dec. 26, 1935, NRA Files; *NP,* July 1, 1933, pp. 17–8; Feb. 17, 1934, p. 21; Oct. 20, 1934, pp. 126–7; New York *Times,* Aug. 21, Oct. 24, 1933.

19. William Green to Gorman, July 13, 1934; W. J. Woolston, Labor Advisory Board, to R. S. Brennan, July 16, 1934; E. J. Tracy, Labor Advisory Board, to Gorman, Dec. 5, 1934; Gorman to Tracy, Dec. 7, 1934, NRA Files; R. Eldridge, "Conference with representative of the Stockyards Labor Council regarding drawing up of a meat packing code, Sept. 25, 1934," NRA Files; John G. Shott, Memorandum, July 17, 1935, NRA Files. The collection of letters and affidavits from AMC local is also deposited in the Packing Industry File, NRA Files.

20. *AMC Executive Board Minutes,* Jan. 25, 1935; Gorman and Lane to Executive Board, May 14, 1935, AMC Files.

21. *AFL Executive Council Minutes,* Sept. 6–15, 1933, p. 56; Jan. 29–Feb. 14, 1935, p. 301; Jan. 15–29, 1936, p. 202; May 5–20, 1936, p. 273; Gorman and Lane to Executive Board, May 14, 1935, AMC Files; AMC, *Proceedings, 1936,* p. 17; *BW,* Dec. 1, 1935, p. 8.

22. Lane to J. E. Bullmaster, Local 23, Jan. 5, 1935, AMC Files.

23. Frank W. Schulte, "Historical Sketches of the Growth of the Packing House Union in Austin, Minnesota," copy in Industrial Relations Library, Harvard Univer-

sity; *AMC Executive Board Minutes,* Jan. 24, 1935; Gorman to John Brophy, Jan. 28, 1936; to John L. Lewis, March 3, 1936, AMC Files; New York *Times,* Nov. 13, 14, 1933; *NP,* Nov. 18, 1933, p. 39; George Mayer, *Floyd B. Olson* (Minneapolis, 1951), pp. 159–61.

24. *Mid-West Union News* (Cedar Rapids), March 1935, pp. 1–3; *BW,* June 1934, p. 1; Sept. 1934, p. 7; Nov. 1934, p. 5; Gorman to John L. Lewis, March 3, 1936, AMC Files; Interview, T. J. Lloyd, Dec. 6, 1963.

25. On Communist activity, see *BW,* Oct. 1933, p. 2; Arthur Kampfert, "History of Unionism in the Meat Packing Industry, 1933 to 1940," III, United Packinghouse Workers of America headquarters; *Daily Worker,* Dec. 4, 1933, in H. R. Cayton, and G. S. Mitchell, *The Black Workers and the New Unionism* (Chapel Hill, 1939), pp. 260–2; Local 28 Correspondence, AMC Files.

26. *In the Matter of G. H. Hammand & Co. and the Stockyards Labor Council, March 2 1934,* Chicago Regional Labor Board, deposited in NRA Files; H. T. McCarthy, Chicago Regional Labor Board, Memorandum, Nov. 30, 1934, NRA Files; *CIO News, Packinghouse Edition,* Nov. 21, 1938, p. 7; Walter Galenson, *The CIO Challenge to the AFL* (Cambridge, Mass., 1960), pp. 360–1.

27. AMC, *Proceedings,* 1922, pp. 18, 35, 81–2; *BW,* Aug. 1922, p. 5.

28. Lane to John Brophy, March 18, 1936, AMC Files; AMC, *Proceedings,* 1936, pp. 8–9, 34, 66; William Green to Gorman, March (?), 1936 (Reel 23, Letter 265) William Green Papers, Cornell University.

29. Lane to John Brophy, March 18, 1936; Brophy to Lane, March 9, 26, 1936; Gorman to Brophy, March 3, 12, 19, 1936, AMC Files; *BW,* May 1, 1936, p. 6; AMC, *Proceedings,* 1936, p. 9.

30. Gorman to John Brophy, Feb. 26, 1936, April 19, 1937; Brophy to Gorman, Feb. 28, 1936, AMC Files; William Green to Gorman, March 9, 1937, Green Papers; *Union News Service,* March 9, 1936, quoted in Galenson, *The CIO Challenge to the AFL,* p. 14; *BW,* Dec. 1, 1936, p. 4; March 1, 1937, p. 5.

31. John Brophy to Lane, March 3, 9, 1936; Brophy to Delegates of Austin and Cedar Rapids Unions, Austin, Minn., Sept. 15, 1936, copy in AMC Files.

32. John Brophy to Lane, Dec. 17, 1936; Lane to John L. Lewis, March 15, 1937, AMC Files.

33. AMC, *Proceedings,* 1936, p. 9; Gorman to John Brophy, April 19, 1937, AMC Files.

34. This is based on Gorman's account of the meetings to the executive board. Lewis' statements are paraphrased or quoted by Gorman. *Minutes,* May 10, 1937.

35. *AMC Executive Board Minutes,* May 10, 1937; Lane to John Brophy, May 12, 1937, AMC Files.

36. For example, Lane to John L. Lewis, March 15, 1937, AMC Files.

37. *AMC Executive Board Minutes,* May 10, 1937.

38. There is no direct evidence to support the analysis in these paragraphs. The Amalgamated position was that it acted solely on the principle of opposition to secession. Patrick Gorman still holds that view. On the other hand, A. A. Myrup of the Bakery Workers, whose situation was very similar to that of the Amalgamated, admitted that it could not enter the CIO "because we felt that the status of our organization in the general labor movement as well as our make-up, especially the importance of our label, could not afford to invite or bear the effect of any possible reprisal." The logic of the situation did not differ materially for the Amalgamated in the retail field. Interview, Gorman, Nov. 17, 1960; Hilton E. Hanna and Joseph Belsky, *Picket and the*

Pen (Yonkers, 1960), pp. 291–5; A. A. Myrup to Charles Howard, March 13, 1936, quoted in Galenson, *The CIO Challenge to the AFL,* p. 5.

39. Gorman to John L. Lewis, June 16, 1937; Lewis to Gorman, June 21, 1937, AMC Files.

40. John Brophy to Lane, March 9, 1936, AMC Files.

41. Iowa Meat Packers' Federation, *Organization Meeting Minutes,* Dec. 5, 1936, copy in possession of Irwin Nack.

42. Kampfert, "History of Unionism in the Meat Packing Industry," IV; *CIO News, Packinghouse Edition,* Nov. 21, 1938, p. 7; Cayton and Mitchell, *The Black Workers,* pp. 273–4.

43. 3 NLRB 103, 895 (1937), p. 7 NLRB 287 (1938); *NP,* June 12, 1937, p. 20.

CHAPTER 9.   THE THREE–CORNERED CONTEST

1. Quoted in James R. Holcomb, "Union Policies of Meat Packers, 1929–1943," unpub. diss., University of Illinois, 1957, p. 173.

2. *NP,* April 17, 1937, p. 17.

3. 7 NLRB 275 (1938).

4. E. A. Cudahy, Jr., 15 NLRB 682 (1939), italics by NLRB.

5. *NP,* May 23, 1937, p. 23.

6. 29 NLRB 853–62 (1941); *NP,* May 29, 1937, p. 18.

7. *NP,* May 15, 1937, p. 23.

8. 5 NLRB 472 (1938), 15 NLRB 992 (1939), 31 NLRB 452 (1941); Holcomb, "Union Policies of Meat Packers," pp. 101–2, 124, 139, 161–2; New York *Times,* Oct. 10, 1939; *CIO News, Packinghouse Edition,* Jan. 9, 1939, p. 2; June 26, 1939, p. 1; Jan. 8, 1940, pp. 1, 8, cited hereafter as *PH News.*

9. 31 NLRB 967–8 (1941).

10. In fact, unions pushed discrimination cases as a deliberate organizing tactic. See Lewis J. Clark to Arthur Kampfert, May 15, 1940, UPWA Files.

11. New York *Times,* Oct. 28, 1936; March 10, 1937; *NP,* May 15, 1937, p. 23.

12. Kampfert, "History of Unionism in the Meat Packing Industry," IV; *CIO Executive Board Minutes,* June 3–5, 1940, p. 251.

13. Holcomb, "Union Policies of Meat Packers," pp. 140–1; T. V. Purcell, *Blue Collar Man* (Cambridge, Mass., 1960), pp. 19–21.

14. Gorman to R. S. Brennan, Oct. 1, 1938, AMC Files; also, *BW,* Jan. 1, 1939, p. 4.

15. *AFL Executive Council Minutes,* Jan. 24–Feb. 8, 1938, p. 240; Aug. 22–Sept. 2, 1938, pp. 146–7; Jan. 29–Feb. 9, 1940, p. 185; *AMC Executive Board Minutes,* March 23, 1938, Feb. 23, 1940; *CIO Executive Board Minutes,* June 3–5, 1940, p. 12; Nov. 15–23, 1940, p. 224; *PH News,* Aug. 7, 1939, p. 2; Dec. 11, 1939, p. 1; Galenson, *CIO Challenge,* p. 600. On PWOC inability to meet organizing demands, see L. J. Clark to Arthur Kampfert, Nov. 19, 1940; Clark to Martin Lane, Dec. 11, 1940, UPWA Files.

16. *AFL Executive Council Minutes,* Jan. 30–Feb. 14, 1939, p. 279.

17. 11 NLRB 950, 953 (1939), 14 NLRB 287 (1939), 16 NLRB 334 (1939), 21 NLRB 1189 (1940). On Teamsters' power see A. J. Shippey to Brother (?) Gebhart, Sept. 13, 1941; to J. C. Lewis, Sept. 14, 1941, UPWA Files.

18. *AMC Executive Board Minutes,* May 11, 1937.

19. *BW,* July 1, 1937, p. 1; Jan. 1, 1939, p. 8.

20. *PH News,* Sept. 18, 1939, p. 7; M. Kennedy, Local 23, to Lane, May 1, 1937, AMC Files.

21. Lane to M. Kennedy, Local 23, Aug. 31, 1938; Bittner to Gorman, Dec. 5, 1938,

AMC Files; L. J. Clark to Van Bittner, April 10, May 15, 1940; to Joe Wheeler, Local 94, June 23, 1941, UPWA Files.

22. U.S. Bureau of Census, *Census*, 1930, V, ch. 7; Taylor, *Mexican Labor in Chicago*, p. 46; Cayton and Mitchell, *Black Workers and New Unions*, pp. 262–73, 277–8; AMC, *Proceedings*, 1936, p. 111.

23. *PH News*, Jan. 2, 1939, p. 2; Feb. 17, 1941, p. 2; Kampfert, "History of Unionism in Meat Packing," IV; H. R. Ballard to J. C. Lewis, April 3, 1942; G. R. Hathaway to J. C. Lewis, Aug. 22, 1941, UPWA Files.

24. See *PH News*, March 20, 1939, p. 2; April 3, 1939, p. 2; May 15, 1939, p. 7; July 10, 1939, pp. 2, 9; June 10, 1940, p. 7. See also, Barbara Warne Newell, *Chicago and the Labor Movement* (Urbana, Ill., 1961).

25. James P. Dean to J. C. Lewis, July 28, 1941, UPWA Files.

26. Joint Executive Board Meeting of Oklahoma City and Ft. Worth Locals, Aug. 9, 1942, UPWA Files; *District 2 Conference Minutes*, Jan. 14, 1940, UPWA Files; *People's Press*, July 23, 1938, quoted in Kampfert, "History of Unionism in Meat Packing," IV.

27. *District 2 Conference Minutes*, Jan. 14, 1940; Jan. 12, 1941, UPWA Files; Kampfert, "History of Unionism in Meat Packing," IV, *PH News*, May 1, 1939, p. 1.

28. *PH News*, Nov. 5, 1938, p. 8; Nov. 14, 1938, p. 2; Dec. 5, 1938, p. 2; Jan. 9, 1939, p. 1.

29. 21 NLRB 1169 (1940); also, New York *Times*, March 29, 1940.

30. Lane to Anthony Lester, Local 5, June 27, Nov. 14, 1939, AMC Files; *BW*, July 1, 1939, p. 1; Feb. 1, 1940, p. 1.

31. *PH News*, Jan. 23, 1939, p. 1; J. C. Lewis to C. L. Hodgert, Cudahy Packing Company, July 30, 1941, UPWA Files. Lewis thereafter denounced several strikes as unauthorized, for instance, at the Cudahy plant several weeks later, and in Sept. 1941 against three Chicago independents. Kansas City *Star*, Aug. 22, 23, 1941; New York *Times*, Sept. 5, 1941.

32. *BW*, Feb. 1, 1940, p. 2. Also, AMC circular printed in *PH News*, Dec. 19, 1938, pp. 1–2.

33. For an account of this strike, see especially the file, including newspaper clippings, in Record Group 280, NMCS Files; also, *PH News*, Nov. 28, 1938, p. 1; Dec. 5, 1938, pp. 1–2; Dec. 12, 1938, pp. 1–2. It does not appear that the national AMC officers were directly involved in this episode. Their names were conspicuously absent from newspaper accounts; President Gorman denied that such measures would have Amalgamated support; and Van Bittner himself told the CIO Executive Board that "the A.F. of L. directly, not through the Amalgamated Meat Cutters, . . . through the personal representatives of the A.F. of L. organized a strike breaking agency and attempted to break the strike." A joint AFL drive was then being attempted in the stockyards, and this episode seems connected with that campaign. On the other hand, there is no evidence that the international tried to exert its authority over one of its locals during the incident. *CIO Executive Board Minutes*, June 3–5, 1940, p. 249; Bittner to Gorman, Dec. 5, 1938, AMC Files; *BW*, Jan. 1, 1939, p. 4.

34. *AFL Executive Council Minutes*, Aug. 22–Sept. 2, 1938, pp. 146, 147; Gorman to John Holmes, Swift & Co., Nov. 3, 1938, AMC Files; Chicago *Tribune*, Nov. 22, 1938; *PH News*, Dec. 5, 1938, p. 8.

35. 8 NLRB 1100 (1938), 10 NLRB 891 (1938); New York *Times*, Sept. 17, 1938; Gorman to Don King, Swift & Co., Feb. 14, 1939, AMC Files; *BW*, Feb. 1, 1940, p. 2. On PWOC propaganda on the company union issue, see *PH News*, Dec. 5, 1938, p. 8; Oct. 30, 1939, p. 8.

36. Gorman to R. S. Brennan, Feb. 15, 1940, AMC Files; W. T. Haywood to J. C.

Lewis, Dec. 30, 1941; Lewis to Haywood, Jan. 5, 1942; Herbert Vogt to Edward Fitzpatrick, May 19, 1942; Roy Franklin to J. C. Lewis, Sept. 21, 1941, UPWA Files. Two distinctions, however, should be drawn here. First, there was a difference between affiliating a company union intact and winning over plant leaders who were company union officers. Several such men—for instance, Walter Piotrowski at the Wilson-Chicago plant—headed aggressive PWOC locals. The second distinction involves the Swift independents in the later period. Once permanently established, they would have been welcomed into either the AMC or the PWOC.

37. Roy Franklin, *District 2 Conference Minutes,* Jan. 12, 1941, UPWA Files.

38. *District 2 Conference Minutes,* Jan. 14, 1940, UPWA Files.

39. *District 2 Conference Minutes,* Jan. 12, 1941, UPWA Files.

40. D. W. Chappell to John R. Steelman, April 25, 1939, NMCS Files.

41. *Locals 42 and 44 Joint Executive Board Minutes,* June 6, 1941; Glen Weidenheimer to Jimmie (James Cunningham?), Dec. 5, 1939, UPWA Files.

42. Clark to Arthur Kampfert, Nov. 5, 1940, UPWA Files.

43. Kampfert to A. T. Stephens, March 2, 1941, UPWA Files.

44. For example, Nicholas Fontecchio to L. J. Clark, April 17, 1940, UPWA Files.

45. Johnson quoted, Nels Peterson, Omaha Joint Council, to Kampfert, n.d.; Clark to Kampfert, Nov. 7, 1940, UPWA Files.

46. *Kansas City Locals' Joint Executive Board Minutes,* June 22, 1941, UPWA Files. On impact of concluded negotiations, see, e.g., Roy Franklin to J. C. Lewis, Sept. 11, 1941, UPWA Files.

47. Gorman to I. W. Ringer, June 14, 1940, AMC Files.

48. *NP,* June 26, 1937, p. 14; Whiting Williams, "IAMP Proceedings, Oct. 25–7, 1937," in *NP,* Nov. 6, 1937, pp. 146–7; Thomas Creigh, Cudahy Packing Co., to Gorman, Dec. 1, 1938, AMC Files.

49. Lane to Anthony Lester, Local 5, June 27, 1939; Lane to Lester, Nov. 14, 1939, AMC Files.

50. Gorman to R. E. Yokum, Cudahy Packing Co., Nov. 16, 1937, July 18, 1938; R. M. Conner to Gorman, July 21, 1938; Gorman to Conner, July 22, 1938; T. J. Lloyd to Gorman, July 25, 1938; C. L. Hodgert, Cudahy Packing Co., to Gorman, Nov. 16, 1939; Gorman to Hodgert, Nov. 17, 1939; Lane to A. Lester, Local 5, Nov. 20, 1939, AMC Files.

51. Gorman to Bob Mythen, Aug. 18, 1939, AMC Files.

52. *CIO Executive Board Minutes,* June 3–5, 1940, p. 251; also, Sam Sponseller to J. C. Lewis, Sept. 11, 1941, UPWA Files.

53. On the impact of the Morrell boycott, see Holcomb, "Union Policies of Meat Packers," pp. 83, 85–7; Galenson, *CIO Challenge,* pp. 356, and 684, note 24; *BW,* Jan. 1, 1936, p. 1; June 1, 1936, p. 3; J. C. S., Director of Sales, Morrell & Co., to Morrell & Co., Oakland, Calif., Dec. 6, 1935, copy in AMC Files.

54. Gorman to Executive Board, June 28, 1935; to G. Steindl, Oct. 27, 1941; to B. W. Compton, Meat Packers, Inc., Sept. 30, 1940, AMC Files; *AMC Executive Board Minutes,* June 20, June 21, 1939; *BW,* March 1939, p. 1.

55. Gorman to William I. Ingram, Safeway Stores, July 6, 1939, May 20, 1941, Jan. 20, 1942; Ingram to Gorman, Jan. 21, 1942; Ingram to Phil Tovrea, Tovrea Packing Co., June 30, 1939, copy; Gorman to Joseph Bappert, Kroger Grocery Co., Jan. 10, 1935, May 27, 1937, AMC Files; *BW,* Sept. 1, 1939, p. 1.

56. For example, Swift obtained an injunction and started a damage suit against the Washington State Federation of Butcher Workmen in 1940, and Wilson took effective legal action against Amalgamated locals in New York and Philadelphia in 1939. R. S.

Brennan to Gorman, March 15, 1940; J. D. Cooney, Wilson & Co., to Gorman, Feb. 15, 1939, AMC Files; New York *Times,* Jan. 15, 1939.

57. Belsky quoted, Gorman to Executive Board, Sept. 17, 1937; J. D. Cooney to Gorman, Dec. 6, 1938; Anthony Lester, Local 5, to Lane, Feb. 2, 1939; Gorman to R. E. Yokum, Cudahy Packing Co., Jan. 4, 1938, AMC Files; New York *Times,* Dec. 16, 1938; *BW,* March 1938, p. 5.

58. *AMC Executive Board Minutes,* June 30, Oct. 9, 1939.

59. Frank Wetterling to Lane, Jan. 21, 1938, AMC Files.

60. *AMC Executive Board Minutes,* May 10, 1937; Gorman to T. M. Sinclair, Kingan & Co., June 16, 1937, AMC Files; Interview, Gorman, Nov. 16, 1960; *BW,* July 1, 1937, p. 4; Aug. 1, 1937, p. 1.

61. *AMC Executive Board Minutes,* March 23, 1938; Gorman to Scott Peterson, Sec., Packers' and Sausage Mfgers.' Association, Chi., July 28, 1938; R. S. Brennan to Gorman, Jan. 17, Feb. 12, 1940, AMC Files. It should be noted that the PWOC also received occasional support from packers. For example, Sam Sponseller to J. C. Lewis, Sept. 11, 1941, UPWA Files.

62. A. F. Hunt, Swift & Co., to Gorman, Sept. 19, 1939; R. S. Brennan to Gorman, Nov. 9, 1939; Gorman to Brennan, Nov. 10, 1939, AMC Files; "Memorandum of conference with Messrs. A. F. Hunt and John Wilson, Swift & Co., relative to Los Angeles situation as well as Portland and Seattle," n.d. (ca. Dec. 1939), AMC Files.

63. J. D. Cooney to Gorman, Dec. 6, 1938; Gorman to Cooney, Dec. 9, 1938; Gorman to C. L. Hodgert, Cudahy Packing Co., June 25, 1941, AMC Files.

64. For example, Gorman to Don King, Armour & Co., Jan. 19, Feb. 10, 14, Nov. 22, 1939; to Frank Green, Armour & Co., June 13, 1939; to C. L. Hodgert, Cudahy Packing Co., April 13, 1939, AMC Files.

65. The boycott was designed to put pressure on employees as well as on their employers, and particularly so in CIO plants. See Circular, March 1938, in Local 5 File; Cudahy Co. Statement to Employees of Los Angeles plant, n.d., in Cudahy File, AMC Files; A. J. Shippey to J. C. Lewis, Aug. 12, Sept. 14, 1941; J. C. Lewis to W. T. Haywood, Jan. 5, 1942; Pete Mosele to J. C. Lewis, Aug. 20, 1941, UPWA Files.

66. Kampfert, "History of Unionism in Meat Packing," IV; *AMC Executive Board Minutes,* Oct. 5, 1937; leaflets from both sides in Local 165 File, AMC Files; *PH News,* May 15, 1939, p. 1. New contract printed in *BW,* June 1, 1938, p. 5.

67. Gorman to Robert Weiner, NLRB, July 12, 21, 1937; Weiner to Gorman, July 15, 23, 1937, AMC Files.

68. 22 NLRB 1019 (1940); T. J. Lloyd to J. W. Madden, Chairman, NLRB, Feb. 26, 1940; Gorman to R. S. Brennan, Jan. 16, 1941; Brennan to Gorman, Jan. 17, 1941, AMC Files; *AMC Executive Board Minutes,* Jan. 22, 1940. On the problem of closed-shop contracts, see Rosenfarb, *National Labor Policy,* pp. 268–70. It should be noted that AMC officials were convinced of NLRB bias in these cases, as well as other matters, and that when Local 207 carried the NLRB decision on the Cudahy-Los Angeles plant to the Federal courts, one of the charges was that the NLRB regional director had acted "with the intent of causing plaintiff to lose members, dues and prestige."

69. Gorman to R. S. Brennan, May 12, 1938; Claude XXX, Local 165, to Lane, Sept. 1, 1938; Jim Whitaker, Local 165, to Gorman, Dec. 12, 1938; Gorman to Whitaker, Dec. 15, 1938, AMC Files.

70. In a few instances, local leaders, attempting to balance conflicting demands of employers and employees, misled the membership about the terms of the agreement. Discovery, of course, was disastrous. See Anthony Lester, Local 5, to Gorman, April 17, 1939, AMC Files; 22 NLRB 1022 (1940); *PH News,* May 13, 1940, p. 1.

71. Gorman to Carl Swenson, July 31, 1941, AMC Files; James Robb to J. C. Lewis, July 31, 1941; Virgil Case to J. C. Lewis, May 1, 1942; Arthur Kampfert to L. J. Clark, Dec. 2, 1943, UPWA Files; *PH News,* May 12, 1941, p. 2; Kampfert, "History of Unionism in Meat Packing," IV.

72. *District 2 Conference Minutes,* Jan. 12, 1941, UPWA Files.

73. See T. V. Purcell, *Blue Collar Man* (Cambridge, Massachusetts, 1960), pp. 22–5. For PWOC explanations for its defeat, see W. T. Haywood to J. C. Lewis, Sept. 25, 27, 1941; Neal Weaver to Lewis, Aug. 21, Sept. 25, 1941, UPWA Files.

74. Melvin J. Humphrey, "Effect of the Government and War Period on Unionism in the Meat Packing Industry," unpub. diss., University of Illinois, 1950, p. 10; Gorman to J. A. Brownlow, NWLB, June 20, 1945, AMC Files.

75. Pete Mosele to Sam Sponseller, Dec. 1, 1942; L. McDonald to J. C. Lewis, May 5, 1942; Murray Cotterill to Sec.-Treas., PWOC, Sept. 25, 1940; F. W. Dowling to Arthur Kampfert, June 23, 1941; J. C. Lewis to Philip Murray, Oct. 6. 1941, UPWA Files; *AMC Executive Board Minutes,* March 28, 1938.

76. *Monthly Labor Review,* Sept. 1943, p. 616; PWOC Financial Statement, June 1, 1943, UPWA Files; *AFL Executive Council Minutes,* Aug. 9–16, 1943, Sec. Report, p. 20.

CHAPTER 10.   THE FIGHT FOR COLLECTIVE BARGAINING

1. For efforts by Wilson & Co. to do this, see 19 NLRB 996 (1940); J. D. Cooney, Wilson & Co., to Van Bittner, June 29, 1939; Maury Hopkins, Wilson & Co., to J. C. Lewis, Aug. 18, 1941, UPWA Files.

2. 19 NLRB 997, 998 (1940), 29 NLRB 764–5 (1941); *PH News,* Oct. 30, 1938, p. 8; Gorman to W. R. Satterlee, Local 1, June 30, 1938, AMC Files.

3. For an extreme case of effective employer resistance, involving the Cedar Rapids plant of Wilson & Co., see 19 NLRB 990 (1940); *PH News,* Feb. 5, 1940, p. 1; March 17, 1941, p. 2; April 28, 1941, p. 8; Lewis Clark to J. C. Lewis, April 24, 1941, UPWA Files; *National Wilson Conference Minutes,* Feb. 14, 1942, UPWA Files.

4. Quoted in Rosenfarb, *National Labor Policy,* p. 197.

5. Documents and conference report in *PH News,* Oct. 1, 1938, pp. 1, 2, 4; Oct. 8, 1938, p. 2.

6. P. W. Chappell to J. R. Steelman, Memorandum, June 30, 1939, FMCS Files.

7. *PH News,* July 24, 1939, pp. 1, 5, 7. For a graphic description, see Benjamin Appel, *The People Talk* (New York, 1940), pp. 176–80.

8. *PH News,* July 10, 1939, p. 1; Aug. 7, 1939, p. 1; New York *Times,* July 18, 1939.

9. Van Bittner, *et al.,* to J. R. Steelman, March 3, 1939; Steelman to Frances Perkins, Memorandum, May 13, 1939; P. W. Chappell to Steelman, April 25, 1939; Steelman to Chappell, May 12, 13, 1939; Don Harris to Henry A. Wallace, June 26, 1939, FMCS Files.

10. John L. Connor to Steelman, Aug. 2, 1939; Memorandum, Aug. 18, 1939; R. H. Cabell, Armour & Co., to Bittner, Aug. 17, 1939; Bittner to Cabell, Aug. 22, 1939; Frances Perkins to H. S. Eldred, Armour & Co., Aug. 18, 1939; Eldred to Perkins, Aug. 22, 1939, FMCS Files; Washington *Times-Herald,* Aug. 19, 1939; New York *Times,* Aug. 17, 19, 1939; *PH News,* Aug. 14, 1939, p. 1; Aug. 21, 1939, p. 1.

11. Gorman and J. J. Keenan to Frances Perkins, Sept. 12, 1939, AMC Files; Gorman to Perkins, Aug. 21, 1939; J. R. Steelman to Gorman, Aug. 21, 1939, FMCS Files.

12. New York *Times,* July 18, 1939; Gorman to H. S. Ellerd (Eldred?), Armour & Co., Aug. 23, 1939, AMC Files.

13. 2 NLRB 159 (1937); Bittner to J. R. Steelman, Aug. 26, 1939; Frances Perkins

to R. H. Cabell, Sept. 8, 1939, FMCS Files; New York *Times,* Sept. 12, 19, 20, 1939; Holcomb, "Union Policies of Meat Packers," pp. 111–2.

14. *PH News,* Oct. 2, 1939, p. 1; New York *Times,* Sept. 21, 1939. The sudden resignation of Armour President Cabell the previous week may have been related to the company's decision.

15. *CIO Executive Board Minutes,* June 3–5, 1940, p. 249.

16. R. S. Brennan to J. A. Padway, Oct. 23, Nov. 4, 1939; Brennan to Gorman, Oct. 23, 30, 1939, AMC Files.

17. *PH News,* March 4, 1940, p. 7. Contracts printed in *PH News* issues, *passim.*

18. J. L. Connor to J. R. Steelman, Jan. 19, 1940, FMCS Files.

19. *CIO Executive Board Minutes,* June 3–5, 1940, p. 251.

20. 6 WLR 427.

21. John Doherty to District Directors, Feb. 8, 1941, UPWA Files. On local negotiations, see L. J. Clark to Bittner, March 10, 1941, UPWA Files; *PH News,* March 3, 1941, p. 1.

22. *National Armour Wage Conference Minutes,* March 23, 1941, UPWA Files.

23. "Armour Bulletin #2," in *PH News,* June 9, 1941, p. 8.

24. J. C. Lewis to Frances Perkins, July 11, 24, 1941; E. J. Cunningham to J. R. Steelman, Memorandum, July 14, 1941, FMCS Files.

25. J. C. Lewis to Armour Locals, Aug. 11, 1941, UPWA Files. The Armour acceptance of a master contract was contingent on a letter from the NLRB stating that such a contract was consistent with the law. This was only a formality and probably a face-saving move by the company.

26. U.S. Dept. of Labor, "Report on the Work of the National Defense Mediation Board," *Bulletin 714* (1942), pp. 202–3.

27. Holcomb, "Union Policies," pp. 149, 159. For typical industry claim concerning the autonomy of plant managers, see Maury Hopkins, Wilson & Co., to J. C. Lewis, Aug. 18, 1941, UPWA Files. The national office of Cudahy was acutely embarrassed by the necessity of reneging on a contract made at its Los Angeles plant in November 1940. C. L. Hodgert, Cudahy Packing Co., to N. Fontecchio, PWOC, Nov. 25, 1940, FMCS Files.

28. Record of Phone Conversation, J. C. Lewis and C. L. Hodgert, Oct. 3, 1941, UPWA Files. In the case of Swift, an additional determinant may have been to help stabilize the Independent Brotherhood, which in early 1942 was certified in six Swift plants. The voluntary offer of a master contract included the Brotherhood as well as the PWOC.

29. Maury Hopkins, Wilson & Co., to J. C. Lewis, Aug. 18, 1941, UPWA Files; *National Wilson Conference Minutes,* Feb. 14, 1942, UPWA Files; 6 WLR War Labor Reports 409; Cooney quoted, Holcomb, "Union Policies," p. 172.

30. L. J. Clark, Joint Executive Board Meeting of Denver Locals, July 26, 1942, UPWA Files. The Wilson contract is printed in 6 WLR 436–41.

31. *UPWA Brief to War Labor Board Panel,* Cases 11–5544, 5760, 6000, 5763, 5914 D, p. 2.

32. National War Labor Board, *Termination Report* (Washington, 1946), I, 1051.

33. NWLB, *Verbatim Transcript of Hearing,* May 17, 1945, pp. 19–20, 31, Industrial Relations Library, Harvard University.

34. See *PH News,* Dec. 12, 1938, p. 8; Kansas City *Star,* Aug. 23, 1941.

35. L. J. Clark to J. E. Hackton, Cudahy Packing Co., May 12, 1942; Clark to Pete Mosele, May 7, 1942, UPWA Files; 6 WLR 421.

36. Sam Sponseller to W. H. Davis, NWLB, Feb. 25, 1943, to NWLB, March 24,

1943, to G. W. Taylor, NWLB, April 7, 1943, to F. D. Green, Armour & Co., April 1, 1943, to Allan Haywood, May 13, 1943, UPWA Files; *Officers' Report,* 2nd Wage and Policy Conference, UPWA, July 8–10, 1943, UPWA Files.

37. For example, R. K. Winkler, Wilson & Co., to G. K. Batt, NWLB, July 7, 1943; Ralph Helstein to Wilson Locals, Sept. 13, 1943, UPWA Files; 21 WLR 690; *Swift Brief Before NWLB Panel,* June 5, 1944, pp. 105–6.

38. The PWOC, for example, claimed that Wilson had offered not to raise the union security issue in exchange for union concessions. *UPWA Brief to NWLB Panel,* n.d. (1944), p. 159. Armour specifically omitted the AMC, which was not involved in wild-cat strikes, from the demand for cancellation of maintenance-of-membership. *Comments of the Amalgamated Meat Cutters to NWLB Panel's Report and Recommendation,* Aug. 11, 1944 (mimeographed), p. 29, Industrial Relations Library, Harvard University.

39. *Swift Presentation Before NWLB Panel,* Aug. 3, 1942, p. 117; *Reply Presentation of Swift & Co.,* Sept. 8, 1942, p. 11; *Swift & Co. Brief Before NWLB Panel,* June 5, 1944, pp. 105–6.

40. Meeting of Joint PWOC Negotiating Committee Representatives, Omaha, March 29, 1942, UPWA Files; *PH News,* May 8, 1942, p. 1.

41. The NWLB panel did reject the union demand for consolidated hearings in 1942, but it submitted a single report, and the board handed down a single decision. 6 WLR 411.

42. A. T. Stephens to Ralph Helstein, Dec. 13, 1943, UPWA Files.

43. 6 WLR 413. Table of company earnings in 21 WLR 709.

44. Roy Franklin to R. B. Fairbanks, April 8, 1943; PWOC Press Release, Feb. 9, 1943, UPWA Files; Jimerson and Gorman to W. H. Davis, NWLB, Feb. 9, 1943; AMC Press Release, Feb. 9, 1943, AMC Files.

45. See *BW,* July 1944, p. 8; Lewis Clark to F. M. Vinson, Director of Economic Stabilization, Jan. 26, 1945, UPWA Files. For summary of arguments, see 31 WLR 664–9.

46. *Officers' Report,* 2nd Wage and Policy Conference, July 8–10, 1943, UPWA Files; *BW,* April 1945, p. 1.

47. NWLB, *Termination Report,* I, 1049; III, 449–50. A classic instance of confusion occurred in the first CIO case involving interplant inequalities. The Vernon group in Los Angeles—the small independents—had given a $7\frac{1}{2}$ cent increase before wage stabilization. The UPWA requested a similar increase for the Swift plant. The permanent company arbitrator, C. O. Gregory, turned this down because the Directive Order permitted adjustments only between plants "in different localities." But, on appeal, the WLB found that its order did not exclude plants in the same labor market. Gregory thereupon granted the increase. Then on Sept. 15, 1944 the board clarified its clarification: inequalities between Big Four and independent plants were *not* grounds for an adjustment. The basis of the UPWA case was thus destroyed. Nevertheless, the Big Four plants in Los Angeles received the increase, for the board simultaneously established a "West Coast rate" of $77\frac{1}{2}$ cents an hour. The outcome doubly chagrined the Amalgamated. Not only did its rival prevail in Los Angeles, but the arbitration award of $2\frac{1}{2}$ cents at its Swift-East St. Louis plant, made on the basis of the first WLB ruling, was reversed as a result of the Sept. 1944 ruling. L. J. Clark to G. W. Taylor, NWLB, July 29, 1944, and other correspondence on this case, UPWA Files; 21 WLR xciv, 1.

48. E. E. Witte, NWLB, to E. W. Jimerson, Aug. 20, 1945; Memorandum, Meeting with Swift representatives, Sept. 12, 1945, AMC Files.

49. NWLB, *Termination Report*, I, 1056–61.

50. *Ibid.*, III, 443 ff.; NWLB, *Transcript of Hearing*, May 17, 1945, p. 23, Industrial Relations Library, Harvard.

51. David Dolnick, Meeting of Swift and AMC representatives, April 17, 1945; AMC to J. E. Wilson, Swift & Co., March 19, 1943, AMC Files; NWLB, *Transcript of Hearing*, May 25, 1945, Industrial Relations Library, Harvard.

52. New York *Times*, Dec. 21, 1939; Sept. 20, 1941; Dec. 25, 1941; *PH News*, Jan. 8, 1940, p. 2; Oct. 14, 1940, p. 2; J. C. Lewis to Philip Murray, Oct. 6, 1941, UPWA Files.

53. *NP*, Nov. 29, 1941, p. 13; "Armour Notice," Dec. 10, 1941, UPWA Files.

54. *Monthly Labor Review*, Jan. 1942, p. 260; Jan. 1946, p. 161. A variety of minor improvements was also forthcoming in such areas as the weekly guarantee and the computation of earnings in piecework and multiple-rate jobs. The only significant exception was Swift's agreement in the second round of negotiations to provide half pay for stipulated periods to employees disabled by sickness or noncompensable accidents.

CHAPTER 11. RIVAL UNIONISM: FROM WARFARE TO INTERDEPENDENCE

1. *CIO Executive Board Minutes*, June 13–5, 1939, pp. 188–9; June 3–5, 1940, pp. 252–3.

2. Gorman to John Holmes, Swift & Co., Nov. 14, 1940 (same letter also to other Big Four packers); Gorman to Executive Board, Dec. 5, 1940, AMC Files.

3. Gorman to A. T. Lester, Local 5, March 11, 1941, AMC Files.

4. *BW*, Sept. 1, 1941, p. 1.

5. *Comments of Swift & Co. on Recommendation of NWLB Panel*, Nov. 13, 1942, pp. 5, 25; *Brief of Swift & Co. Before NWLB Panel*, June 5, 1944, pp. 66–7, Industrial Relations Library, Harvard; 6 WLR 410; E. W. Jimerson to W. H. Davis, NWLB, March 15, 1943; F. D. Green, Armour & Co., to Davis, March 17, 1943, H. J. Meyer, NWLB, to Jimerson, April 6, 1943, AMC Files.

6. M. Guerra to Gorman, April 20, 1943, quoted in Gorman to J. E. Wilson, Swift & Co., April 26, 1943, AMC Files; *BW*, April 1, 1943, p. 3; 21 WLR 712.

7. NWLB, *Termination Report*, I, 1050; *BW*, April 1945, p. 1; AMC, *Proceedings*, 1944, p. 10.

8. Jimerson to J. E. Wilson, Swift & Co., Nov. 8, 1943; Wilson to Jimerson, Nov. 10, 1943, AMC Files. Actually, the AMC tried unsuccessfully to negotiate apart from the UPWA. See Jimerson to J. E. Wilson, March 19, 1945; Gorman to W. H. Davis, NWLB, Jan. 24, 1944, AMC Files.

9. NWLB, *Transcript of Hearing*, May 17, 24, 1945, Industrial Relations Library, Harvard; *BW*, July 1945, p. 1; *Packinghouse Worker*, June 8, 1945, p. 1, cited hereafter as *PW*.

10. *PH News*, Dec. 18, 1942, p. 1; Feb. 26, 1943, p. 1; *BW*, Feb. 1, 1943, p. 2; March 1, 1943, pp. 1, 7.

11. M. J. Osslo to E. E. Walker, Sept. 8, 1944, AMC Files.

12. *BW*, May 1945, p. 1.

13. See *PW*, July 30, 1943, p. 1; July 12, 1944, p. 1; *BW*, April 1945, p. 1.

14. Ralph Helstein to W. E. Wilson, Swift & Co., Dec. 3, 1943, UPWA Files; *UPWA Brief to the NWLB Panel*, n.d. (1944), pp. 157–8, Industrial Relations Library, Harvard.

15. *Officers' Report*, 2nd Wage and Policy Conference, July 8–10, 1943, UPWA Files.

16. *Officers' Report,* PWOC Wage and Policy Conference, Feb. 19, 1943, UPWA Files.

17. *District 3 Conference Minutes,* Aug. 2, 1942; *Districts 3 and 10 Joint Conference Minutes,* Sept. 13, 1942; *PWOC Minutes,* Nov. 12, 1942; Sponseller to H. P. Ballard, Oct. 12, 1942, to Nels Peterson, Oct. 12, 1942, UPWA Files; UPWA, *Proceedings,* 1943, p. 227; *CIO Executive Board Minutes,* Nov. 14, 1942, pp. 354-5.

18. W. T. Haywood to J. C. Lewis, Feb. 4, 1942; Roy Franklin to Lewis Clark, Feb. 28, 1942, to J. C. Lewis, Feb. 11, 1942; Kampfert to Franklin, Feb. 2, 1942; James Robb to J. C. Lewis, Feb. 3, 1942; Frank Ellis to Sponseller, Aug. 17, 1942; Frank McCarty, District 1 Director, Report, 1942, UPWA Files; *PH News,* Aug. 14, 1942, p. 4.

19. Frank McCarty to John L. Lewis, Feb. 9, 1942; J. L. Lewis to J. C. Lewis, Feb. 14, 1942; J. C. Lewis to J. L. Lewis, Feb. 24, 1942; O. E. Gasaway, District 50, to J. C. Lewis, May 15, 1942; J. C. Lewis to G. R. Hathaway, May 23, 1941, UPWA Files; *PH News,* May 15, 1942, p. 1; June 12, 1942, p. 1; *PM,* May 8, 1942.

20. J. C. Lewis to James Dean, July 24, 1941; Lewis Clark to John Doherty, Jan. 20, 1941, to Van Bittner, April 10, 1940; Joe Ollman to Sam Sponseller, March 28, 1943, UPWA Files; *PH News,* Dec. 11, 1939, p. 1; Galenson, *CIO Challenge,* p. 374; Wilson Record, *The Negro and the Communist Party* (Chapel Hill, 1951), p. 180; U.S. House Un-American Activities Committee, *Hearings,* 82 Cong., 2nd sess. (1952), pp. 3814 ff.; Iowa-Nebraska States Industrial Union Council, *Proceedings,* 1939, pp. 93 ff.

21. A. Kampfert to all District 1 Locals, March 29, 1940; Frank McCarty to Herbert March, Oct. 1, 1940; N. Fontecchio to Kampfert, Jan. 30, 1941, UPWA Files; *PH News,* July 7, 1941, p. 2; Sept. 29, 1941, p. 2; Galenson, *CIO Challenge,* p. 376.

22. Roy Franklin to J. C. Lewis, Oct. 21, 1941, UPWA Files; District 1 circular distributed at Darling Rendering Co., Chicago, 1942, UPWA Files.

23. Richard Francis, CIO Regional Director, Seattle, to J. C. Lewis, May 19, 1941; Lewis to David Fowler, July 21, 1941; Roy Franklin to Lewis, Sept. 1, Oct. 1, 1941; Joe Ollman to Sponseller, March 28, 1943, UPWA Files.

24. Roy Franklin to R. B. Fairbanks, Nov. 5, 1942, March 25, 1943, UPWA Files.

25. On the devious politics preceding the convention, see UPWA, *Proceedings,* 1943, pp. 223-4, 233; Allan Haywood to Sponseller, July 16, 1943; Joe Ollman to Sponseller, June 24, 1943; Floyd Brouillard to Sponseller, March 20, 1943, to Lewis Clark, Aug. 4, 1943; Sponseller to T. M. Covey, June 25, 1943, UPWA Files.

26. UPWA, *Proceedings,* 1943, p. 233.

27. *Ibid.,* 52.

28. The national officers had very limited means of acting against district directors. The latter could be removed by the executive board for "an offense against the Constitution and By-Laws of the Union" (Art. 23); and they could be undermined from within their districts through the efforts of the national office, which could place locals in trusteeship and, of course, controlled the staff representatives and the union newspaper. In practice, neither of these tactics was effective except under unusual circumstances.

29. *PW,* March 22, 1946, p. 8; *BW,* April 1946, p. 8; T. V. Purcell, *The Worker Speaks his Mind on Company and Union* (Cambridge, Mass., 1953), p. 65.

30. Chicago *Defender,* May 26, 1944, reprinted in *BW,* June 1944, p. 4; HUAC, *Hearings,* 1952, p. 3810, and *passim.*

31. UPWA, *Proceedings,* 1946, pp. 142-4; *PW,* May 16, 1947, p. 2.

32. *PW,* Aug. 31, 1945, p. 1; Sept. 28, 1945, p. 1; Dec. 14, 1945, p. 1; *BW,* Oct.–Nov. 1945, p. 1; Dec. 1945, pp. 1, 4; Feb.–March 1946, p. 1; New York *Times,* Dec. 1, 11, 1945; Jan. 3, 4, 6, 12, 1946.

33. *NP*, Jan. 5, 1946, pp. 5, 22; Jan. 12, 1946, pp. 4, 48; New York *Times*, Jan. 4, 6, 8, 1946.

34. Quoted, Hanna and Belsky, *Picket and the Pen*, p. 194.

35. New York *Times*, Jan. 17, 18, 23–5, 1946; *BW*, Feb.–March 1946, p. 1; April 1946, p. 1; Hanna and Belsky, *Picket and the Pen*, pp. 194–5.

36. New York *Times*, Jan. 12, 24, 26, 1946; *PW*, Jan. 25, 1946, p. 1; Feb. 8, 1946, p. 6; *NP*, Jan. 26, 1946, p. 11.

37. New York *Times*, Jan. 25–8, 1946.

38. *NP*, Feb. 9, 1946, pp. 13–4; March 9, 1946, pp. 11, 27; April 13, 1946, p. 17; New York *Times*, Feb. 7, 27, 1946; Joel Seidman, *American Labor from Defense to Reconversion* (Chicago, 1953), p. 229.

39. On the Hormel plan, see Fred H. Blum, *Toward a Democratic Work Process* (New York, 1953); Jack Chernick, *Economic Effects of Steady Employment and Earnings: A Case Study of the Hormel Annual Wage System* (Minneapolis, 1942), chs. 2, 3.

40. New York *Times*, Aug. 3, 1946.

41. *PW*, Sept. 20, 1946, p. 1; Oct. 4, 1946, p. 3; Oct. 18, 1946, p. 1; *BW*, Nov. 1946, p. 1.

42. AMC–Swift agreement given, *BW*, Jan. 1947, p. 1; also, New York *Times*, Dec. 5, 1946; *PW*, Dec. 13, 1946, p. 4.

43. Russell Porter, New York *Times*, Dec. 16, 1946. On views of the rival unions, see *PW*, March 22, 1946, p. 4; Feb. 8, 1946, p. 7; *BW*, Feb.–March 1946, p. 1; April 1946, p. 1.

44. AMC, *Proceedings*, 1948, p. 14; UPWA, *Proceedings*, 1947, pp. 189–90; *PW*, May 30, 1947, p. 7; June 27, 1947, p. 1; *NP*, May 10, 1947, p. 13; June 28, 1947, p. 11. It should be noted that the UPWA moves toward a *rapprochement* contained sources of tension, for March was advocating activities at the local and district level, and this of course was opposed by the national AMC leadership.

45. UPWA, *Proceedings*, 1950, p. 231; *PW*, Sept. 5, 1947, p. 8; Chicago *Tribune*, Oct. 10, 1946, Chicago *Daily News*, Oct. 14, 1946, reprinted in *BW* (Nov. 1946); Purcell, *Worker Speaks his Mind*, pp. 64–5; Max M. Kampelman, *The Communist Party vs. the C.I.O.* (New York, 1957), p. 64.

46. UPWA, *Proceedings*, 1948, p. 73; *PW*, Sept. 19, 1947, p. 1.

47. *BW*, March 1948, p. 1; AMC, *Proceedings*, 1948, p. 14; *PW*, Feb. 6, 1948, p. 1; *NP*, Feb. 7, 1948, p. 13.

48. For pre-strike events and documents, see *PW*, March 19, 1948, p. 8; UPWA, *Proceedings*, 1948, p. 43; *NP*, March 6, 1948, pp. 15–6; New York *Times*, March 15, 1948.

49. Quoted, Purcell, *Worker Speaks his Mind*, p. 61. Also, on board of inquiry report, see New York *Times*, April 8–10, 1948; *NP*, April 17, 1948, pp. 11–2; *PW*, April 16, 1948, p. 46.

50. UPWA, *Proceedings*, 1948, pp. 62, 86; *NP*, Jan. 10, 1948, p. 11; Jan. 17, 1948, p. 31.

51. On this incident, see *PW*, May 14, 1948; UPWA, *Proceedings*, 1948, pp. 48, 52, and *passim*.

52. *PW*, May 28, 1948, p. 2.

53. UPWA, *Proceedings*, 1948, pp. 43 ff. and 79; 1949, p. 172; A. B. Held, "The CIO Packers' Convention," *New Leader*, July 10, 1948, p. 3.

54. *PW*, Oct. 29, 1948, p. 4.

55. The decisive act was Armour's decision in November 1948 voluntarily to reinstate the discharged strikers; this move set the pattern for the others. UPWA, *Proceedings*, 1949, p. 104.

56. *PW*, Aug. 20, 1948, p. 7; *NP*, June 5, 1948, p. 19.

57. *PW*, Jan. 14, 1949, p. 2; Jan. 28, 1949, p. 2; April 8, 1949, p. 3; *Armour Magazine*, March–April 1949.

58. *PW*, Sept. 3, 1948, p. 2; Oct. 1, 1948, p. 1; Dec. 10, 1948, p. 1; Dec. 24, 1948, p. 6; March 25, 1949, p. 4; June 24, 1949, p. 7; Aug. 5, 1949, p. 1; UPWA, *Proceedings*, 1949, pp. 61 ff.

59. *PW*, June 11, 1948, p. 1; April 8, 1949, p. 1; UPWA, *Proceedings*, 1948, p. 62; *BW*, June 1948, p. 20; Dec. 1948, p. 9.

60. *PW*, Oct. 7, 1949, p. 1; UPWA, *Proceedings*, 1950, pp. 113, 241; *BW*, Aug. 1950, p. 10; *Monthly Labor Review*, Sept. 1950, pp. 366–7; Feb. 1951, p. 10.

61. *PW*, Dec. 24, 1948, p. 6; Purcell, *Blue Collar Man*, p. 38; D. J. Saposs, *Communism in American Unions* (New York, 1959), pp. 202–4.

62. UPWA, *Proceedings*, 1950, pp. 222 ff.; 1952, pp. 271 ff.

63. *BW*, Jan. 1952, p. 21; Feb. 1952, pp. 12–3, 16–7; UPWA, *Proceedings*, 1952, pp. 261 ff.

64. *BW*, Aug. 1953, p. 19.

CHAPTER 12.   THE NEW EXPANSIONISM

1. John W. Kendrick, *Productivity Trends in the United States* (Princeton, 1961), pp. 162, 483; Solomon Fabricant, *Employment in Manufacturing 1899–1939* (New York, 1942), pp. 68, 273; U.S. Bureau of Labor Statistics, *Productivity and Unit Labor Cost in Selected Manufacturing industries, 1919–1940* (February 1942), p. 100; Milton Derber, "Economic Trends in the Meat Packing Industry," (Report submitted to the Armour Automation Committee, Jan. 15, 1961, mimeographed); Corey, *Meat and Man*, ch. 9; U.S. Bureau of Census, *Census of Manufacturing*, 1954, II, 20A–21; 1958, II, 20A1–A20. The indications of a break from the slow rate of productivity increase actually began before World War II. Between 1937 and 1939 meat packing output per man-hour jumped 20 per cent. From 1939 to 1947, however, productivity became relatively static, no doubt partly reflecting the effect of the war. Beginning in 1947, the industry experienced a productivity increase in the range of 3 per cent a year.

2. *NP*, June 26, 1948, p. 9; *BW*, June 1953, pp. 7–8; AMC, *Officers' Reports*, 1960, pp. 114, 126–7, 134, 163–4.

3. *BW*, July 1955, p. 1.

4. UPWA, *Officers' Reports*, 1958, pp. 30, 33, 34; *PW*, March 1959, pp. 6–7, 11; March 1960, p. 11.

5. Jimerson and Gorman to the Executive Board, July 13, 1945, AMC Files.

6. UPWA, *Proceedings*, 1950, p. 114; UPWA, *Officers' Reports*, 1957, pp. 38–9, 41.

7. AMC, *Proceedings*, 1956, p. 112.

8. AMC, *Proceedings*, 1956, pp. 161–5; AMC, *Officers' Reports*, 1956, pp. 57, 68–9, 92; 1960, pp. 61–2, 91–3; *BW*, Aug. 1955, p. 30; Dec. 1955, p. 28; Interview, Harry Poole, Dec. 6, 1963.

9. *BW*, June 1949, p. 6; Nov. 1949, p. 10; April 1954, p. 8.

10. UPWA, *Officers' Reports*, 1957, pp. 41, 44.

11. AMC, *Proceedings*, 1940, p. 137; 1948, p. 51; *AMC Executive Board Minutes*, June 13, 1936.

12. *AFL Executive Council Minutes*, Aug. 9–16, 1943, p. 74; Aug. 21–Sept. 2, 1937, pp. 181–2.

13. See AMC, *Proceedings*, 1917, pp. 111, 139; 1920, pp. 7, 114, 145; *BW*, Dec. 1919, p. 3; Feb. 1920, p. 3; Gompers to John F. Hart, March 9, 1918, Gompers Letterbooks.

14. AMC, *Proceedings,* 1944, p. 46.

15. *BW,* Oct. 1, 1936, p. 6; Feb. 1, 1937, p. 1; June 1, 1937, p. 3; Nov. 1, 1937, p. 7; Feb. 1, 1938, p. 7; Interview, Harry Poole, Dec. 6, 1963.

16. *AMC Executive Board Minutes,* June 13, 1936; Oct. 5, 1937; Nov. 24, 1940.

17. *AFL Executive Council Minutes,* Feb. 8–19, 1937, p. 74; May 22, 26–30, 1937, p. 113; May 13–21, 1940, p. 89; Sept. 30–Oct. 10, 1940, pp. 40–1; Aug. 4–13, 1942, pp. 88–9; AMC, *Proceedings,* 1948, pp. 132–33.

18. *AFL Executive Council Minutes,* Feb. 10–17, 1918, pp. 29–30, 54; AMC, *Proceedings,* 1917, p. 64; 1922, p. 45; *Monthly Labor Review,* May 1926, p. 208.

19. *AFL Executive Council Minutes,* Jan. 23–Feb. 8, 1938, p. 132.

20. See AMC, *Proceedings,* 1940, pp. 48, 51; H. F. Koerble, Milwaukee RMDA, to I. W. Ringer, copy, AMC Files. On the internal weakness of the Clerks, see United Retail and Wholesale Employees, *Proceedings,* 1939, p. 62; *Retail Clerks International Advocate,* May–June 1937, pp. 20–2.

21. Memorandum of Agreement, Dec. 15, 1937, included in letter, T. M. Finn to J. R. Steelman, Dec. 16, 1937, Federal Mediation and Conciliation Service Files; *AFL Executive Council Minutes,* Jan. 24–Feb. 8, 1938, p. 133; Gorman to E. J. McCarthy, March 28, 1939, AMC Files.

22. *RCI Advocate,* Nov.–Dec. 1937, p. 8; RCIA, *Proceedings,* 1939, p. 42.

23. *BW,* Jan. 1933, p. 3; Jan. 1, 1938, p. 3; *AMC Executive Board Minutes,* March 23, 1938; AMC, *Proceedings,* 1940, pp. 51, 68.

24. Gorman to H. J. Mullally, Local 2, Feb. 7, 1938, AMC Files; Gorman to Green, in *AFL Executive Council Minutes,* Jan. 24–Feb. 8, 1938, pp. 132–4.

25. Gorman to Schimmat, Nov. 29, 1938, AMC Files.

26. C. C. Coulter to AFL and RC Bodies, Dec. 6, 1938; Green to Coulter, Dec. 5, 1938, in *RCI Advocate,* Sept.–Oct. 1940, p. 10; Gorman to Green, Feb. 9, 1939, in *AFL Executive Council Minutes,* Jan. 30–Feb. 14, 1939, p. 311, and Green's remark, *ibid.,* p. 310; AMC, *Proceedings,* 1940, p. 11.

27. Entries, Nov. 2, 16, 1937, *Minute Book,* AMC Local 358, Wisconsin State Historical Society; *RCI Advocate,* March–April 1940, p. 22.

28. Lane to H. J. Mullally, Local 2, Feb. 7, 1938; Gorman to C. C. Randall, Local 2, July 1, 1938, AMC Files; AMC, *Proceedings,* 1944, pp. 43–4.

29. On policy of First National, see C. E. L. Gill to J. R. Steelman, Jan. 17, 26, 1940, Federal Mediation and Conciliation Service Files.

30. W. A. Ingram, Safeway, to Gorman, Aug. 19, 1941; Joseph Bappert, Kroger, to Gorman, Aug. 25, 1941; Gorman to Bappert, Aug. 25, 1941, AMC Files. Fear of adverse consequences particularly affected A & P, since it would have been self-defeating to alienate the Clerks (who, moreover, had the sympathy of the AFL leadership) when the first purpose of dealing with organized labor was to win support against the Patman Bill.

31. Gorman to F. H. Massman, National Tea, Nov. 9, 1938; Massman to Gorman, Nov. 10, 1938; W. I. Ingram, Safeway, to C. N. Saunders, Aug. 19, 1941, to Gorman, Aug. 19, 1941, to C. C. Coulter, Aug. 19, 1941, copy, AMC Files. The Wagner Act had little effect in this area. The NLRB followed a policy of not intervening in jurisdictional disputes between affiliates of the same federation, although it did order an election between the Clerks and the Amalgamated in one exceptional case in 1944, involving Safeway clerks in San Diego. The labor legislation, in fact, sometimes made the Amalgamated more conciliatory, since it was foolhardy to enter a jurisdictional fight while facing a contest with the CIO. The New York State Labor Relations Board actually ruled that the two unions would not be placed on the ballot in an election involv-

ing the CIO in the A & P Stores in New York City unless the jurisdictional fight was privately resolved. 59 NLRB 936 (1944); Rosenfarb, *National Labor Policy,* pp. 318–22; Jay Oliver, NLRB, to Gorman, Sept. 17, 1941, AMC Files; 3 NYLRB 933 (1940).

32. Green to Gorman, Aug. 29, 1940, in *RCI Advocate* (Sept.–Oct. 1949), p. 11; Coulter to AFL Executive Council, Dec. 31, 1941, in *AFL Executive Council Minutes,* Jan. 12–7, 1942, p. 42, also, pp. 53–4; Gorman to Executive Board, Jan. 13, 1942, AMC Files.

33. Soderstrom to Green, April 21, 1942, in *AFL Executive Council Minutes,* May 13–22, 1942; *AMC Executive Board Minutes,* Oct. 6, 1942; *BW,* Aug. 1, 1942, p. 4; Nov. 1, 1942, p. 3; RCIA, *Proceedings,* 1947, pp. 5, 8.

34. Agreement printed in RCIA, *Proceedings,* 1951, pp. 5–6; *BW,* April 1949, p. 10; Aug. 1949, p. 16.

35. "Minutes of Proceedings of 20th Convention of AMC," 2nd day, June 28, 1960, p. 18; Interview, Harry Poole, Dec. 6, 1963.

36. For Amalgamated nervousness on this point, see *BW,* Feb. 1954, p. 16; Dec. 1955, p. 16; AMC, *Officers' Reports,* 1956, pp. 15–9; 1960, pp. 45–6, 93–4, 128–9. One solution would have been to acquire retail jurisdiction through a merger with the Wholesale and Retail Workers, whose jurisdiction came from its CIO charter. Actually, such a merger was considered in the early 1960's, but had not been achieved at this writing (1963).

37. AMC, *Proceedings,* 1948, pp. 52–3; *BW,* March 1949, p. 10. On the RCIA in the 1950's, see Martin S. Estey, "Patterns of Union Membership in the Retail Trades," *Industrial and Labor Relations Review,* VIII, 557–64 (July 1955); Ralston G. Zundel, "Conflict and Cooperation among Retail Unions," *Journal of Business,* XXVIII, 301–11 (Oct. 1954).

38. AMC, *Proceedings,* 1940, pp. 114, 119; *AMC Executive Board Minutes,* June 21, 1939; Interview, T. J. Lloyd, Dec. 6, 1963.

39. For a history of the IFLWU, see Philip Foner, *The Fur and Leather Workers' Union* (Newark, 1950).

40. Interview, Patrick Gorman, Nov. 17, 1960; Interview, Harry Poole, Dec. 6, 1963; New York *Times,* Aug. 29, 30, Sept. 1, Oct. 18, 25, 1953; April 3, May 1, 31, Oct. 3, 1954.

41. AMC, *Proceedings,* 1956, p. 43.

42. "Merger Agreement," Dec. 28, 1954, p. 4.

43. *BW,* March 1955, pp. 6–8; Philip Taft, *The A.F. of L. from the Death of Gompers to the Merger* (New York, 1959), pp. 434–5.

44. *BW,* Oct. 1955, p. 37; July 1956, pp. 53–4.

45. *BW,* Dec. 1955, p. 3; AMC, *Officers' Reports,* 1956, p. 84; Taft, *The A.F. of L.,* p. 435.

46. *AFL Executive Council Minutes,* Feb. 3–11, 1958, pp. 88–9; Interview, Harry Poole, Dec. 6, 1963; Interview, Patrick Gorman, Dec. 6, 1963.

47. Author's notes at Fur and Leather Caucus, June 28, 1960.

48. AMC, *Proceedings,* 1956, p. 243; *PW,* Feb. 1955, p. 3.

49. *BW,* May 1952, p. 3; Corey, *Meat and Man,* pp. 302–3.

50. *BW,* April 1954, p. 17.

51. *BW,* May 1952, p. 3.

52. *PW,* April 1955, p. 13; May 1955, p. 1; Oct. 1955, p. 4.

53. *PW,* Dec. 1955, pp. 1–2; Joel Seidman, "Unity in Meat Packing: Problems and Prospects," *New Dimensions in Collective Bargaining,* eds., H. W. Davey *et al.* (New York, 1959).

54. Helstein ruefully acknowledged the fact when a suggestion was made that the UPWA should embark on an organizing drive that would make it as large as the

Amalgamated. "Then we will revise the [proposed] Constitution," retorted the UPWA president. UPWA, *Proceedings of the 2nd Special Convention*, 1956, p. 32.

55. AMC, *Proceedings*, 1956, p. 242.

56. The Amalgamated calculated that the annual income from the UPWA membership would total $1,260,000 for usable purposes, while salaries alone for UPWA officials would reach more than $800,000. *Butcher Workman Supplement* (n.d., probably end Oct. 1956).

57. *Work*, Jan. 1957; *BW*, May 1956, pp. 2–4; March 1956, pp. 6–7.

58. Seidman, "Unity in Meat Packing," note 12, p. 42.

59. AMC, *Proceedings*, 1956, pp. 159, 219; *BW*, June 1956, pp. 34 ff.; John Jurkanin, "Open Letter to Delegates of 1956 AMC Convention, March 9, 1957" (mimeographed).

60. *BW*, Oct. 1956, p. 3.

61. UPWA, *Proceedings of 2nd Special Convention*, 1956, pp. 8, 10, and *passim; Butcher Workman Supplement;* Interviews, Gorman, T. J. Lloyd, Harry Poole, Dec. 6, 1963.

62. UPWA, *Proceedings of 2nd Special Convention*, 1956, pp. 8 ff.

63. *Butcher Workman Supplement.*

64. *BW*, Oct. 1956, p. 2; Nov.–Dec. 1956, pp. 2–3; *Butcher Workman Supplement;* Interview, T. J. Lloyd, Dec. 6, 1963.

65. *BW*, March 1952, p. 16.

66. *Butcher Workman Supplement.*

67. *BW*, Nov.–Dec. 1956, p. 2. On neither side did men consider the failure to reflect irreconcilable differences. Rather, the mutual feeling was that the other camp did not really want the merger, or did not want it badly enough to make the necessary concessions. In private interviews several years later, the two principals assigned personal motives to the other for scuttling the merger: Helstein said that Gorman was aging and did not want to undergo the uncomfortable adjustments involved in merger; Gorman said that Helstein had an ambitious wife who did not want her husband to occupy less than a top union post. Interviews, Helstein and Gorman, Nov. 17, 1960. These statements were significant, not because they were necessarily accurate, but because they revealed a common assumption that basic differences did not divide the two unions.

68. John Jurkanin, "Open Letter #2, May 7, 1957" (mimeographed).

INDEX

Wertheim Publications in Industrial Relations

PUBLISHED BY HARVARD UNIVERSITY PRESS

J. D. Houser, *What the Employer Thinks*, 1927
Wertheim Lectures on Industrial Relations, 1929
William Haber, *Industrial Relations in the Building Industry*, 1930
Johnson O'Connor, *Psychometrics*, 1934
Paul H. Norgren, *The Swedish Collective Bargaining System*, 1941
Leo C. Brown, S.J., *Union Policies in the Leather Industry*, 1947.
Walter Galenson, *Labor in Norway*, 1949
Dorothea de Schweinitz, *Labor and Management in a Common Enterprise*, 1949
Ralph Altman, *Availability for Work: A Study in Unemployment Compensation*, 1950
John T. Dunlop and Arthur D. Hill, *The Wage Adjustment Board: Wartime Stabilization in the Building and Construction Industry*, 1950
Walter Galenson, *The Danish System of Labor Relations: A Study in Industrial Peace*, 1952
Lloyd H. Fisher, *The Harvest Labor Market in California*, 1953
Theodore V. Purcell, S.J., *The Worker Speaks His Mind on Company and Union*, 1953
Donald J. White, *The New England Fishing Industry*, 1954
Val R. Lorwin, *The French Labor Movement*, 1954
Philip Taft, *The Structure and Government of Labor Unions*, 1954
George B. Baldwin, *Beyond Nationalization: The Labor Problems of British Coal*, 1955
Kenneth F. Walker, *Industrial Relations in Australia*, 1956
Charles A. Myers, *Labor Problems in the Industrialization of India*, 1958
Herbert J. Spiro, *The Politics of German Codetermination*, 1958
Mark W. Leiserson, *Wages and Economic Control in Norway, 1945–1957*, 1959
J. Pen, *The Wage Rate Under Collective Bargaining*, 1959
Jack Stieber, *The Steel Industry Wage Structure*, 1959
Theodore V. Purcell, S.J., *Blue Collar Man: Patterns of Dual Allegiance in Industry*, 1960
Carl Erik Knoellinger, *Labor in Finland*, 1960
Sumner H. Slichter, *Potentials of the American Economy: Selected Essays* edited by John T. Dunlop, 1961
C. L. Christenson, *Economic Redevelopment in Bituminous Coal: The Special Case of Technological Advance in United States Coal Mines, 1930–1960*, 1962
Daniel L. Horowitz, *The Italian Labor Movement*, 1963
Adolf Sturmthal, *Workers Councils: A Study of Workplace Organization on Both Sides of the Iron Curtain*, 1964
Vernon H. Jensen, *Hiring of Dock Workers and Employment Practices in the Ports of New York, Liverpool, London, Rotterdam, and Marseilles*, 1964

STUDIES IN LABOR-MANAGEMENT HISTORY

Lloyd Ulman, *The Rise of the National Trade Union: The Development and Significance of Its Structure, Governing Institutions, and Economic Policies*, 1955.

Joseph P. Goldberg, *The Maritime Story: A Study in Labor-Management Relations*, *1957*, 1958

Walter Galenson, *The CIO Challenge to the AFL: A History of the American Labor Movement, 1935–1941*, 1960

Morris A. Horowitz, *The New York Hotel Industry: A Labor Relations Study*, 1960

Mark Perlman, *The Machinists: A New Study in Trade Unionism*, 1961

Fred C. Munson, *Labor Relations in the Lithographic Industry*, 1963

Garth L. Mangum, *The Operating Engineers: The Economic History of a Trade Union*, 1964

David Brody, *The Butcher Workmen: A Study of Unionization*, 1964

PUBLISHED BY McGRAW–HILL BOOK CO., INC.

Robert J. Alexander, *Labor Relations in Argentina, Brazil, and Chile*, 1961

Carl M. Stevens, *Strategy and Collective Bargaining Negotiations*, 1963